Lecture Notes in Computer Science

Edited by G. Goos, J. Hartmanis, and J. van Lee

Springer
Berlin
Heidelberg
New York
Barcelona
Hong Kong
London
Milan
Paris
Tokyo

Tony Marsland   Ian Frank (Eds.)

Computers and Games

Second International Conference, CG 2000
Hamamatsu, Japan, October 26-28, 2000
Revised Papers

 Springer

Series Editors

Gerhard Goos, Karlsruhe University, Germany
Juris Hartmanis, Cornell University, NY, USA
Jan van Leeuwen, Utrecht University, The Netherlands

Volume Editors

Tony Marsland
University of Alberta, Department of Computer Science
Edmonton, Alberta, Canada
E-mail: tony@cs.ualberta.ca

Ian Frank
Future University - Hakodate
Hakodate, Hokkaido, Japan
E-mail: ianf@fun.ac.jp

Cataloging-in-Publication Data applied for

Die Deutsche Bibliothek - CIP-Einheitsaufnahme

Computers and games : second international conference ; revised papers /
CG 2000, Hamamatsu, Japan, October 26 - 28, 2000. Tony Marsland ; Ian Frank
(ed.). - Berlin ; Heidelberg ; New York ; Barcelona ; Hong Kong ; London ;
Milan ; Paris ; Tokyo : Springer, 2001
    (Lecture notes in computer science ; Vol. 2063)
    ISBN 3-540-43080-6

CR Subject Classification (1998): G. I.2.1, I.2.6, I.2.8, F.2, E.1

ISSN 0302-9743
ISBN 3-540-43080-6 Springer-Verlag Berlin Heidelberg New York

Springer-Verlag Berlin Heidelberg New York
a member of BertelsmannSpringer Science+Business Media GmbH

http://www.springer.de

Typesetting: Camera-ready by author, data conversion by Olgun Computergrafik
Printed on acid-free paper   SPIN 10781624   06/3142   5 4 3 2 1 0

Preface

This book contains the papers presented at CG 2000 – the Second International Conference on Computers and Games – held at the CURREAC Center in Hamamatsu, Japan, on October 26–28, 2000.

The CG conferences provide an international forum for researchers working on any aspect of computers and games to meet and exchange information on the latest research. CG 2000 was attended by 80 people from over a dozen different countries, thus building on the success of the inaugural Computers and Games conference, held in 1998. The third conference in the series is scheduled to take place alongside the AAAI conference in Edmonton, Alberta, Canada in 2002. The interests of the conference attendees and organizers cover all issues related to game-playing; for instance, the implementation and performance of programs, new theoretical developments in game-related research, general scientific contributions produced by the study of games, social aspects of computer games, cognitive research on how humans play games, and issues related to networked games.

This book contains all the new developments presented at CG 2000. The CG 2000 technical program consisted of 23 presentations of accepted papers and a panel session. In addition there were invited talks by Michael Littman of AT&T Labs, Kei-ichi Tainaka of Shizuoka University, and Nob Yoshigahara, noted inventor, collector, and popularizer of puzzles. The conference was preceded by an informal workshop on October 26, 2000.

The 23 accepted papers were selected by the Program Committee from a total of 44 submissions. Each paper was sent to three referees, who were chosen on the basis of their expert knowledge. A total of 18 papers were clearly acceptable, and 16 papers were not yet ready for publication. Two other submissions were accepted as papers for the panel session (for which an extra three papers were solicited) and the remaining eight works were returned to the authors with the request to address the substantial concerns of one referee, and with the statement that they would be thoroughly reviewed again. Finally, with the help of many referees (see the end of this preface), the Program Committee accepted an additional five of the eight re-reviewed papers for presentation and publication.

The international and varied nature of the papers presented at CG 2000 reflects the diversity of the backgrounds and the many different views brought to "computers and games" (both the conference and the research field) by researchers around the globe. Unfortunately, this diversity also makes it difficult to group the papers neatly into a small number of categories. To provide an ordering for including the papers in this book, we somewhat arbitrarily impose the following structure: Search and Strategies (four papers), Learning and Pattern Acquisition (seven papers), Theory and Complexity Issues (six papers), Further Experiments with Games (six papers), Invited Talks (two papers), and Panel Session (five papers).

Acknowledgements

CG 2000 was made possible by the efforts of many people and organizations. On the following pages, we give a list of names ordered by function and by support.

Special thanks are due in a number of areas. First and foremost, neither CG 2000 nor this volume would have been possible without the generous support of our sponsors. We express our gratitude to: the Japanese Society for the Promotion of Science (JSPS), the Electrotechnical Laboratory (ETL), Shizuoka University, the Hamamatsu Visitors and Convention Bureau, The Shizuoka Organization for Creation of Industries, and the University of Alberta (Computing Science).

Further, the editors gratefully acknowledge the expert assistance of the Program Advisory Committee and the time and expertise given by the referees. We would also like to express our sincere gratitude to Ms. Anne Nield who assisted us during the reviewing phase of the conference.

During the conference itself, the schedule and events organized by the Local Arrangements Committee were much appreciated by all. We thank all the members of the Local Arrangements Committee for the enthusiasm and attention to detail they put into ensuring the smooth running of the conference (looking after 80 researchers and theoreticians for a few days isn't easy...). We especially thank Ms. Akiko Taniguchi and Ms. Karen Ohara for managing the registration desk and Uri & Sandra Globus and Markian Hylnka for organizing the excursions. Thanks are also due to numerous other volunteer helpers, including Zahidul Haque, Tsuyoshi Hashimoto, Atushi Hondoh, Tatsuya Irisawa, Takashi Sugiyama, Tsuyoshi Suzuki, Katsutoshi Usui, and Satoshi Watanabe.

October 2001

Tony Marsland
Ian Frank

Organization

We thank all the following for their support in organizing CG 2000.

Organizing Committee

Conference Chairs:	Hiroyuki Iida (University of Shizuoka, Japan)
	Yoshiyuki Kotani (Tokyo University of Agriculture and Technology, Japan)
	Hitoshi Matsubara (Future University Hokodate, Japan)
	Takenobu Takizawa (Waseda University, Japan)
	Atsushi Yoshikawa (NTT Basic Research Laboratories, Japan)
Program Chairs:	Tony Marsland (University of Alberta, Canada)
	Ian Frank (ETL, Japan)
Workshop:	Hitoshi Matsubara (Future University Hakodate, Japan)

Local Arrangements Committee

Chair:	Hiroyuki Iida (Shizuoka University, Japan)
	Uri Globus (Shizuoka University, Japan)
	Reijer Grimbergen (ETL, Japan)
	Jin Yoshimura (Shizuoka University, Japan)
	Markian Hylnka (University of Alberta, Canada, and Shizuoka University, Japan)
Secretary-Treasurer:	Akiko Taniguchi (Shizuoka University, Japan)

Program Committee

Program Chair: T. Anthony Marsland (University of Alberta, Canada)
Program Co-chair: Ian Frank (ETL, Japan)

Don F. Beal (Queen Mary and Westfield College, UK)
Michael Buro (NEC Institute, USA)
Keh-hsun Chen (UNC Charlotte, USA)
Susan Epstein (Hunter College, USA)
Aviezri Fraenkel (Weizmann Institute, Israel)
Reijer Grimbergen (ETL, Japan)
Ernst Heinz (MIT, USA)
Jaap van den Herik (Universiteit Maastricht,
                        The Netherlands)
Shun-chin Hsu (National Taiwan University, Taiwan)
Takuya Kojima (NTT Basic Research Laboratories,
                        Japan)
Richard Korf (UC Los Angeles, USA)
Shaul Markovitch (The Technion, Israel)
Martin Müller (ETL, Japan)
Monty Newborn (McGill University, Canada)
Jurg Nievergelt (ETH Zurich, Switzerland)
Kohei Noshita (University of Electro-Comm., Japan)
Wim Pijls (Erasmus University Rotterdam,
                        The Netherlands)
Jonathan Schaeffer (University of Alberta, Canada)
Gerald Tesauro (IBM Watson Labs, USA)
Janet Wiles (Queensland University, Australia)

Referees (In addition to those on the Program Committee)

D. Avis	G. Kaminka	M. Tajima
A. Beacham	D. Khemani	T. Takizawa
D. Billings	Y. Kotani	K. Thompson
Y. Björnsson	H. Matsubara	T. Uehara
A. Blair	D. Nau	J. Uiterwijk
B. Bouzy	K. Nakamura	P. Utgoff
T. Cazenave	I. Noda	R. Wu
M. Crasmaru	J. Richardson	X. Yao
J. Culberson	J. Romein	A. Yoshikawa
A. Junghanns	S.J. Smith	

Sponsoring Institutions

Electrotechnical Laboratory (ETL)
Hamamatsu Visitors and Convention Bureau
The Japan Society for the Promotion of Science (JSPS)
Shizuoka University
The Shizuoka Organization for Creation of Industries
University of Alberta (Computing Science)

Cooperative Organizations

Computer Shogi Association (CSA)
Computer Go Forum (CGF)
International Computer Chess Association (ICCA)
IEEE, Tokyo

Table of Contents

Part 4: Further Experiments with Games

Part 5: Invited Talks and Reviews

A Least-Certainty Heuristic for Selective Search

Paul E. Utgoff and Richard P. Cochran

Department of Computer Science
University of Massachusetts
Amherst, MA 01003

{utgoff,cochran}@cs.umass.edu

Abstract. We present a new algorithm for selective search by iterative expansion of leaf nodes. The algorithm reasons with leaf evaluations in a way that leads to high confidence in the choice of move at the root. Performance of the algorithm is measured under a variety of conditions, as compared to minimax with α/β pruning, and to best-first minimax.

Keywords: Selective search, evaluation function error, misordering assumption, certainty, confidence, swing, evaluation goal, swing threshold, LCF, random trees, artificial time, Amazons, Othello.

1 Introduction

It has been recognized for quite some time that some moves in a game are hopelessly suboptimal, and that search effort should not be expended in exploring lines of play that emanate from them. Several methods have been devised to grow the tree selectively, as discussed below. We present a new algorithm of this kind, and investigate its characteristics.

That some positions are obviously worse than others rests on an evaluation mechanism that recognizes apparently debilitating or fatal flaws in the position. It is pointless to investigate the nuances of the poor play that will ensue. Less obvious are those positions in which lesser strengths and weaknesses are evident. Nevertheless, issues of search control remain paramount. It is best to invest effort where it will help the decision process. This calls for a more thorough expansion of lines of play in which the relative advantage of a position is less certain, and a less thorough expansion of lines in which the relative advantage is more certain.

2 Evaluation Function Error

The purpose of search is to obtain information that enhances the decision process at the root. With an evaluation function that is perfectly correlated with the true game value, there is no information to be gleaned from searching. Similarly, with an evaluation function that has no correlation with the game value, search is equally uninformative. Search is useful when it serves to overcome error in an evaluation function that is only imperfectly correlated.

T.A. Marsland and I. Frank (Eds.): CG 2000, LNCS 2063, pp. 1–18, 2001.

An evaluation function assesses present utility by measuring indicators of future merit. When these indicators predict imperfectly, as will typically be the case, search can reduce the effect of prediction errors. For example, one can either estimate the weight of an object (heuristic evaluation), or instead actually weigh it (search). The estimate is imperfectly correlated with the actual weight, and weighing the object obviates the need to use a predicted value.

Because the evaluation function assesses a position imperfectly, it would be best to search those positions for which the evaluation errors are most likely to affect the decision at the root. To the extent that the distribution of errors is known, it is possible to improve the evaluation function itself. The very nature of heuristic evaluation is that the error distribution of the evaluation function cannot be known.

To direct search effort in a useful manner, it is necessary to make at least some weak assumptions about the error distribution. We make three assumptions, all of which are quite common. The first is that the evaluation function is the best available estimator of the game value of a node without searching below it. The second is that the variance in the error of the evaluation function is not so large that it renders comparison of evaluations meaningless. Finally, we assume that the evaluation of a position is on a linear scale, such that the size of the difference in two evaluations is a useful measure of the difference in quality of the two positions.

Minimax search has the strength that its brute force search control is not guided by the evaluations of the states, making the error distribution irrelevant. However, the algorithm pays the price of searching to a uniform depth, wasting search effort on many suboptimal lines of play. Selective search has the weakness that its search control is guided by the state evaluations, making it sensitive to evaluation errors and therefore making it susceptible to being misled. However, the approach can search apparently promising lines to much greater depth by not squandering time on those of less merit. There is a classic tradeoff of being unguided and shallow (risk averse) versus being guided and deep (risk accepting).

Although we do not address evaluation function learning here, it is quite common to use an automatic parameter tuning method such as temporal difference learning [11] to adjust the coefficients of the functional form of the evaluation function. When doing so, the error distribution changes dynamically over time. One needs a decision procedure that is sensitive to the error distribution under these circumstances too.

3   Related Work

There is much appeal to the notion of expending search effort where it has the highest chance of affecting the decision at the root. Numerous approaches have been developed that grow the game tree by iteratively choosing a leaf, expanding it, and revising the minimax values for the extant tree. An alternative is to retain the depth-first search of minimax, but search to a variable depth, such as singular extensions [2] and ProbCut [5]. The discussion here focuses on iterative tree growing.

Berliner's [3] B* algorithm attempts to bound the value of a node from below and above by a pair of admissible heuristic functions, one pessimistic and the other optimistic. The algorithm repeatedly selects a leaf node to expand that is expected to drive the pessimistic value of one node above the optimistic value of all the siblings. Palay [8] improved the algorithm by providing it with explicit reasoning about which child of the root to descend. This was possible by assuming a uniform probability distribution of the range of values in the node's admissible interval of values. More recently Berliner & McConnell [4] simplified the procedure for backing up probability distributions to the root.

McAllester [7] proposed a search algorithm based on conspiracy numbers, further illustrated by Schaeffer [10]. The algorithm counts the number of leaves whose values would need to change by a given amount to affect the value of the root by a given amount. As the number of such leaves rises, the likelihood of all the values changing as needed drops. When the algorithm becomes sufficiently confident that the best node will remain so, search terminates.

Rivest [9] suggested guiding the iterative growth process by a penalty-based heuristic. His algorithm expands the leaf node to which the root value is most sensitive. The edge penalty is a function of the derivative of the min or max operator when represented as a generalized mean. The cost associated with a node is the minimum path cost to a leaf below. Starting at the root, the algorithm can follow the path of least total cost to a leaf, which is then expanded. For Connect-4, the algorithm searches fewer nodes than minimax, but with higher total cost due to overhead.

Proof-number search [1] of Allis & van den Herik is similar in spirit to conspiracy numbers. Instead of exploring various candidate game values, the algorithm temporarily equates all values below a candidate threshold with a loss, and the rest with a win. With just two possible values, the algorithm searches selectively by following a path that requires the fewest leaf values to change in order to change the decision at the root. This search mechanism is contained within an outer binary search on candidate games values, allowing the algorithm to home in on the game value at log cost in candidate game values. The algorithm depends on a variable branching factor.

Korf & Chickering [6] explored best-first minimax search, which always expands the leaf of the principal variation. The algorithm depends on error in the evaluation function, and it also depends on tempo effects to cause its opinion of the principal variation to change as search proceeds. Expansion of a leaf can make it look less attractive because the opponent has had a following opportunity to influence the position in its favor.

These algorithms all endeavor to spend search effort expanding those nodes that are needed to reach the best decision at the root, given the error in the evaluation function. We offer the LCF algorithm, which is based on a different heuristic for selecting the node to expand next. The algorithm is closest in spirit to conspiracy numbers, but discards buckets of supposed target values in favor of a real-valued approach.

4 A Least-Certainty Heuristic

The purpose of search is to establish a high-confidence best choice at the root. How can this top-level goal be realized as a search strategy? Our approach is patterned as a proof by contradiction, but instead of obtaining an absolute guarantee of contradiction, we accumulate as much evidence as possible. Assume that the actual evaluations of the first and second best children at the root misorder them with respect to the unknown true

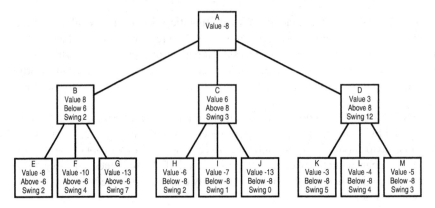

Fig. 1. Example NegaMax Search Tree

evaluations. Under this *misordering assumption*, the evaluations of these two children have at least as much error collectively as the difference in their evaluations. Define *swing* of a node to be the minimum amount of change needed in its actual evaluation to alter the decision at the root. Now the search strategy can be oriented to achieving this minimum change at a child of the root.

We associate a larger swing with a lower probability that the original misordering assumption is correct. To the extent that this probability can be driven near zero, one can reject the misordering assumption with increasing confidence. We do not compute confidence levels explicitly, but instead work directly with swing values. As the amount of minimum swing (needed to change the decision) grows, so too does the confidence that the misordering assumption is false. This approach is motivated in much the same way as conspiracy numbers, but we depart by considering the minimum amount of real-valued swing, not the coarser number of leaf evaluations that would need to change by some fixed amount. When considering the swing for each of the evaluations, one should search a line of play in which the swing is least. Our algorithm for doing this is called LCF because is uses best-first search to investigate the line of least certainty (confidence/swing) first.

Consider the example search tree in Figure 1, which assumes that the value of a node is the negative of the maximum (negamax) of its children. Although the figure shows a fixed depth tree, the search tree will typically vary in depth. At any node, the best move is the one that produces the highest-valued child. Node B, with value 8, is currently the highest-valued child of the root. The second best child is node C, with value 6. To change the move selection at the root, node B must come not to have the highest value. One can attempt to accomplish this in a variety of ways.

One can infer a target value for a node based on a value that, if achieved, would change the decision at the root. The only means of changing a value of a node is to expand a leaf at or below the node such that the backed-up move value at the node will be different. So, in this example, one goal for changing the decision at the root would be to drive the value of node C above 8. Another goal would be to drive the value of node B below 6. Yet another goal would be to drive the value of node D above 8. Generally, one can attempt to drive the best value down below the second best value, or one can attempt

select_move()
    1. allocate time for this move
    2. update_all_swing()
    3. while (time remains and tree incomplete and swing below threshold) grow_tree()
    4. return best move

grow_tree()
    1. descend path of least non-zero swing to leaf
    2. expand leaf
    3. backup negamax value and swing value
    4. if new or revalued first or second child then update_all_swing()

update_all_swing()
    1. for first child, update_swing_below(first,second→lookahead)
    2. for each non-first child, update_swing_above(non_first,first→lookahead)

update_swing_above(node,val)
    1. if node is a leaf then set node→swing to max(0,val-node→lookahead)
    2. otherwise, for each child of node, update_swing_below(child,-val)
        and set node→swing to sum of swings of children

update_swing_below(node,val)
    1. if node is a leaf then set node→swing to max(0,node→lookahead-val)
    2. otherwise, for each child of node, update_swing_above(child,-val)
        and set node→swing to minimum of swings of children

Fig. 2. Move selection for LCF

to drive a non-best value above the best value. There is no need to consider simultaneous goals, such as to drive B down to 6.8 and drive C up to 6.9 because one can achieve the same effect by proceeding one goal at a time, which is necessarily enforced when expanding one leaf at a time.

Given the three goals, one each for each child of the root, which should be pursued first? If these three nodes were leaves, then one could reason that to change the value 3 of node D to its goal of 8 would require a swing of 5 in the valuation. However, to change the value 6 of node C to 8 would require a smaller swing of 2. Similarly, to change the value 8 of node B to 6 would also require only a swing of 2. Assuming that the size of swing is related to the size of the implied error in the evaluations, smaller swings should be easier to achieve than larger swings. Hence, there is more promise in descending node B or C than for node D.

Consider how evaluation goals and swing values are computed when searching deeper than one ply. How are these values managed in the negamax representation? How is a goal to raise an evaluation handled differently from a goal to lower an evaluation?

Suppose that each of node B, node C, and node D have been expanded previously, and that their children evaluate as indicated in the figure. To drive the value of node D up to 8, one would need to drive the value of *every* child below -8. This is because the negamax formulation assigns node values from the point of view of the player on move, and negates the maximum when backing up to the parent. One negates the evaluation subgoals in a corresponding manner. For example, all values need to be driven below

-8, so that the maximum will be below -8, which will push the negated backed-up value above 8.

Given that the goal at each of nodes K, L, and M is to drive its value below -8, one can determine the size of the swing in evaluation needed at each. For node K it is $-3 - (-8) = 5$, for node L it is $-4 - (-8) = 4$, and for node M, it is $-5 - (-8) = 3$. Because all these swings need to occur, the total amount of swing needed below node D is now $5 + 4 + 3 = 12$. The same reasoning applies below node C, giving a total swing of $2 + 1 + 0 = 3$. Notice that node J has value -13, which is already below the goal of -8, so a swing of 0 (none) is needed to meet that goal.

For node B, the goal is to drive its value below 6, which means that the goal for each of nodes E, F, and G is to drive it above -6. For node E, the required swing is $-6 - (-8) = 2$, for node F it is $-6 - (-10) = 4$, and for node G it is $-6 - (-13) = 7$. If any of these goals were achieved, there would be a new maximum value among them, so it is necessary to achieve just one to change the decision at the root. The minimum of the swings is ascribed to the parent node B, giving it a swing of 2. Of course it already had a swing of 2. The example would become more interesting after expanding node E, because the swing needed at E would become the sum of the swings of its children.

The LCF algorithm is shown in Figure 2. It is best-first search, with its metric based on total swing required to achieve the goal of changing the decision at the root. One grows the tree by repeatedly starting at the root, and selecting the node with the least non-zero amount of swing required. Note that a swing of 0 is acceptable for children of the root because it indicates two or more top-rated children. Further below however, a 0 means that there is nothing to achieve by searching below the node. Ties by swing value are broken randomly and uniformly. When a leaf is reached, it is expanded, and the normal backing up of negamax values to the root is performed.

The swing values are updated during the backup to the root. If the child selected at the root retains its value, then all the swing values are correct, and nothing more need be done. However, if the value has changed and the child was or is one of the first or second-best values, then all the goals and swing values have been invalidated. In this case, the tree is traversed with the new pair of best and second best values, updating the goals and swing values. From an efficiency standpoint, one would hope for no change in best/second values at the root, but this would be shortsighted since the objective is to determine the best choice at the root.

Finally, how does one know when to stop searching? Of course one needs to stop when the time allocation runs short. However, when the best node is highly likely to remain the best node, the utility of additional searching diminishes. One measure of diminishing utility is the minimum amount of swing needed to change the decision at the root. As this number grows, the chances of achieving it presumably shrink. Depending on the unit of evaluation, one can set a fixed threshold on total swing that, if exceeded, would cause the search to terminate.

5   Experimental Comparison

How does LCF compare to other known algorithms? In this section, we compare three algorithms in three domains. In addition to LCF, we include best-first minimax (BFMM) and minimax with α/β pruning (AB). The first domain is the class of random-tree games. The second is the game Amazons, and the third is the game Othello.

5.1   Random-Tree Games

The characteristics of a two-person zero-sum perfect-information game can be summarized reasonably well by a small set of parameters. The *branching factor* of the game tree is important because it dictates how quickly the number of states multiplies as a function of search depth. For a fixed amount of search effort, one can search more deeply in a tree that branches less. The length of the game affects the *depth* of the search that is needed to analyze the game tree. For average branching factor b and average depth d, the game tree will contain $O(b^d)$ states. For non-trivial games, one searches a relatively small portion of the game tree when selecting a move. The *total time* available affects how much search effort can be expended during play. More time enables more searching, which improves play in non-pathological games. The *node generation* cost, including node evaluation, also impacts the amount of search that can be accomplished in a fixed amount of time. Finally, the *error distribution* of the evaluation function will mislead search algorithms that are guided by functions of node evaluation.

For our purposes, we vary average branching factor and total time available, holding the others fixed. We follow Korf & Chickering, and Berliner, by using an artificial class of games modeled as random trees. These games can be played at low cost, making a large exploration feasible.

For a random tree, an integer index is associated with each node. For a complete $O(b^d)$ tree, it is straightforward to map each possible state to a unique integer by computing the breadth-order index of the node. The index of each node indexes a random number sequence. This means that a random number is associated with each node, but this indexing method ensures that the mapping of nodes to random numbers is reproducible, no matter what part of the tree is being traversed at any time. The random value associated with each node is the incremental change in the evaluation of that node, called the edge value. The value of the root, which has no edge to a parent, is 0. The heuristic value of a leaf node is the sum of the edge values from that leaf node up to the root.

For our experiments, we follow K&C, using a random number sequence whose period is 2^{32}. Every node index is taken as the modulus of this period length, ensuring that it falls within the bounds of the sequence. The random number that is indexed in this way is further modulated to fall within $[-2^{14}, 2^{14} - 1]$. The purpose of this indexing scheme is to determine a reproducible random edge value in the evaluation. For any particular branching factor b, game depth d, and random number sequence r, one obtains a particular game. Changing any of these parameters defines a new game. For our experiments, we fixed the depth at 64.

There are two attractive aspects of using random trees. The first is that the computational requirements for node generation and evaluation are only slight. Second, the evaluation function is imperfectly correlated with the true game value. For example, the sum of the edge values from the root to a final game state gives the game value of the final position exactly. Backing up one step, the sum of the edge values from the root to this previous state is well correlated with the game value, with the last unspecified incremental change introducing some variability. This models very well the typical situation

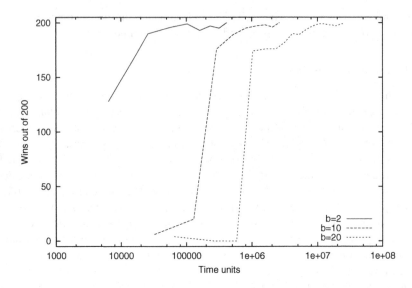

Fig. 3. LCF versus AB

in which an evaluation function is better at evaluating positions near the end of the game than near the beginning.

Because node generation is inexpensive, we simulate this cost in artificial time units. This is a departure from previous studies in which time is measured by the number of nodes generated. By using artificial time units, one can charge for a variety of different operations as deemed appropriate. For most of the experiments here, we charge one (1) time unit for traversing a node pointer for any reason, and 100 time units for generating and evaluating a node. The swing cutoff threshold was not used for these experiments because we wanted to minimize confounding time management with heuristic search control. It would help to stop searching when the choice is clear because that time would be available for a subsequent move.

Figure 3 shows the number of wins for LCF when pitted against AB. Artificial time units were charged identically for the AB algorithm, which used static node ordering. Time management for AB was implemented by computing the maximum depth k that could be searched within the time allotment, assuming that $b^{0.75k}$ nodes will be generated. The 200 games played for each branching factor b and time allotment t, consisted of 100 pairs of games. In each pair, the same random number sequence r was employed, with each player going first in one of the games. The random number sequence was varied from one pair to the next.

In the figure, one can observe a variety of combinations of branching factor and time allotment. The lines connecting these points help visually with grouping, but are only suggestive of results for intervening time allotments. For average branching factor $b = 2$, LCF won all 200 games for all time allotments. For $b = 10$, LCF lost a majority of the games at the smallest time allotment, but won all 200 games for the remaining

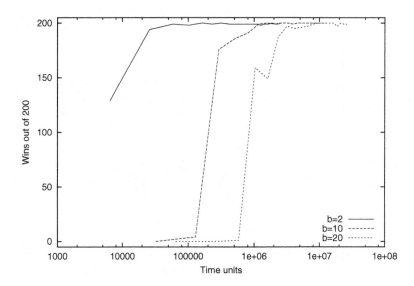

Fig. 4. BFMM versus AB

allotments. Similarly, for $b = 20$, LCF is at a disadvantage only for the two smallest allotments.

K&C's best-first minimax (BFMM) was implemented by modifying the LCF code to expand the principal leaf, and to skip computation of swing values. Figure 4 shows how BFMM fared in the identical experimental setup that was used for the LCF comparison. One can observe the same pattern, that the leaf expander is at a disadvantage for the smaller time allotments, and at an advantage for the larger allotments.

Which of LCF and BFMM is stronger under various conditions? Figure 5 shows a comparison following the same experimental design as before. In this context, as the time allotment is increased, the LCF algorithm becomes weaker than the BFMM algorithm. We note that for smaller allotments, LCF is favored. We ran a version of LCF that did not charge for the overhead of maintaining the swing values, and saw the same fundamental pattern. The explanation rests somewhere within the actual node selection strategy.

Figure 6 shows the search activity of the LCF algorithm for the first move of a particular random tree game. Figure 7 shows the search activity of the BFMM algorithm in the same setting. LCF expands 74 leaves in the time that BFMM expands 88 leaves. This difference is most likely due to the overhead cost of LCF in computing swing values. An attractive property of BFMM is that it has no extra measures to compute for the game tree.

Each figure consists of three columns of information. The middle column depicts information regarding the children of the root. A lower triangle indicates a node with lowest swing value, and an upper triangle indicates a node with highest move value. A box identifies a node that has both lowest swing value and highest move value. A circle denotes a node that is suboptimal in both senses. If the symbol is filled (solid black), then that is the child of the root that was selected for descent to a leaf to be expanded.

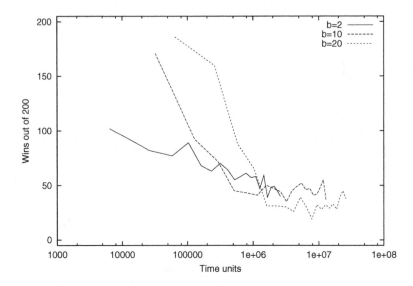

Fig. 5. LCF vs BFMM

Notice that LCF explores several moves, principally $child_0$ and $child_4$. BFMM spends most of its time below $child_4$, though at expansion 48 (not shown), it switches to $child_0$ and sticks with it to the end of the search at expansion 88. LCF also settled fundamentally on $child_0$ at expansion 44.

The lefthand column of the figure shows the proportion of nodes in each subtree of the root after each expansion. The values are connected to improve readability. At each expansion (row in the figure), observe the distance between each of the lines. It is evident that for LCF the growth is mostly below $child_0$ and $child_4$, whereas for BFMM the growth is mostly below $child_4$.

The righthand column shows after each expansion (row) the proportion of nodes at each depth of the tree. Initially, all nodes are at depth 1, just below the root. Subsequently, there are expansions at depth 2, and later at various depths. The proportion of nodes at depth 0 is the leftmost area (between lines) in the column. The BFMM algorithm tree becomes narrower and deeper than the LCF tree during this particular search.

5.2 Amazons

Amazons is played on a 10x10 board. Each player has four amazons, which move like chess queens. To take a turn, the player-on-move selects one of his/her amazons, moves it like a chess queen to an empty square, and from its resting point throws a stone as though it were a second chess queen move. Where the stone lands is permanently blocked. The last player to move wins. The game has elements of piece movement and territory capture.

Three versions of an Amazons program were implemented, an LCF version, a BFMM version, and an AB version. The LCF version generates the children of a node in a

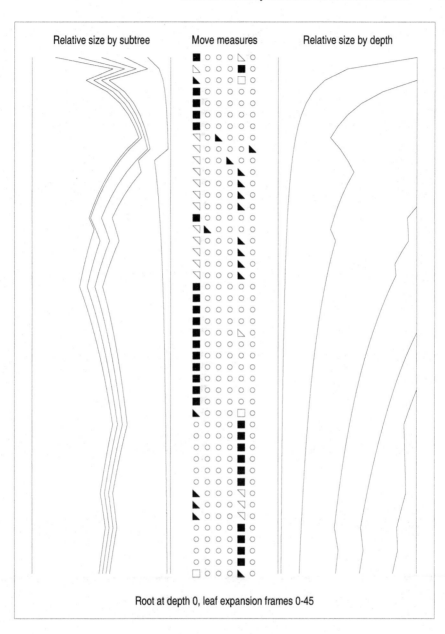

Root at depth 0, leaf expansion frames 0-45

Fig. 6. LCF Expansions for Random Tree

particular way, keeping up to the ten best. The BFMM version is identical except for its search method as described above. Similarly, the AB version differs in just its search method and time management policy. It uses iterative-deepening negamax search with static node ordering. The three programs were pitted against one another in four round-

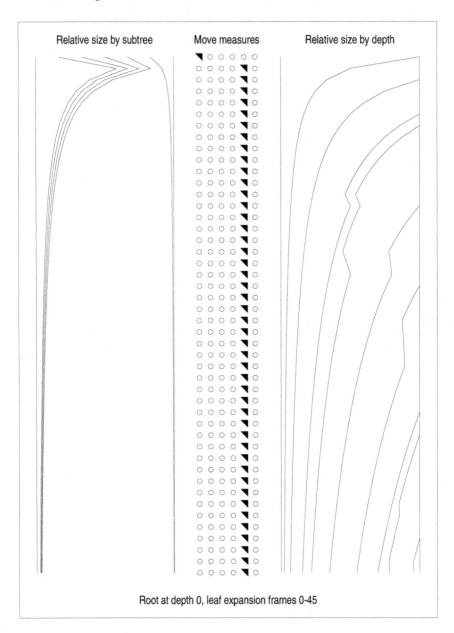

Fig. 7. BFMM Expansions for Random Tree

robin tournaments of various time allocations. The results are summarized in Table 1. AB won a total of ten (10) games, BFMM nine (9) games, and LCF five (5) games. Again, LCF is weaker than BFMM. Remarkably however, AB is not dominated by LCF or BFMM, as it was in the Random-Trees case. Variable branching factor, an issue

Table 1. Amazons Tournaments Results

5 mins		10 mins		20 mins		30 mins		Total	
LCF	0	LCF	0	LCF	0	LCF	2	LCF	2
BFMM	2	BFMM	2	BFMM	2	BFMM	0	BFMM	6
LCF	2	LCF	0	LCF	1	LCF	0	LCF	3
AB	0	AB	2	AB	1	AB	2	AB	5
BFMM	1	BFMM	1	BFMM	0	BFMM	1	BFMM	3
AB	1	AB	1	AB	2	AB	1	AB	5

Table 2. Othello LCF versus BFMM Results

1 min		10 mins		30 mins		60 mins		Total	
LCF	5	LCF	8	LCF	3	LCF	5	LCF	21
BFMM	5	BFMM	2	BFMM	7	BFMM	4	BFMM	18

discussed by Korf & Chickering, is not an issue here because it was uniformly ten during the decisive portion of the game.

Figure 8 shows the search behavior of LCF during the first move of a game of Amazons. Similarly, Figure 9 shows the same information for BFMM. LCF expanded 250 leaves, and BFMM expanded 245 leaves, which is virtually identical. LCF explores principally $child_2$ and $child_3$ throughout this search, settling on $child_3$. BFMM explores a large variety throughout its 245 expansions, settling on $child_0$. These behaviors have 'swapped' in some sense from random trees, where LCF was the more varied of the two.

5.3 Othello

For the game of Othello, we also prepared LCF, BFMM, and AB versions that were identical except for the method of search. At a time allocation of one minute total per player for all moves of the game (very quick game), the LCF version won two of ten games against the AB version, and at all large time allocations lost all games. We did not pit BFMM against AB, but Korf & Chickering report their BFMM Othello program doing poorly against their AB Othello program. Regarding our LCF version against our BFMM version, we ran a large number of games at various time allocations, as summarized in Table 2. The programs appear to be of fundamentally equal strength in this setting. A tenth game at the 60 minute time allocation was a draw, and is not included in the table.

Figure 10 shows the search behavior of LCF during the second move of a game of Othello, as does Figure 11 for BFMM. LCF explores the two openings that are well-regarded and shuns the third. BFMM latches onto one of the openings, exploring it at great depth. Examination of traces of this kind shows that both algorithms conduct searches that are quite narrow and deep.

6   Discussion

It is disappointing that LCF or BFMM do not produce play that is stronger than AB. (It appears that superior performance for random tree games is a special case.) It is

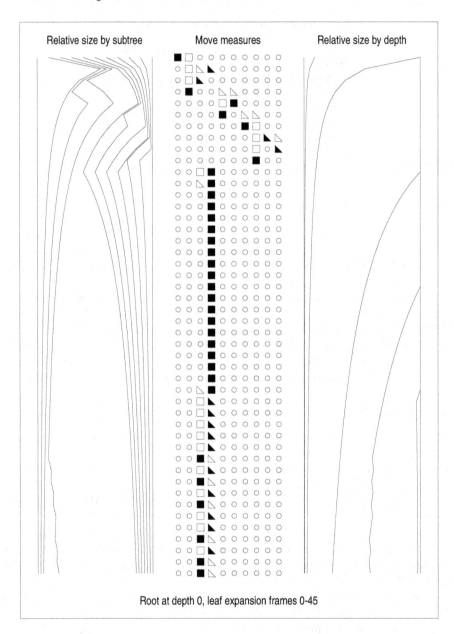

Fig. 8. LCF Expansions for Amazons

naturally appealing to want to search those parts of the tree that are most likely to be informative. However, evaluation functions are imperfect, encoding a variety of strengths and weaknesses that collectively bias position assessment. To use such a function to guide search has a circular dependency. Imperfect value assessment implies imperfect

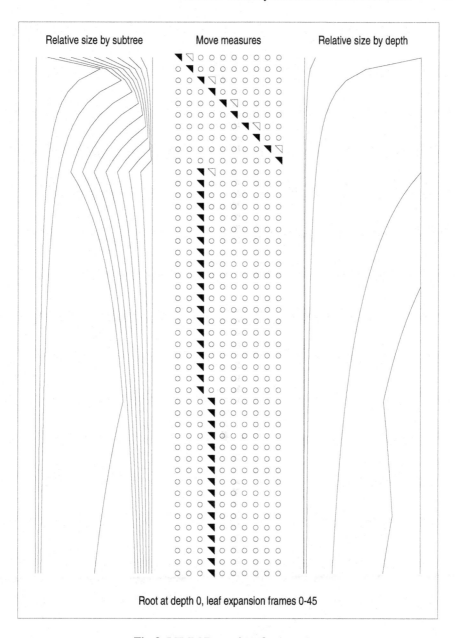

Fig. 9. BFMM Expansions for Amazons

relevance assessment. Searching those nodes that appear to bear on (are relevant to) the choice at the root is subject to the blind spots of the evaluation function, yet that is exactly what the search was intended to overcome.

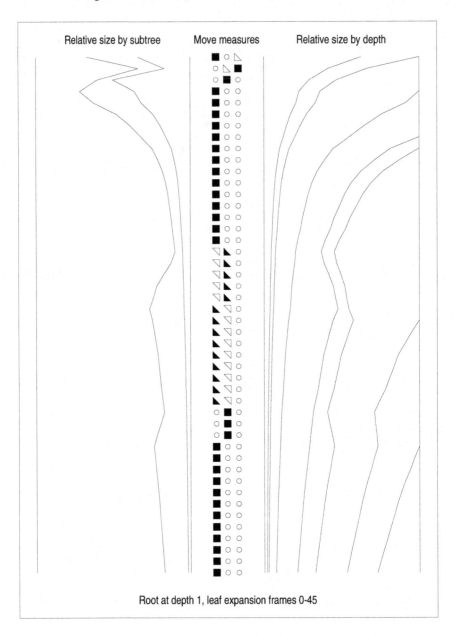

Fig. 10. LCF Expansions for Othello

7   Summary

We have presented a new heuristic for guiding best-first adversary search. The misordering assumption provides a basis for achieving goals in real-valued node evaluation. The

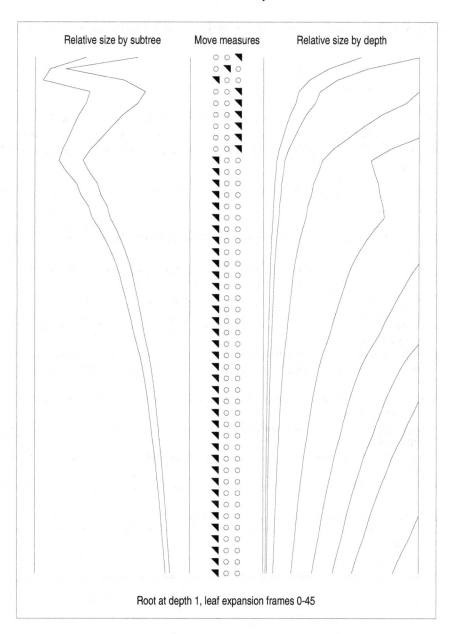

Fig. 11. BFMM Expansions for Othello

notion of swing, and its relation to certainty, provides a new search heuristic. This approach was motivated by the work on conspiracy numbers, with the goal of eliminating the need to compute certainty with respect to an assortment of proposed game values. Although this goal has been achieved, the only evidence we have so far is that a player using the LCF algorithm in a real game (Amazons, Othello) will be weaker than a player using a minimax variant.

Acknowledgments

This material is based on work supported by the National Science Foundation under Grant IRI-9711239. Rich Korf shared his code for efficient indexing of a random number sequence. Gang Ding, David Stracuzzi, and Margaret Connell provided helpful comments.

References

1. L. V. Allis, M. van der Meulen, and H. J. van den Herik. Proof-number search. *Artificial Intelligence, 66*, 91-124, 1994.
2. T. Anantharaman, M. S. Campbell, and F-h. Hsu. Singular extensions: Adding selectivity to brute-force searching. *Artificial Intelligence, 43*, 99-109, 1990.
3. H. Berliner. The B* tree search algorithm: A best-first proof procedure. *Artificial Intelligence, 12*, 23-40, 1979.
4. H. Berliner, and C. McConnell. B* probability based search. *Artificial Intelligence, 86*, 97-156, 1996.
5. M. Buro. Experiment with multi-Probcut and a new high-quality evaluation function for othello. *Proceedings of the Workshop on Game-Tree Search*. Princeton: NEC Research Institute, 1997.
6. R. E. Korf, and D. M. Chickering. Best-first minimax search. *Artificial Intelligence, 84*, 299-337, 1996.
7. D. A. McAllester. Conspiracy numbers for min-max search. *Artificial Intelligence, 35*, 287-310, 1988.
8. A. J. Palay. The B* tree search algorithm - new results. *Artificial Intelligence, 19*, 145-163, 1982.
9. R. Rivest. Game tree searching by min/max approximation. *Artificial Intelligence, 34*, 77-96, 1988.
10. J. Schaeffer. Conspiracy numbers. *Artificial Intelligence, 43*, 67-84, 1990.
11. R. S. Sutton. Learning to predict by the method of temporal differences. *Machine Learning, 3*, 9-44, 1988.

Lambda-Search in Game Trees – with Application to Go

Thomas Thomsen

Stockholmsgade 11, 4th.
DK-2100 Copenhagen, Denmark
thomas@t-t.dk

Abstract. This paper proposes a new method for searching two-valued (binary) game trees in games like chess or Go. Lambda-search uses null-moves together with different orders of threat-sequences (so-called lambda-trees), focusing the search on threats and threat-aversions, but still guaranteeing to find the minimax value (provided that the game-rules allow passing or zugzwang is not a motive). Using negligible working memory in itself, the method seems able to offer a large relative reduction in search space over standard alpha-beta comparable to the relative reduction in search space of alpha-beta over minimax, among other things depending upon how non-uniform the search tree is. Lambda-search is compared to other resembling approaches, such as null-move pruning and proof-number search, and it is explained how the concept and context of different orders of lambda-trees may ease and inspire the implementation of abstract game-specific knowledge. This is illustrated on open-space Go block tactics, distinguishing between different orders of ladders, and offering some possible grounding work regarding an abstract formalization of the concept of relevancy-zones (zones outside of which added stones of any colour cannot change the status of the given problem).

Keywords: binary tree search, threat-sequences, null-moves, proof-number search, abstract game-knowledge, Go block tactics

1   Introduction

λ-search is a general search method for two-valued (binary) goal-search, being inspired by the so-called null-move pruning heuristic known from computer chess, but used in a different and much more well-defined way, operating with direct threats, threats of forced threat-sequences, and so on.

To give a preliminary idea of what λ-search is about, consider the goal of mating in chess. A direct threat on the king is called a *check*, and we call a sequence of checks that ends up mating the king a forced mating check-sequence. However, a *threat* of such a forced mating check-sequence is not necessarily a check-move itself, and such a meta-threat – being of a second order compared to the first-order check moves in the check-sequence – can be a more quiet, enclosing move.[1]

---

[1] Note: for readers not familiar with Go, the appendix contains a chess-example similar to the Go-example given below. In chess, goals other than mating could be, for instance, trapping the opponent's queen, creating a passed pawn, or promoting a pawn.

T.A. Marsland and I. Frank (Eds.): CG 2000, LNCS 2063, pp. 19–38, 2001.
© Springer-Verlag Berlin Heidelberg 2001

Similarly, in Go, a direct threat on a block of connected stones is called an *atari*. A forced atari-sequence resulting in the capture of the block is called a working *ladder* (Japanese: *shicho*). Hence, in this context, a (meta-)threat of a forced threat-sequence would be a ladder-threat, which does not have to be an atari itself. Consider figure 1: Black cannot capture the white stone *a* in a ladder, because of the ladder-block (the other white stone). An example of such a failed attempt is shown in figure 2. Now, providing that black gets a free move (a move that is answered with a pass by white), there are a number of ways in which black can make the ladder work. One of the possibilities could be playing a free move at *c* in figure 3. This threatens the ladder shown in figure 4. Another possibility could be playing at *d* in figure 3, threatening the ladder shown in figure 5.

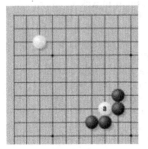

Fig. 1. How to capture a?

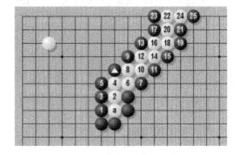

Fig. 2. A failing ladder ($\lambda^1=0$)

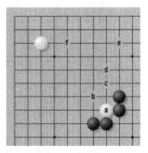

Fig. 3. Black λ^2-moves?

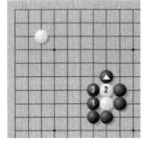

Fig. 4. Working ladder ($\lambda^1=1$) after black *c*

Fig. 5. Working ladder ($\lambda^1=1$) after black *d*

The moves *c* and *d* in figure 3 are thus more indirect threats on the white stone *a*, compared to a direct (first-order) atari, and this possible generalization of the concept of threats is the main theme of the present paper.

To introduce the terminology used in this paper, we denote a direct threat-sequence a λ^1-tree, consisting solely of λ^1-moves (checks/ataris for the attacker and moves averting these for the defender). The λ^1-tree is a standard search tree solvable by means of any minimax search technique, with value 1 (success for the attacker) or 0 (success for the defender).

At the next level of abstraction, a λ^2-move for the attacker is a move that threatens a λ^1-tree with value 1, if the attacker is allowed to move again. Figure 6 depicts all possible black λ^2-moves given unlimited search depth; i.e., all black moves threatening a working ladder (or a direct capture; hence the two atari-moves on white *a* are also λ^2-moves) if left unanswered. If e.g. 1 in figure 7 (= c in figure 3) is chosen, black threatens the ladder depicted in figure 4. Figure 7 then shows all possible white moves averting that threat. In the general λ^2-tree, these two white moves in Figure 7 are hence λ^2-children of black 1, and in figure 8, the two black moves are λ^2-children of white 2 (again reviving the ladder-threat).

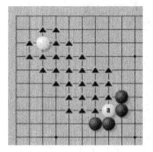

Fig. 6. Black λ^2-moves to kill a

Fig. 7. White λ^2-moves after black 1

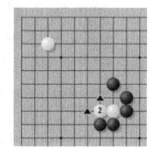

Fig. 8. Black λ^2-moves after white 2

In this way, a λ^2-tree can be built out of λ^2-moves, these being threats of threat-sequences (ladder-threats) for the attacker, and aversions of threat-sequences (ladder-breakers) for the defender.[2] See the appendix, figures A1-A3, for a similar chess-example.

In the context of Go block tactics, a λ^2-tree can be interpreted as a net or loose ladder (Japanese: *geta* or *yurumi shicho*), but the order of λ can easily be > 2. For instance, loose loose ladders (λ^3-trees) are quite common in open-space Go block tactics, the λ^3-trees being built up of λ^3-moves, originating from λ^2-trees. Tsume-go problems (enclosed life/death problems) or semeai (race-to-capture) usually involves high-order λ^n-trees (mostly $n \geq 5$), and connection problems (trying to connect two blocks of stones) often features high λ-orders, too. For example, in figure 9, white a can be killed in a λ^7-tree (the solution is black b, 21 plies/half-moves), whereas in figure 10, black a and b can be connected into one block in a λ^3-tree (the solution is black c or d, 15 plies).[3]

As it is shown in section 3, compared to standard alpha-beta, λ-search can often yield large reductions in the number of positions generated in order to solve a given problem, especially if the search tree is highly non-uniform in a way exploitable by λ-

---

[2] As most readers with a basic knowledge of Go would know, the best black λ^2-move is b in figure 3, after which there are no white λ^2-moves. Another way of stating this is that after black b, there is no way for white to avoid being killed in a subsequent ladder (λ^1-tree), and so white can be declared unconditionally dead.

[3] Note generally that if $\lambda^n = 1$, the attacker has to spend $n + 1$ extra moves (moves that can be answered by a pass/tenuki by the defender) in order to "execute" the goal. Hence, in figure 9, white a can be conceived of as having $n + 1 = 8$ effective liberties, instead of only the 5 real/visible ones. If the black block surrounding white was vulnerable, such knowledge would be crucial in for instance semeai problems (cf. [10]), or regarding the difficult problem of "opening up" tsume-go problems (cf. [12]).

search. Hence, the λ-search methodology could probably prove useful for goal-search in almost any two-player, full-information, deterministic zero-sum game.[4]

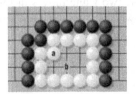

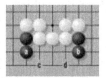

Fig. 9. Black to kill white a (λ^7)   **Fig. 10.** Black to connect a and b (λ^3)

The point about λ-search is that in each λ-tree, all legal moves are not just being blindly generated and ploughed through, but instead the search focuses on those making and averting (more or less direct) threats. This implicit move-ordering directs the search towards its goal, and even though the λ-search methodology does not in itself contain any knowledge of the concrete game-problem analyzed, λ-moves often seem to have at least some minimum of meaning or purpose (cf. figures 6-8, or figures A1-A3 in the appendix), in contrast to the often random-looking move sequences generated by no-knowledge alpha-beta search.

In addition to this implicit threat-based move-ordering, the λ-search approach can often ease and inspire the implementation of abstract game-specific knowledge. For instance, regarding λ^1-moves in chess mating problems, there are only three ways of averting a check (capturing the threatening piece, putting a piece in between, or moving the king). Similarly, regarding λ^1-moves in Go block tactics (ladders), the attacker must always play on one of the defender's liberties, whereas the defender must always play on his remaining liberty, or capture a surrounding block. Section 4 describes an attempt at formulating some abstract/topological Go-tactical knowledge regarding general λ^n-trees, these abstract rules relying both upon the value of the λ-order n, and on information generated during search of lower-order λ-trees.

2   Formalizing the λ-Search Method

In this section, the λ-search method is formalized in its most general form, not tied to the context of any specific game. The λ^n-tree is defined as follows:

Definition 1: A λ^n-tree is a search tree for trying to achieve a single well-defined goal, the search tree consisting solely of λ^n-moves (defined in definition 2), and a λ_a^n-tree being a λ^n-tree where the attacker moves first. The minimax value of a λ^n-tree is either 1 (success for the attacker) or 0 (success for the defender).[5] A node is a leaf

---

[4] Provided that the game-rules allow passing, or zugzwang (move-compulsion) is not always a motive for obtaining goals.

[5] The terms "attacker" and "defender" are arbitrary and could just as well be "left" and "right". For instance, in Go, "attacking" could just as well be trying to connect two blocks of stones, and "defending" trying to keep them disconnected.

(terminal node) if the node has no children because there are no legal λ^n-moves following it. If this is so, the value of the leaf is 1 if the defender is to move, and 0 if the attacker is to move.

The λ^0-tree is particularly simple: $\lambda_a^0 = 1$ if the attacker's goal can be obtained directly by at least one of the attacker's legal moves, and $\lambda_a^0 = 0$ otherwise.

Definition 2: A λ^n-move is a move with the following characteristics. If the *attacker* is to move, it is a move that implies – if the defender passes – that there exists at least one subsequent λ_a^i-tree with value 1, $0 \leq i \leq n-1$. If the *defender* is to move, it is a move that implies that there does not exist any subsequent λ_a^i-tree with value 1, $0 \leq i \leq n-1$.

Example (Go block tactics): Regarding the construction of a λ^2-tree, cf. figure 3. Here, a black play at *c* is a λ^2-move. After *c*, provided that white passes, $\lambda_a^0 = 0$ (the block cannot be captured directly), but $\lambda_a^1 = 1$ (the block can be captured in a ladder, cf. figure 4).

After black plays this λ^2-move at *c,* we have the situation in figure 7. Here, a white move at any of the two marked points is a λ^2-child of the black λ^2-move, since after any of these two white moves, $\lambda_a^0 = \lambda_a^1 = 0$ (black cannot follow up with neither a direct capture nor a working ladder).

The λ-method can be illustrated as in figure 11. In this figure, black *a* is a λ^n-move, because – if white passes – it can be followed up with a λ_a^{n-1}-tree with value 1, cf. the small tree to the left. Considering the white move *b*, this is likewise a λ^n-move, since it is not followed by a λ_a^{n-1}-tree with value 1, cf. the small tree to the right.[6] Other legal black and white moves are pruned off, and hence the λ^n-tree generally has a much smaller average branching factor than the full search tree. Note that in all the three depicted λ-trees the attacker moves first, and note also that the algorithm works recursively in the "horizontal" direction, so that the moves in the small λ^{n-1}-trees to the left and right are in turn generated by λ^{n-2}-trees, and so on downwards, ultimately ending up with λ^0-trees. Note finally that a λ^{n-1}-tree generally contains much fewer nodes than a λ^n-tree. Having made the above definitions, it is time to introduce an important theorem:

Theorem 1 (confidence): If a λ_a^n-tree search returns the value 1 (success for the attacker), this is the minimax value of the position – the attacker's goal can be achieved with absolute confidence.

Proof: For a λ_a^n-tree to return the value 1, the attacker must in each variation be able to force the defender into situations where he has no λ^n-moves (cf. definition 1). The defender not having any λ^n-move means that no matter how the defender plays, the attacker can follow up with a lower-order λ-tree with value 1: $\lambda_a^i = 1, 0 \leq i \leq n-1$. By the same reasoning as before this in turn means that the attacker can force the defender into situations where he has no λ^i-moves. By recursion, this argument can be

---

[6] In fact, small λ_a^{n-2}-trees, λ_a^{n-3}-trees, and so on, should also be included at the left and right (cf. definition 2), but these are omitted here for the sake of clarity.

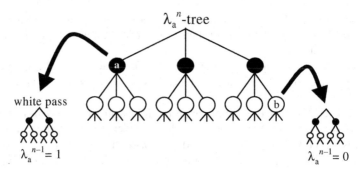

Fig. 11. The λ^n-tree

repeated for lower and lower orders of λ-trees, until we (in at most n steps) end up with a λ^0-tree. And if a λ_a^0-tree has value 1, the goal is obtained directly (cf. definition 1). □

This theorem is crucial for establishing the *reliability* of λ-search. As theorem 1 shows, if a λ_a^n-tree returns the value 1, the attacker's success *is* a proven success no matter how the defender tries to refute it, so if a λ-search finds a win for the attacker, one can have absolute confidence in it.

Theorem 1 is obviously not symmetric. If a λ_a^n-tree search returns the value 0, this does not entail that there could not be a forced attacker's win in the full game-tree. It only means that the defender cannot lose by means of attacker's λ^n-moves, but there might very well be a forced win for the attacker lying in wait in trees of order $n+1$ or higher.

Example (Go block tactics): If black plays a working ladder ($\lambda^1=1$), he is certain to kill the white block: the defender cannot avoid being killed. But the other way round, if a white block cannot be killed in a ladder, this does not, of course, imply that it could not, for instance, be killed in a loose ladder.

Seen in the light of the above, given some specific goal for the attacker to achieve, and provided that the attacker cannot achieve the goal directly within the first move ($\lambda_a^0 = 0$), the strategy will be to first try out a λ_a^1-tree. If $\lambda_a^1 = 1$, we are finished, cf. theorem 1. If $\lambda_a^1 = 0$, we continue with a λ_a^2-search, and so on until time runs out. If we end up with $\lambda_a^n = 0$ for some value of n, we cannot say anything authoritative on the status of the problem, other than that the defender survives a λ_a^n-attack.

The next question is what happens asymptotically as λ^n-trees of higher and higher order get searched? Provided that we know a priori that the attacker's goal *can* be reached within at most d plies by optimal play (d being an odd number), can we be sure that the value of λ_a^n will converge to 1 for some n, as n tends towards infinity?

If the game-rules allow passing, the answer is yes, and in that case it is possible to prove that $n \leq (d-1)/2$, as shown in theorem 2 below. First a lemma:

Lemma 1. Consider a full search tree consisting of all legal moves for both players. In this search tree, consider a node x where the attacker is on the move. Assume that we are not considering the root node (move #1), but some attacker's move deeper down the tree (i.e., some odd move number ≥ 3). Consider the sub-tree unfolding from node x. For the attacker to be able to force the defender into a λ_a^n-tree with value 1 expanding from node x, the attacker must have been *threatening* $\lambda_a^n = 1$ when he was on the move last time. Had he not been threatening $\lambda_a^n = 1$, the defender could have passed, after which (cf. definition 2) the value of a subsequent λ_a^n-tree would have been 0.

Example (Go block tactics): in order to be able to force the defender into a working ladder ($\lambda_a^1 = 1$), the preceding attacker's move must have been a ladder-threat. If this had not been so, the defender could have passed, and the ladder would still not work.

Theorem 2 (convergence). Consider again the full search tree consisting of all legal moves for both players. Assume that we know a priori that there exists a forced win for the attacker in at most d plies, d being some odd number ≥ 1. Consider a node y at depth $d - 1$ (with the attacker to move) after which the forced win is executed at depth d. This means that the tree expanding from y is a λ_a^0-tree with value 1. Using Lemma 1, this again means that the attacker must have been *threatening* $\lambda_a^0 = 1$ when he played last time (i.e., played a move at depth $d - 2$). Threatening $\lambda_a^0 = 1$ is the same as playing a λ^1-move; hence the attacker must have played a λ^1-move at depth $d - 2$.

In order for the attacker to be able to force the defender into this λ^1-move at depth $d - 2$, it must have been part of a λ_a^1-tree with value 1. The first attacker's move in this tree can be at any depth d_1, where $1 \leq d_1 \leq d - 2$. If $d_1 = 1$, we are finished (hence establishing $n = 1$). Else, if $3 \leq d_1 \leq d - 2$, in order to force the defender into this λ_a^1-tree with value 1, the attacker must have been threatening $\lambda_a^1 = 1$ (= playing a λ^2-move) at some depth d_2, where $1 \leq d_2 \leq d_1 - 2$.

This goes on recursively for λ-trees of higher and higher order (each time moving at least two plies upwards in the full search tree), so in the worst case we end up with a tree of order $n = (d-1)/2$. This reasoning can be repeated for all nodes y at depth $d - 1$ after which the forced win is executed at depth d, and hence it is proved that $n \leq (d-1)/2$. $\square$

Example (Go block tactics): Assume that we know a priori that the attacker can kill the defender in at most 5 plies. The attacker's move #5 must have been the capture (taking the stones off the board), so the attacker's move #3 must have been an atari (λ^1-move) – otherwise the defender could have passed, and the stones could not have been taken off the board at move #5. Regarding the attacker's move #1, it can have been a(nother) λ^1-move if the whole killing sequence is a working ladder (λ^1-tree with value 1). Otherwise, the first attacker's move could have been a λ^2-move (loose ladder move), threatening $\lambda_a^1 = 1$ (an ordinary ladder). But if the attacker's move #1 had not been either a λ^1-move or a λ^2-move, the defender could have passed at move #2, and there would have been no way for the attacker to establish a working ladder (λ^1-tree) at move #3, and hence no way to kill the defender at move #5. Thus, if $d = 5$, the λ-search can only be of order 1 or 2, meaning that if $d = 5$ and a λ^2-search returns value 0, the problem cannot be solved in 5 plies. This corresponds to the inequality, since the formula states that $n \leq (d-1)/2 = 2$.

Considering games like chess where passing is not allowed, theorem 2 does not hold, since the ability to pass is a prerequisite for Lemma 1. However, apart from zugzwang-motives (e.g. in late chess endgames), this is not a problem if the defender has always at least some harmless/non-suicidal move to play instead of a pass move. Hence, for practical purposes, theorem 2 also applies to goals such as e.g. mating in middle-game chess positions. And it should be noted that theorem 1 applies whether passing is allowed or not.

Thus, theorems 1 and 2 show that λ-search is not to be interpreted as some heuristic forward-pruning technique, since a λ-search – in the worst case searched to λ-order $n = (d-1)/2$ – returns the minimax value, provided that passing is allowed or zugzwang is not a motive. Some pseudo-code containing the λ-search methodology can be found in [11].

3   Comparing λ-Search with Other Search Methods

Having offered the proofs of confidence and convergence, the question is how effective the λ-search method really is compared to other techniques, such as, e.g., standard alpha-beta, alpha-beta with null-move-pruning, or proof-number search?

3.1   Standard Alpha-Beta

Given some simplifying assumptions it is possible to indicate the effectiveness of λ-search compared to standard alpha-beta. Below, it is shown that this effectiveness generally depends on the branching factor of the λ-search trees compared to the branching factor of the full search tree, combined with the λ-order n.

Consider a general λ-search tree of some order n, with the attacker moving first, and assume that an attacker's win exists in at most d plies (d being uneven). Also assume that passing is allowed or zugzwang is not a motive. We will now try to count the total number of positions generated in order to search such a λ^n-tree, including the positions generated in the lower-order λ-trees. It is assumed that minimax is used in both the λ-tree, and as the standard reference algorithm, for reasons to be explained.

Assume furthermore that B is the average branching factor of the full search tree (= the number of legal moves in each position: in chess, around 30-40, and in Go around 200-250). Assume furthermore that all λ-trees have the same average branching factor, b (b being $\leq B$). In order to generate the attacker's moves at depth 1 in the λ^n-tree, all legal attacker's moves must be tried out, to see whether or not they can be followed by some λ^i-tree ($i \leq n-1$) with value 1 (cf. a in figure 11). This amounts to Bn λ^i-tree searches (the λ^i-trees being of order $i = 0, 1, \ldots, n-1$). And at depth 2 in the λ^n-tree, for each of the defender's λ^n-nodes, all legal defender's moves must be tried out, to see whether or not they are followed by λ^i-trees ($i \leq n-1$) with value 0 (cf. b in figure 11). This amounts to a total of bBn λ^i-tree searches (the λ^i-trees being of order $i = 0, 1, \ldots, n-1$).

Hence, in order to generate all λ^n-moves at depths 1 and 2, corresponding to the three black moves and the nine white answers in the λ^n-tree in figure 11, $(1+b)Bn$ number of λ^i-trees must be analyzed (the λ^i-trees being of order $i = 0, 1, \ldots, n-1$), all

```
lambdaMovesGenerated(n,d) {
  if(n == 0)return B;
  sum = 0;
  for(i=2; i<=(d-1)-2*(n-1); i=i+2) {
    for(j=0; j<=n-1; j=j+1) {
      sum = sum + (b**(i-2)+b**(i-1))
                  *B*lambdaMovesGenerated(j,d-i);
    }
  }
  return sum;
}
```

Fig. 12. Pseudo-code for counting the total number of positions generated by a λ^n-search to depth d.

of them to depth d–2. By a similar argument, the λ^n-moves at depths 3 and 4 in the λ^n-tree require that $(b^2+b^3)Bn$ number of λ^i-trees (the λ^i-trees being of order $i = 0, 1, \ldots ,$ n–1) are searched to depth d–4, and so on.

Now the question is how deeply the λ^n-tree needs to be searched? This depends on circumstances, and the depth will generally be between 2 and $(d–1) – 2(n–1)$ plies. To simplify, we assume the worst case, namely that the λ^n-tree needs to be searched to its maximum depth, i.e., $(d–1) – 2(n–1)$ plies.

The above reasoning leads to the recursive algorithm shown in figure 12 – stated in pseudo-code – for the total number of positions generated in order to search a λ^n-tree with search depth d.

The recursion ends with the number of generated moves in the λ^0-tree, since the algorithm knows that a λ^0-tree demands B positions to be evaluated. For given values of B and b, the above-mentioned recursive formula is thus capable of estimating the total number of positions generated in a λ^n-search to depth d, where all the λ^n-trees are searched by means of minimax. This estimate can be compared to $(B^{d+1}–1)/ (B–1)$, being the estimated total number of positions generated in a minimax-search. Thus, for given B, b, n and d, we define the reduction factor as follows: $r = [(B^{d+1}–1)/(B–1)] /$ lambdaMovesGenerated(n, d).

The reader might now ask why alpha-beta is not used instead of minimax? This is only because assuming minimax makes the reduction factor easier to compute – the point being that using alpha-beta (or any other search technique) instead of minimax would not alter the reduction factor, assuming that alpha-beta (or any other search technique) works equally efficiently in the λ-trees and in the standard (full) search tree. The efficiency of alpha-beta-pruning thus cancels out in the numerator and denominator of the reduction factor, so that the reduction factor can also be interpreted as the efficiency of the λ-search technique (using standard alpha-beta in the λ-trees) over standard alpha-beta.

To provide an idea of the magnitude of the reduction factor, we show two tables below: namely $(b, B) = (4, 40)$ and $(b, B) = (8, 40)$, calculated by means of the algorithm in figure 12.

Table 1. Reduction factor, λ-search relative to standard alpha-beta, $(b, B) = (4, 40)$

	$n=1$	$n=2$	$n=3$	$n=4$	$n=5$
$d=3$	8				
$d=5$	772	65			
$d=7$	76.942	3.174	482		
$d=9$	7.692.425	209.432	16.192	3.328	
$d=11$	769.231.503	15.625.505	812.550	88.201	21.791

Table 2. Reduction factor, λ-search relative to standard alpha-beta, $(b, B) = (8, 40)$

	$n=1$	$n=2$	$n=3$	$n=4$	$n=5$
$d=3$	4,6				
$d=5$	112	20			
$d=7$	2.804	251	76		
$d=9$	70.112	4.173	666	254	
$d=11$	1.752.804	78.158	8.530	1.810	781

In order to have something to compare with, the reduction factor of standard alpha-beta relative to minimax is shown in the table below.[7]

Table 3. Reduction factor, standard alpha-beta relative to minimax, $B = 40$

	good move order	average move-order	bad move-order
$d=3$	15	3,9	1,8
$d=5$	60	7,8	2,4
$d=7$	238	15,6	3,1
$d=9$	952	31,2	4,2
$d=11$	3.810	62,4	5,6

Tables 1 and 2 show that the largest reductions are found for $n = 1$, i.e., if the problem can be solved by means of a number of *direct* attacker's threats (chess mating: checks – Go block tactics: ataris). For growing depth, this renders a huge reduction factor relative to standard alpha-beta, both if the threats and threat-aversions on average make up 10% of the legal moves (table 1), or 20% (table 2). This corresponds to the full search tree being highly non-uniform in a way exploitable by λ-search. But even for more "saturated" problems; i.e., problems needing a larger value of n, the reduction factor is still impressive. For instance, with b/B being 20% as in table 2, a λ^2-search to a depth of 5 plies would still – on average – require only about 1/20 of the positions generated with standard alpha-beta to depth 5, this

---

[7] Note: In the calculation of these reduction factors it is assumed that the best attacker's move is found after 10 tries (good), 20 tries (average), or 30 tries (bad). So for instance, for $t = 3$, the reduction for good move-ordering is calculated as follows: $(1+40+40^2+40^3)/(1+10+10\cdot40+10\cdot40\cdot10)$. It should also be noted that the move-ordering can often be improved upon in a number of ways, but here we focus on *standard* alpha-beta only.

reduction factor being comparable to the reduction factor of good or average move-ordering alpha-beta over minimax at depth 5 (cf. table 3).[8]

As it is seen, the relative reduction in search space by using λ-search instead of standard alpha-beta is comparable to the relative reduction in search space by using alpha-beta instead of minimax, especially for low saturation problems with $n < (d-1)/2$. It is possible to prove that – for large B and b, and for n relatively small compared to $(d-1)/2$ – the reduction factor, r, can be approximated by:

$$ r \approx \frac{\left(\dfrac{B}{b}\right)^{d-n-1}}{\left(\begin{array}{c}(d-3)/2\\n-1\end{array}\right)}, \qquad \left(\begin{array}{c}x\\y\end{array}\right) = \frac{x!}{y!\,(x-y)!} \ . \tag{1} $$

In that case, the reduction factor depends on the size of B compared to b in the $(d-n-1)$'th power, divided by a binomial coefficient. A possible interpretation: Assume that we had some divine knowledge, knowing in each position of the full search tree which b out of the total of B legal moves contain the best move (thus in each position having only to consider b instead of B moves). In that case, the reduction factor would be $r = [(B^{d+1}-1)/(B-1)] \,/\, [(b^{d+1}-1)/(b-1)] \approx (B/b)^d$. However, since we do not have access to such divine knowledge, lower-order λ-trees have to be searched in order to find the b interesting moves (this being more costly for higher n), explaining the binomial coefficient and the fact that the exponent is $(d-n-1)$ rather than just d.

From the formula it is seen that the crucial factor for the reduction factor is the *relative* size of b in relation to B. Thus, we would expect similar reduction factors (tables like table 1 and 2) for $(b, B) = (8,40)$ and, e.g., $(b, B) = (50,250)$, whereas the reduction factors for, e.g., $(b, B) = (8,250)$ would be much larger than for $(b, B) = (8,40)$. It should, however, be emphasized that all the above calculations must be taken with a large pinch of salt, since the calculations just yield some theoretical and not empirical indications of the strength of λ-search over standard alpha-beta. A more authoritative estimate of the efficiency of λ-search would imply analyzing a large number of realistic game-problems with different search techniques, including λ-search, thus being able to offer some real-world statistics on the relative merits of the different approaches.

3.2   Null-Move Pruning

Alpha-beta augmented with null-move pruning (see e.g. [5] for an overview) might seem similar to λ-search, but even though the underlying idea is quite similar, there are a number of important differences – apart from the fact that alpha-beta augmented

---

[8] This could, for instance, be a model of those mate-in-two problems in chess where the first attacker's move is not a check (whereas the next is the checkmate). In that case – a λ^2-tree searched to depth 5 – a reduction factor of about 20 would be expected. Alternatively, if the first attacker's move is a check (and the next the checkmate) – a λ^1-tree – the reduction factor would be expected to be about 112.

with null-move pruning can be directly applied to non-binary search trees (in contrast to λ-search):[9]

First, in λ-search, only the defender makes null-moves, and λ-search does not use depth-reduction when searching the lower-order λ-trees.

Second, a λ^n-search controls the total number of admissible defender's null-moves in any branch between some node and the root, this number always being $\leq n$. Knowing the concrete λ-order of a problem (i.e., distinguishing clearly between different orders of threats) often eases the implementation of abstract game-specific knowledge.

Third, because null-move pruning operates with a depth reduction factor for searching the sub-tree following a null-move, an erroneous result might occasionally be returned. Null-move pruning with depth-reduction works extremely well in many applications (for instance computer chess), but it cannot offer the same absolute reliability as λ-search (cf. the theorems of confidence and convergence).

Finally, λ-search is not tied to any specific search-technique for searching the λ-trees, thus opening up the possibility of easily combining null-moves with for instance proof number search.

3.3   Proof-Number Search

A recent and very popular search technique for searching two-valued search trees is the so-called proof-number search (see e.g. [1] and [2]). Like λ-search, proof-number search exploits non-uniformity of the search tree, trying to search the thinner parts of the search tree first. Still, the differences between the two approaches should be pointed out – apart from the fact that proof-number search has no problems with zugzwang-motives (in contrast to λ-search):

First, proof-number search uses a large working memory overhead, since the whole (or at least much of the) search tree needs to be stored in memory. In contrast, λ-search using, for instance, standard alpha-beta (without transposition tables etc.) as the "search-engine" has negligible working memory requirements.

Second, it is not an easy task to incorporate transposition tables into proof-number search. In contrast, if alpha-beta is used to search the λ-trees, transposition tables can be used with as little difficulty as in standard alpha-beta.

Third, λ-search depends on some technique for searching the λ-trees. But the concrete "search-engine" could be anything as long as it works – including proof-number search. Hence, proof-number search could be combined with λ-search with very little effort (thus combining null-moves and proof-numbers), and even if some new power-

---

[9] The reader might ask whether λ-search could somehow be used for non-binary search trees? A possibility would be to define the goal as follows: If a leaf-node has evaluation value larger than some threshold, the value of the node is 1, otherwise 0. This way, λ-search could be used in the same way as a null-window alpha-beta-search. However, a general non-binary search tree is usually very uniform (unless, for instance in chess, a hidden mate exists), implying $n = (d-1)/2$ and hence a limited reduction factor (full "saturation"; cf. the diagonals of tables 1 and 2).

ful tree-searching technique should be invented in the future, λ-search would most likely benefit from it as well.

Finally, and very importantly, as mentioned in the comparison with null-move pruning, the possibility of distinguishing clearly between different orders of λ^n-trees – i.e., in each position knowing the context of the value of n (and in addition: having access to information generated in lower-order λ-trees) – often makes it easier to implement abstract game-specific knowledge. Additionally, the distinction between different orders of λ-trees also makes it possible to concentrate time- and/or memory-expensive techniques on the highest-order λ-trees. Time-expensive techniques could e.g. be elaborate move-ordering schemes or pattern-matching, and memory-expensive techniques could be the use of transposition tables, or the use of proof-number search.

4   Implementing Abstract Game-Knowledge (Go Block Tactics)

4.1   The Relevancy-Zone

One of the hardest problems in Go is how to limit the search in open-space Go block tactics by means of characterizing some a priori zone of *relevance*; that is, some area of the board outside of which the addition of any number of black or white stones in any configuration cannot alter the status of the problem. For instance, white *a* in figure 3 cannot be caught in a normal ladder, an example of which is shown in figure 2. Adding a free black stone at *e* in figure 3 will not alter the status of the ladder (still not working), whereas a free black stone at e.g. *f* will make the ladder work. Actually, adding a free black stone at any of the marked points in figure 6 makes the ladder work; hence each of these points is a possible λ^2-move (loose ladder move) for the attacker.

However, it does not take long looking at figures 2 and 6, before one realizes that the true relevancy-zone seen in figure 6 should be somehow derivable from or dependent upon the "shadow" of the stones played in figure 2. So the idea is that the true relevancy-zone, called the R-zone, is situated inside another zone, called the R*-zone, consisting of all stones (= "shadow stones") played to prove that a lower-order λ-search (an ordinary ladder) cannot kill the white stone in figure 3. If all the moves of these failing ladders (of which one of them is shown in figure 2) are recorded (circles), and all points adjacent to the circles (liberties of the shadow stones) are added as well (squares), the result is seen in figure 13:[10]

Definition 3. A *shadow stone* corresponding to a λ^n-tree is a stone that is played in this λ^n-tree, or in one of the lower-order λ-trees called from the λ^n-tree.

Comparing figures 6 and 13, we see that the R-zone of figure 6 is contained within the R*-zone of figure 13. This seems simple enough, reducing the points considered for finding the attacker's λ^2-moves from 355 to 36. Using this methodology, a point

---

[10] Note that in this and the following examples, there is no limit on the maximal search depth. With limited search depth, the relevancy zones would be smaller. Also, note in figure 13 that the liberties of the white ladder-breaker are added (more on this in section 4.2 on inversions).

around *e* in figure 3 *will* never be considered as a λ^2-move, whereas a point such as *f* would. There is a problem, however, namely what to do with so-called *inversions*?

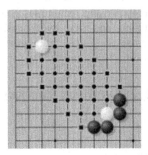

Fig. 13. Relevancy-zone for black?

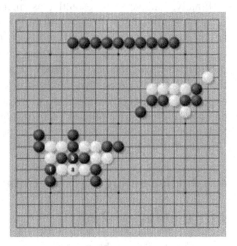

Fig. 14. Is white *a* dead after black 1?

4.2   Inversions

In Go block tactics, the attacker tries to confine the defender, whereas the defender tries to break loose or make two eyes. The defender can break loose by either moving/extending out, or trying to capture some of the attacker's surrounding stones. We denote the defender's trying to catch some attacker's stones an *inversion*, since the roles are switched.

To yield an example, we consider the defender disturbing/averting the threat of a loose ladder (λ^2-tree) by means of an (inverted) ordinary ladder (λ^1-tree). An example is shown in figure 14.

Black has just played 1, a λ^3-move threatening to kill white *a* in a λ^2-tree (loose ladder). Black 1 is a standard tesuji for capturing a block like *a,* but before dooming *a* dead, we need to consider the fact that the surrounding black block *b* is vulnerable due to a limited number of liberties. Hence, the question is: what are the possible white λ^3-moves following black 1?

In order to answer this, we play out the threatened λ^2-tree, killing white. This is shown in figure 15.[11] These moves can also be found as circles in figure 16.[12] Since the black block *b1* has three liberties only, these liberties are also added to the R˙-

---

[11] The sequence in figure 15 contains both λ^2-, λ^1- and λ^0-moves (black 1 is a λ^2-move, black 3 is a λ^1-move, and black 5 is a λ^0-move), since all moves in any lower-order λ-tree are "recorded" to yield the shadow stones.

[12] The reason why there are 8 circled moves in figure 16 compared to the 5 shown moves in figure 15 is that the shadow stones in figure 16 (circled moves) also contain white moves played at the three extra points. These moves are (non-working) white escape-attempts.

zone. The general rule for doing this makes use of the concept of quasi-liberties, defined below.

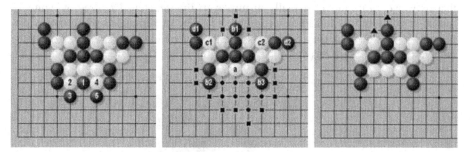

Fig. 15. Killing white in a λ^2 **Fig. 16.** Relevancy-zone   **Fig. 17.** White λ^3-moves
                                                    for white

Definition 4. A *quasi-liberty* corresponding to a λ^n-tree is a liberty that is not coincident with a shadow stone corresponding to the λ^n-tree. (The number of quasi-liberties will always be $\leq$ the number of real liberties)

Example: In figure 16, the block $b1$ has 3 quasi-liberties (coinciding with its real liberties since none of the points surrounding $b1$ are circles (shadow stones)). The blocks $b2$ and $b3$ have 3 and 2 quasi-liberties, respectively.

Using the concept of quasi-liberties, the inversion rule can be stated:

Inversion rule: Consider a λ_a^n-tree with value 1 or 0. Call the defender's block a 0-surrounding block. Now, repeat the following for $m = 1, 2, 3, \ldots , \infty$: Find all blocks of opposite colour touching the already found $(m-1)$-surrounding block(s). Of these new potential m-surrounding blocks, only keep those that have not been found already for smaller m, and for which the following is observed: $q < n - m + 3$, where q is the number of quasi-liberties of the block.

Given this, the R^*-zone corresponding to the λ_a^n-tree can now be defined. First, take the shadow stones and points adjacent to the shadow stones. Next, if $\lambda_a^n = 1$, all quasi-liberties of all surrounding blocks from the above list of the *same* colour as the attacker are added to the R^*-zone. Conversely, if $\lambda_a^n = 0$, all quasi-liberties of all surrounding blocks of the *opposite* colour of the attacker are added to the R^*-zone.

Example: All attacker's blocks touching the defender's block and obeying the equality $q < n - m + 3$ are called 1-surrounding. In the example, we are analyzing the relevancy zone of a λ_a^2-tree with value 1 (in order to find λ^3-moves), and hence $n = 2$. Thus, in order for black blocks to be 1-surrounding, they must have fewer than $n - m + 3 = 2 - 1 + 3 = 4$ quasi-liberties. In figure 16, it is seen that $b1$ has 3 quasi-liberties, whereas $b2$ and $b3$ have 3 and 2 quasi-liberties, respectively. Hence, both $b1, b2$ and $b3$ are 1-surrouding blocks.

The blocks $b1, b2$ and $b3$ touch three white blocks, a already being in the list, and $c1$ and $c2$ being new. In order for a white block to be 2-surrounding, it must have less than $n - m + 3 = 2 - 2 + 3 = 3$ quasi-liberties. Since $c1$ and $c2$ have 3 and 4 quasi-liberties, respectively, these blocks are not 2-surrounding. As there are no 2-surrounding blocks, the search for m-surrounding blocks is terminated, not worrying about the potential 3-surrounding blocks *d1* and *d2* (even if they had had less than 2 quasi-liberties).

The true R-zone corresponding to the λ_a^2-tree is shown in figure 17, where it is seen that only two white moves actually disturb/avert the threatened λ_a^2-tree of figure 15 (all other white moves fail to disturb figure 15). The two triangled white moves disturb the λ_a^2-tree because they render possible the inverted ladder shown in figure 19 – an inverted ladder black has no way of escaping.

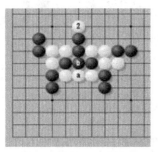

Fig. 18. Does white 2 save white *a*?

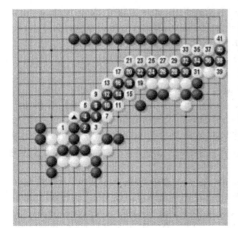

Fig. 19. The threatening inverted ladder

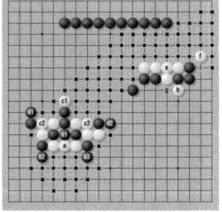

Fig. 20. Zone made by the inversion rule

Now consider figure 18, where white has just played the λ^3-move white 2, disturbing the threatened λ_a^2-tree of figure 15. To find candidates for the next black λ^3-move, the now failing λ_a^2-tree is played out, and all those λ^2-moves (and the λ^1- and λ^0-moves generating those λ^2-moves) are recorded. The shadow of these stones is shown as circles in figure 20.

Using the inversion rule, the relevancy zone would look like figure 20. This seems to be a safe bet for a R^*-zone containing the true R-zone of black disturbance-moves (black λ^3-moves following white 2 in figure 18). However, the inversion rule is not 100% safe, which is seen by the fact that a black stone at *g* is actually a disturbance threatening $\lambda_a^2 = 1$. This is so, because after black g, the inverted ladder shown in figure 19 no longer works – and the example is constructed in such a way that after black g there is no other ladder that works for white. Thus, black g should in principle be counted among the possible black λ^3-moves in this position, and thus be part of the R^*-zone.

Obviously, g is not the best move for black to play in order to fend off the threatened ladder of figure 19, but we are interested in ensuring that *all* possible black λ^3-moves are generated, leaving it up to some successive forward-pruning heuristic to cut some of them off afterwards, if they can safely be judged inferior to other black moves. But at least we should know that moves like black g exist.

A much better black λ^3-move would e.g. be a move at 1 in figure 19. After such a move, white is close to being dead, since white will soon run out of (inverted) ladder-threats on the black block $b1$. But the interesting thing about a black move at g is that it encloses the territory below g at the same time as threatening to kill a. Hence, black g could be used as a forcing territory-enclosing move or as a ko threat.[13]

The question is how the R^*-zone catches a move like black g? The answer is that the white block e is actually surrounding the black block $b1$ in much the same way as $c1$, $c2$ and $c3$, even though it is, of course, much more indirectly. It *is* possible to formalize this with help of additional concepts of so-called quasi-surroundedness (white e quasi-surrounding $b1$) and so-called shadow blocks, but space does not permit going further into this here.[14]

Apart from the remaining problems of the inversion rule, however, this rule at least seems to contain some of the right abstract concepts for the construction of reliable relevancy-zones, even though it may still overlook a few exotic moves. Such exotic moves are typically easy for the opponent to fend off and usually not relevant to the solution of the problem.[15]

More work needs to be done in this field, and until some algorithm for the construction of R^*-zones can be proved to contain the true R-zones in all cases of Go block tactics, an algorithm like the proposed inversion rule would just have to be seen as providing some useful foundations for the implementation of abstract topological knowledge regarding open-space Go block tactics. Perhaps some kind of automated theorem-proving could be of use here; cf. the interesting approach in [4]. And for another example illustrating the inversion rule, the reader is referred to [11].

5   Conclusions and Scope for Further Work

To conclude briefly, the λ-search method seems to be very well suited for obtaining well-defined goals in two-player board games like chess or Go, provided that passing

---

[13] Interestingly, a white λ^3-answer to a possible black λ^3-move at g in figure 20 could be a move directly to the left of or below g (giving an atari on the black stone at g). This would be yet another inversion.

[14] With these additional concepts it is also possible to justify the inclusion of the liberties of the white ladder-breaker in figure 13 in the R^*-zone.

[15] As noted before, such moves could still be relevant as forcing moves or ko threats. Also, in the context of finding double threats, such moves can be highly interesting.

is allowed or zugzwang is not a motive. As shown in section 2, λ-search offers the theorems of confidence and convergence, and the algorithm is simple and requires negligible working memory in itself. As section 3 indicates, in many cases λ-search is capable of offering a relative search space reduction over standard alpha-beta comparable to the relative reduction of standard alpha-beta over minimax. In addition, the λ-search method often eases the implementation of abstract game-specific knowledge, and λ-search can be combined with any search method for searching the λ-trees, including proof-number search (thus rendering possible an easy combination of null-moves and proof-numbers).

In Go, the λ-search algorithm is capable of solving open-space tactical problems (tesuji) from e.g. volume 4 of *Graded Go Problems* [6] with relative ease.[16] Regarding the scope for further work, the following points could be considered:

Generally:
- Can the algorithm be taught to find at least one of two (or more) goals by means of identifying double-threat-moves?
- Which kinds of search techniques are well-suited for λ-trees? How to manage the use of transposition tables in the different orders of λ-trees? Which kinds of move-ordering techniques could be useful (iterative deepening with transposition tables, history heuristic etc.)? Use of proof-number search to solve the λ-trees?
- Investigate a large number of realistic game-problems with λ-search and other search methods, in order to provide some real-world statistics on the efficiency of the different approaches.
- Can zugzwang-motives somehow be incorporated into λ-search?
- How to implement parallelism (multiple processors) most conveniently into λ-search?

Go-specific
- Go block tactics: Solving *inversions* (cf. section 4.2) as independent sub-problems (local games) if possible. Perhaps some of the tools described in [8] could be of use. Try to reduce the size of the R^*-zone as much as possible by means of solving all λ^n-trees two times – the second time using the best moves (stored in a transposition table) from the first search. Try to classify certain kinds of λ^2-moves, see e.g. [7].
- Implementing abstract knowledge regarding tsume-go, semeai and connection problems in the context of the λ-search methodology.
- Using λ-search to search for eyes or Benson-immortality, see [3] and [9].
- How to handle seki or ko?

---

[16] As an example, the following numbers have been tried out successfully: 2, 27, 28, 29, 30, 31, 32, 33, 34, 35, 36, 37, 38, 39, 41, 103, 104, 105, 107, 108, 110, 113. The program uses iterative deepening alpha-beta with transposition tables for searching the λ-trees, together with the inversion rule described in section 4. In addition, moves in all λ-trees are ordered by counting the number of liberties and meta-liberties (points adjacent to liberties) of the attacked block after each possible move (minimizing these if the attacker moves, and maximizing these if the defender moves, and also prioritizing moves with low Manhattan distance to the attacked block).

References

1. Allis, L.V: Searching for Solutions in Games and Artificial Intelligence. PhD Thesis, University of Limburg, Maastricht (1994)
2. Allis, L.V., M. van der Meulen, and H.J. van den Herik: Proof-Number Search. Artificial Intelligence, Vol. 66, ISSN 0004-3702 (1994) 91-124
3. Benson, D.B.: Life in the Game of Go. Information Sciences, vol. 10 (1976) 17-29
4. Cazenave, T.: Abstract Proof Search. In I. Frank and T.A. Marsland (eds.): Computers and Games 2000, Lecture Notes in Computer Science, Springer-Verlag (2001)
5. Heinz, E.A.: Adaptive null-move pruning. ICCA Journal, Vol. 22, No. 3 (1999) 123-32
6. Kano, Y: Graded Go Problems for Beginners, Volume Four, The Nihon Ki-in, Tokyo, Japan (1990)
7. Kierulf, A.: Smart Go Board: Algorithms for the Tactical Calculator. Diploma thesis (unpublished), ETH Zürich (1985)
8. Müller, M.: Computer Go as a Sum of Local Games: An Application of Combinatorial Game Theory. PhD thesis, ETH Zürich, 1995. Diss. ETH Nr. 11.006 (1996)
9. Müller, M.: Playing it safe: Recognizing secure territories in computer Go by using static rules and search. In H. Matsubara (ed.): Game Programming Workshop in Japan '97, Computer Shogi Association, Tokyo, Japan (1997) 80-86
10. Müller, M.: Race to capture: Analyzing semeai in Go. In Game Programming Workshop in Japan '99, volume 99(14) of IPSJ Symposium Series (1999) 61-68.
11. Thomsen, T.: Material at http://www.t-t.dk/go/cg2000/index.html (2000)
12. Wolf, T.: About problems in generalizing a tsumego program to open positions, 3rd Game Programming Workshop in Japan, Hakone (1996)

Appendix: λ-Search Illustrated on Chess

Consider the mating problem in figure A1. In chess mating problems, a λ^1-move is a check, but it is easily seen that the black king cannot be mated with a sequence of white check-moves. Hence, in this problem, $\lambda_a^1 = 0$. However, the black king can be mated in a λ^2-tree, as shown below.

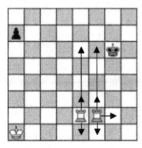 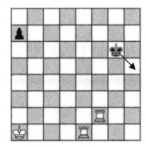 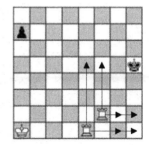

Fig. A1. How to mate black? White λ^2-moves

Fig. A2. Black λ^2-move following Re2-e1

Fig. A3. White λ^2-moves following Kg6-h5

In figure A1, all the white λ^2-moves (moves threatening a forced mating check-sequence) are shown. Of these 7 white moves, three of them are direct checks, in contrast to the four other more "quiet" moves (Re2-e3, Re2-e1, Rf2-f3 and Rf2-f1). For instance, after white Re2-e1, if black passes, a mate results from Re1-g1+, Kg6-h7/h6/h5, Rf2-h2++. Now, consider Re2-e1. As shown in figure A2, there is only one black λ^2-move averting the threatened mating check-sequence: Kg6-h5. After Kg6-h5, white has no mating check-sequence. His best try would be Re1-h1+, Kh5-g4, Rh1-g1+, but it peters out after Kg4-h3. In figure A3, the 6 white λ^2-moves following Kg6-h5 are shown. Of these, four are direct checks, while the more quiet Re1-g1 and Rf2-g2 threaten mating check-sequences. After either Re1-g1 (or Rf2-g2), black cannot avoid a mating check-sequence no matter where he moves, implying that there are no λ^2-moves following Re1-g1 (or Rf2-g2). Hence black can be declared dead.

Abstract Proof Search

Tristan Cazenave

Laboratoire d'Intelligence Artificielle
Département Informatique, Université Paris 8,
2 rue de la Liberté, 93526 Saint Denis, France.
cazenave@ai.univ-paris8.fr

Abstract. In complex games with a large branching factor such as Go, programs usually use highly selective search methods, heuristically expanding just a few plausible moves in each position. As in early Chess programs, these methods have shortcomings, they often neglect good moves or overlook a refutation. We propose a safe method to select the interesting moves using game definition functions. This method has multiple advantages over basic alpha-beta search: it solves more problems, the answers it finds are always correct, it solves problems faster and with less nodes, and it is more simple to program than usual heuristic methods. The only small drawback is the requirement for an abstract analysis of the game. This could be avoided by keeping track of the intersections tested during the search, maybe with a loss of efficacy but with a gain in generality. We give examples and experimental results for the capture game, an important sub-game of the game of Go. The principles underlying the method are not specific to the capture game. The method can also be used with different search algorithms. This algorithm is important for every Go programmer, and is likely to interest other game programmers.

Keywords: Computer Go, Search, Theorem Proving, Capture Game.

1   Introduction

It is very important in complex games where search trees have a large branching factor to safely select the possible moves worth trying. Finding the moves worth trying and the moves that can be eliminated, drastically reduces the search trees [1]. It is important to select the moves safely, which includes not forgetting a possible refutation and not considering as a refutation a useless move. Abstract Proof Search uses game definition functions to safely select complete and minimal sets of moves worth trying. The capture game is used as an illustration of the algorithm, experimental results for this sub-game of Go show that Abstract Proof Search is very efficient: it is more accurate, more safe and faster than basic alpha-beta search for this kind of problems.

The capture game is a fundamental sub-game of the game of Go. All the non-trivial computer Go programs use it. A Go proverb says "If you don't read ladders,

T.A. Marsland and I. Frank (Eds.): CG 2000, LNCS 2063, pp. 39–54, 2001.

don't play Go", its equivalent in computer Go is "if you don't program ladders, don't program Go" as Mark Boon pointed it. The capture game is important by itself, but it is also an important sub-game of other useful sub-games such as the connection, eye and life sub-games.

Proving theorems on the capture game is important because most or even all the other sub-games of Go rely on it. False results of the capture game can invalidate a connection or a life and death analysis, and it often results in the program losing a group or being under severe attack. It is responsible for many lost games.

In our experiments we use a variant of Alpha Beta Null Window Search. However, our method works with other search algorithms, it has also been successfully tested with Proof Number search for example.

Abstract Proof Search improves the speed and the accuracy of Go programs, it is likely that it can also be used to improve search in other games. The difference between our algorithm and other planning approaches to game playing using abstraction [3, 9] is that we concentrate on the classes of states that are worthwhile searching (ip1, ip2 and ip3 states at AND nodes) instead of identifying abstract operators. The word abstract in our algorithm means that the moves are selected using abstract properties of the objects of the games, such as the liberties of the strings.

The second section describes the capture game and its relation to other sub-games of Go. The third section uncovers our search algorithm. The fourth section explains what is the abstract analysis of games that enables Abstract Proof Search. In the fifth section we invalidate the widely accepted knowledge among Go programmers that the number of liberties is a good heuristic for the capture game, we show that the capture game is more subtle and that using too simple heuristics can be harmful. We propose a more accurate classification of situations worthwhile searching as well as a more selective move generator. The sixth section details experimental results and compares Abstract Proof Search to usual alpha-beta search for the capture game on standard test sets.

2   The Capture Game

The capture game is the most fundamental sub-game of Go. It is usually associated to deep and narrow search trees. It has strong relations with connections, eyes, life and death, safety of groups and many important Go concepts.

Figure 1 gives some examples of the capture game. The first example is called a geta, a white move at A captures the black stone marked with an x, it can be found in 5 plies. The second example is an illustration of the capture game as a sub-game of the connection game, a white move at B captures the marked black stone and connects the two white strings, it requires 9 plies. The third example shows the capture game as a sub-game of the life and death game, white at C can make two eyes by capturing the marked black stone in a simple but 15 plies depth ladder. Note that move B is harder to find than move C, the depth of a problem is not always a good measure of its complexity in Go.

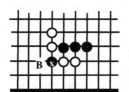
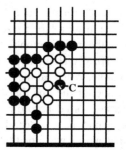

Fig. 1. Examples of captures

3   The Search Algorithm

We use Null Window Search [7] with some modifications tailored to computer Go. We do not use forward pruning with null move search because we are looking for exact results, however some of our experimental results show that null move pruning can speed-up the algorithm with very little drawbacks or even improve it. We use iterative deepening, transposition tables, quiescence search, null-window search when not at the root and the history heuristic. We stop search early when the goal is reached. We search all the moves at the root, even if a previous move at the root has solved the problem, in order to find all the moves that reach the goal. Because it is useful for a Go program to know more than one way to accomplish its goals. Especially when it is useful for the program to reach multiple goals with one move.

To make the explanations easier, from now on, the friend color is black and the enemy color is white. The string that is under attack is black.

In the capture game, the evaluation can take three values : -INFINITY if the string has more than 5 liberties and it is white to play or the string is captured and it is black to play; +INFINITY if the string has more than 5 liberties and it is black to play or the string is captured and it is white to play; 0 when the state of the string is unknown. At the beginning of each node, the evaluation function is called, and the value is returned if it is different from 0.

We also use incrementality so as not to recalculate all the abstract properties of the strings after each move. We keep track of the liberties of the strings, and of the adjacent strings of each string. Each intersection is associated to a bit in a bit array so as to optimize checking of liberties, and the same is done for adjacent strings numbers.

Transposition Tables are used to detect identical positions and return the associated value if the search depth of the stored position is greater than the depth of the node or if the value is +INFINITY or –INFINITY. Transpositions are also used to recall the best move from previous search in the position and try it first when searching deeper so as to maximize cut-off. The size of the transposition table is set to 16384, a larger table could easily contain in memory, but the time to initialize the table before each search becomes too important for large tables. Given the simplicity of some problems and the number of different problems that have to be solved a small table is enough as the threshold for the number of nodes is set to 10 000. Another interesting possibility would be to set a larger size for the table, and to switch it off for the first 100 nodes, keeping the initialization for harder problems that potentially require many nodes.

The History Heuristic is used to order the moves that are not given by the transposition table. When all the moves at a node have been tried, the move that returned the best value, or the one that caused a cut-off, is credited with 2^{Depth}. At each node, the moves are sorted according to their credit, and tried in this order. For the capture game, it may be a better idea to order moves also taking into account simple heuristics that works well for this game: trying the liberties of the string first (ordered by number of neighbor second order liberties), then the liberties of the liberties (ordered by the number of neighbor liberties), and then other moves sorted by the distance to the string. The History Heuristic is a general domain independent heuristic, but it can be improved by using domain-dependent knowledge such as trying the check moves first (playing the liberties first in the Capture game is equivalent to playing the check moves first in Chess).

A Quiescence Search is performed at leaf nodes. The quiescence search alternatively calls two function QSCapture() that plays on the liberties of the string to capture if it has 2 liberties, and QSSave() that plays the liberty of the string to capture and the liberties of the adjacent strings in atari[1], if the string to capture is in atari. This ensures that the Quiescence search sends back correct results on the capture status of the string and quickly reads simple ladders.

Iterative deepening does not stop after the first winning move, it continues two more plies to find some other working moves. There are multiple stopping criteria to iterative deepening: the time allotted to the search, the number of visited interior nodes, the depth of the search, and the comparison between the depth of the first solution found and the current depth.

In order to find the games status and the moves associated to goals in the test positions, one or two searches may be performed. The first search is made with the player trying to capture playing first. If the goal is to prove that the string can be captured, no more search is performed. Otherwise another search is made with the player trying to save the string playing first, it is useful to know when the string is captured, and when it can be saved and which moves save it.

4   Abstract Analysis of Games

The possible moves that can modify the outcome of the search can be easily found when the goal is almost reached. However, when the goal is not one or two moves away, it becomes less clear. This section deals with the selection of a complete set of worthwhile possible moves, when the goal cannot be directly reached.

We try to find the complete set of abstract moves that can possibly change the outcome of a game a given number of moves in advance. For example, given that a string can be captured in 5 plies, we want to find all the abstract moves than can pos-

---

[1] atari means only one liberty left

sibly prevent it to be captured in 5 plies. An abstract move is a move that is defined using abstract properties either of the strings or of the board. An example of an abstract move is 'a liberty of the string to capture'.

The set of possible moves that can modify the outcome of the search could be found dynamically by simply recalling the intersections tested during the search. The only moves that can modify the issue of the search are the moves that modify one of the tested intersections. Selecting forced moves in this way may be more general than an abstract analysis. It is done in [8], and it is similar to keeping an explanation of the search to find the forced moves as in early learning versions of Introspect [1]. Abstract analysis is more related to pre-computation of some parts of the search tree in order to be more efficient, such as in the partial evaluation version of Introspect [2]. Some more tests need to be performed to compare the two approaches and decide which one is the most efficient.

In the following we will use names for the different games states. The names of the games are usually followed by a number that indicate the minimal number of white moves in order to reach the goal. A game that can be won if white moves is called 'gi', a game where black has to play otherwise white wins the game by playing in it is called 'ip', it is the almost the same as 'gi' except that it is associated to black moves. A game that is won for white whatever black plays is called 'g'. A game is always associated to a player, the g and gi games are associated to the player that can reach the goal, the ip games are associated to the player that tries to prevent to reach the goal. Here the goal is to capture strings, it can be easily defined as: removing the last liberty of the string to capture. A forced move is a move associated to an ip game. For example, when the program checks whether a game is ip2, it begins with verifying that white can capture in two moves if it plays first (a gi2 game). The forced ip2 moves are the black moves that prevent white from capturing the string in two moves once one of the black ip2 moves has been played (we can say that the gi2 game has been invalidated by the black move).

It is quite simple to find forced moves, one move from the goal: when the string has only one liberty, the only moves to save the string are the moves that directly increase the number of liberties. There are only two ways to increase the number of liberties of a string: play one of its liberties, or remove an adjacent string in atari. These moves are associated to the ip1 game.

Figure 2 gives the dependencies between games. A game can be defined using the games for the lower number of plies, for example, the g1 game for white is defined as: the game is ip1 for black, and all the forced black moves lead to a gi1 game for white after the black move. So the g1 game is defined using the definitions of the gi1 and of the ip1 games, as it is shown in figure 2 where the g1 game depends on the gi1 and ip1 games. Another example is the gi3 game for white: a white move leads to a g2 game for white. So the gi3 game depends on the g2 game only. Some more detailed game definition functions are given in the next section on the selection of moves. In order to make things clear some examples of games are given in figure 3.

The only abstract moves that can change an ip1 game are the liberty of the string and the liberties of the adjacent strings in atari. A g1 game for white is defined as an ip1 game for black that is still gi1 for white after each of the forced black move is played. The set of intersections that are responsible for the state of a g1 game are the intersections involved in the corresponding ip1 game, and the intersections involved in the gi1 games following each forced black move. But as we know the abstract set of intersections for the ip1 and the gi1 games, we can deduce that the intersections responsible for a g1 game are the liberty of the string, the liberties of the adjacent strings in atari, the liberty after a black move is played on the liberty of the string, or on the liberties of the adjacent strings in atari.

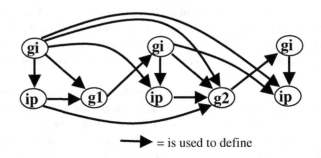

➤ = is used to define

Fig. 2. The dependencies between games

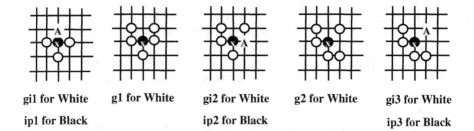

| **gi1 for White** | **g1 for White** | **gi2 for White** | **g2 for White** | **gi3 for White** |
| **ip1 for Black** | | **ip2 for Black** | | **ip3 for Black** |

Fig. 3. Examples of games

Another example of how the abstract sets of moves can be calculated is the transition from a gi set of moves to an ip set of move: the only move that can modify an empty intersection is to play on this intersection, therefore if some empty intersections are involved in the definition of a gi game, the set of moves that can prevent it (the corresponding possible ip moves) contains all these empty intersections.

More detailed explanations of how this knowledge can be automatically generated can be found in [2], but more formal and easy to use tools for analyzing games need to be investigated.

A more practical example of such a set of abstract moves is the function that finds the complete set of abstract moves for the ip2 game. The function begins with adding the liberties of the string to the set of moves to prevent gi2, then for each liberties, it plays a black move on it, and adds the liberties of the string after the move. It also adds the liberties of the white strings adjacent to the string to capture that have strictly less than three liberties after the black move. Then it adds the liberties of all the adjacent string that have strictly less than four liberties (because a gi2 string has two liberties and the only adjacent strings that can be captured to save it have strictly less than four liberties: any four liberties adjacent string cannot be captured before the two liberties gi2 string). The code for CompleteSetOfMovesToPreventgi2 is quite simple, requiring only 16 lines of C. Similarly the code for CompleteSetOfMovesToPreventgi3 is 23 lines of C.

Here is the CompleteSetOfMovesToPreventgi2 function in pseudo-code that finds the complete set of abstract moves for the ip2 game:

```
CompleteSetOfMovesToPreventgi2(S) {
   for each liberty l {
      add l to S // add the liberty to the set of moves
      if (LegalMove (l,StringColor)) {
         MakeMove (l,StringColor);
         // add the liberties of liberties
         add the liberties after the move to S;
         // liberties of adjacent strings after the move
         for each adjacent string adj
            if (number of liberties of adj < 3)
               add the liberties of adj to S;
         UndoMove();
      }
   }
   // liberties of adjacent strings < 4 liberties
   for each adjacent string adj
      if (number of liberties of adj < 4)
         add the liberties of adj to S;
}
```

A property of the game of Go is that the minimum number of moves to take a string is its number of liberties. As a consequence, it is often useless to try to increase the number of liberties of a string by capturing an adjacent string that has more liberties or to play the external liberties of an adjacent string that has many liberties trying to make a seki[2] with it. There are exceptions to this rule when the adjacent string and the string to capture share some liberties or when the string to save has protected liberties and a sufficient number of liberties. Only in this case, it can be useful to fill the external liberties of the adjacent string in order to obtain a seki or to capture it so as to save the string under attack.

The figure 4 gives illustrations of these two cases. In the left position, playing at A, one of the three liberties of a string adjacent to the string to save that has two liberties, enables to save it by capturing the adjacent string in 5 plies. The reason is that the string to save has two protected liberties after the move. In the right position, playing at B saves the marked string by making a seki between the black and the white strings. In the cases of ip2 and ip3 games, the string to capture has only two or three liberties and can be captured in 3 or 5 plies. These limitations ensure that looking at

---

[2] Seki: Two strings that are mutually alive. One string cannot capture the other by playing a common liberty because it will be captured itself first. However as the pass move is legal in Go, the two strings of a seki are safe provided all the adjacent strings are also safe.

the adjacent strings that have less than one or two liberties more than the string to capture, is enough. Strings that become adjacent after a move can also be taken into account as shown in the CompletSetOfMovesToPreventgi2 function, where the abstract properties of the string are taken into account after some black moves are tried. Improvements could be made by also counting shared liberties between the string and its adjacent strings so being more selective on the adjacent strings to consider.

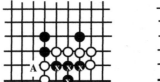 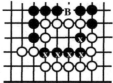

Fig. 4. Playing liberties of adjacent strings with more liberties

5   Selection of Moves

The functions that safely select moves use practically very little knowledge, they are quite simple to program and are based on the abstract analysis of the game and the definition of games values. This way of coding the functions is simpler than explicitly programming all interesting cases. Experiments with coding the cases related to the Preventip3 knowledge show that it needs 22034 lines of C for the Preventip3 function, and some more lines to write the functions associated to the high level definitions and abstract concepts which are called by the main function [1, 2].

Instead of explicitly coding all the cases, either in patterns or in complex programs, it is better to rely on the definition of games, and to rely on simple concepts only, simulating the playing of moves.

At each node and at each depth of the Abstract Proof Search, the game definition functions are called, they are equivalent to the development of small search trees. So Abstract Proof Search is a search algorithm that can be considered as developing small specialized search trees at each node of its search tree. At OR nodes, the program first checks if the position is gi1, if it is not, it checks if it is gi2 (equivalent to a depth 3 search tree), and if it is not, it checks if it is gi3 (equivalent to a depth 5 search tree). As soon as one of the gi games is checked, the program stops searching and sends back Won. Otherwise it tries the OR node moves associated to the position. At AND nodes, the same thing is done for ip1, ip2 and ip3 games, if none of them is verified, the programs sends back Lost, otherwise it tries the moves associated to the verified ip1, ip2 or ip3 game. Note that the game definition functions are equivalent to the programs generated by the Introspect system to safely select moves in games search trees [1, 2].

For example the pseudo-code that finds whether the string can be captured in 3 plies at each OR node is:

```
Capturegi2 () {
  res = 0;
  if (number of liberties == 2)
    for each liberty l
      if (res == 0)
        if (LegalMove (l, Opposite(StringColor))) {
          MakeMove (l, Opposite(StringColor));
          if (Capturegl())
            res = 1;
          UndoMove ();
        }
  return res;
}
```

It relies on the Capturegl function as shown by the arrow between gi1 and gi2 in the figure 2. The functions begins with verifying that the string to capture has two liberties. Then for each of the two liberties, and if the results has not been proved yet (res==0), it tries to fill the liberty, and verifies that the game is g1 after the liberty is filled, using the Capturegl game definition function.

The function defining the ip2 game and its associated moves is equivalent to find the forced moves that prevent the string to be captured in 3 plies. It is checked at every AND nodes of the Abstract Proof Search tree provided the ip1 function has not been verified before:

```
Captureip2 (S) {
  res = 0;
  if (Capturegi2()) {
    res = 1;
    CompleteSetOfMovesToPreventgi2 (S1);
    for each move m of S1
      if (LegalMove (m, StringColor)) {
        MakeMove (m, StringColor);
        if (!Capturegi1())
          if (!Capturegi2())
            add move m to S;
        UndoMove ();
      }
  }
  return res;
}
```

Again it is defined using simple concepts and the functions corresponding to other games. Here again, as shown in the figure 2, the ip2 game definition function relies on the functions defining the gi1 and gi2 games. The function adds the forced moves to prevent capture in 3 plies (the ip2 moves). The function begins with verifying that the string can be captured in two moves if white plays first, by calling the Capturegi2 game definition function. If it is the case, the function finds the complete set of black moves that may change the issue of a gi2 game by calling the function Complete-SetOfMovesToPreventgi2. Then, for each move of this set, it plays it and verifies that the game is not gi1 and not gi2 after the move. If it is the case, then the move has been successful in preventing the gi2 game, and is therefore an ip2 black move, so it adds the move to the set of forced ip2 moves.

In order to give an example for each kind of game, here is the pseudo-code that detects situations won 4 plies ahead:

```
Captureg2 () {
   res = 0;
   if (Captureip1(S)) {
      res = 1;
      for each move m of S
         if (LegalMove (m, StringColor)) {
            MakeMove (m, StringColor);
            if (!Capturegi1())
               if (!Capturegi2())
                  res = 0;
            UndoMove();
         }
   }
   else if (Captureip2(S))
      res = S is empty;
   return res;
}
```

The Captureg2 game definition is a little more complex than the previous ones because there are two possibilities:

- Either the black string can be captured in one move by white, so it has only one liberty, and the Captureip1 function fills the set S with it. And after playing on its liberty the string can still be captured in two white moves (the Capturegi2 function matches).
- Or the function Captureip2 is verified, but all the moves that could prevent the game to be gi2 do not work, so the Captureip2 function sends back an empty set in S for the preventing moves. In that case, the game is won for white because none of the black moves to prevent gi2 works.

Again, as shown in the figure 2, the g2 game is defined using the ip1, ip2, gi1 and gi2 games.

Figure 5 gives a part of an Abstract Proof Search tree, some moves at OR nodes (white moves) have been omitted. Each move is labeled with its color and for forced moves (black moves) with the name of the game that found it.

A widely accepted knowledge among Go programmers is that the number of liberties is a good heuristic for the capture game. In this paper we show that the capture game is more subtle and that using too simple heuristic can be harmful. We propose a finer classification of situations worthwhile searching, by considering forced moves only when a position can be proved to be winning a given number of plies in advance (gi games that enable to define ip ones).

Forgetting a move at an OR node can lead a program to miss a winning move, however it does not invalidate the result of the search: the result will be Unknown (0) instead of Won (+INFINITY). In the capture game OR node moves are moves that try to capture the string. On the contrary, forgetting an AND node move can make the

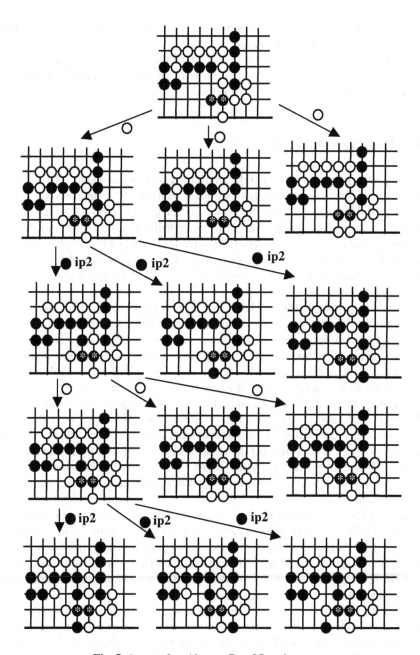

Fig. 5. A part of an Abstract Proof Search tree

result of a search wrong by missing a refutation. Our approach to games enable to be sure of not forgetting any move. Moreover it also enables to select only a subset out of all the possibly refuting moves. Selecting the minimal number of moves is as important as selecting all the necessary moves. Because, if a move does not interfere with the result, but the associated search returns Unknown or Lost, it is considered as a refutation and the program gives a false result.

Finding the complete set of forced moves, enables to prove theorems about games by not forgetting to consider some moves, and also by not considering moves that are proved not to have influence on the result of the game.

6   Experimental Results

This section gives the results and the analysis of some experiments on a standard test set. We begin with describing how we have managed to compare basic alpha-beta search and Abstract Proof Search. At the end of the section some of the results are detailed and discussed. We give experimental results on a standard test set for capturing strings in Go: we call them ggv1 [4], ggv2 [5] and ggv3 [6]. We have selected all the problems involving a capture of a string, including semeai[3] and some connection problems. There are 114 capture problems in ggv1, 144 in ggv2 and 72 in ggv3. Experiments were performed with a K6-2 450 MHz microprocessor.

In order to compare Abstract Proof Search with the basic alpha-beta search usually performed in Go programs, we choose to use as basic alpha-beta search the same search algorithm with a different move generation function. The basic alpha-beta search calls the function CompleteSetOfMovesToPreventgi3 to generate the set of possible moves at AND nodes. It uses the same function for move selection as Abstract Proof Search at OR nodes. This way, we are fair to basic alpha-beta search, as it uses exactly the same move generation function as Abstract Proof Search, except that Abstract Proof Search uses games definition functions so as to be more selective and to ensure the validity of the results of the search. So our basic alpha-beta search is already an improvement over the usual alpha-beta search as it never overlooks a five plies deep refutation. The CompleteSetOfMovesToPreventgi3 function verifies that a string has strictly less than four liberties, and if it is the case, it returns the liberties of the string, the liberties of the string if black moves are played on its liberties, the liberties of the adjacent strings that have strictly less than four liberties after the black moves on a liberty of the string to capture is played, the liberties of the string and the liberties of the adjacent strings that have strictly less than three liberties if two black moves are played on the liberties of the string, and finally all the liberties of the adjacent strings that have strictly less liberties than the number of liberties of the string to capture plus two.

In the tables below, the basic alpha-beta search that stops whenever the time allotted to the search exceeds 1 second or the number of interior nodes exceeds 10,000 is called Preventip3-1s-10000N. Similarly, Preventip3-1s is the basic alpha-beta search that stops when the time exceeds 1 second. The Abstract Proof Search, using moves that prevent a goal up to 5 plies in advance is called ip3-1s-10000N. We do not give the results for ip3-1s since they are the same as for ip3-1s-10000N. In the number of nodes, we only count the interior nodes where some moves have been played. We do

---

[3] Semeai: race to capture between two or more strings.

not count the leaf nodes (nodes where a transposition has occurred and has directly returned +INFINITY or –INFINITY are considered as leaf nodes).

The problem number 172 in volume 1 is not solved with our algorithm but is solved with the basic alpha-beta algorithm. Problem ggv1_172 can be considered as a mix of capture and life and death. It involves a nakade[4] shape that cannot be a part of the capturing game three moves ahead. The basic problem solver continues to play AND nodes moves even if the moves are not forced, provided the number of liberties is small enough. For these kinds of problems only, the basic algorithm can give better results. However, given the experimental results, the drawbacks of the basic algorithm are more important than its gains. As we can see with the different tables, the basic alpha-beta search method is not selective, and spends more time in useless branches of the tree.

Table 1. Results for ggv1

Algorithm	Total time	Number of nodes	% of problems
Preventip3-1s-10000N	19.79	109117	99.12%
Preventip3-1s	19.79	109117	99.12%
ip3-1s-10000N	11.82	10340	99.12%

Table 2. Results for ggv2

Algorithm	Total time	Number of nodes	% of problems
Preventip3-1s-10000N	113.20	836387	78.47%
Preventip3-1s	118.60	870938	77.78%
ip3-1s-10000N	34.13	42382	88.19%

Table 3. Results for ggv3

Algorithm	Total time	Number of nodes	% of problems
Preventip3-1s-10000N	65.61	449987	65.28%
Preventip3-1s	74.25	483390	65.28%
ip3-1s-10000N	21.13	27283	73.61%

One of the proposed metrics for performance is the percentage of solved problems, this percentage corresponds to the number of problems with a correct game value and a correct move, however on some problems, the basic alpha-beta algorithm sometimes also gives moves that do not work. They are not counted as wrong answers, so the metric favors the basic algorithm.

Surprisingly, there is one more solved problem in the ggv2-Preventip3-1s-10000N test than in the ggv2-Preventip3-1s, this is due to the complexity of some problems. When trying to find a move that saves the black stones, the algorithm does not stop until the Alpha-Beta returns –INFINITY or +INFINITY or one of the stopping criterion is met. So a move that returns 0, can be considered as a move that saves the en-

---

[4] Nakade: a shape of string that makes an unsettled life shape when captured.

dangered stones if the search threshold corresponding to the number of nodes is passed over. However, if more search is performed and the algorithm does not send back correct results, as Preventip3 does, the saving move can then be associated to – INFINITY, in other words as not saving the string. This is what happens here for one problem in ggv2, where more search with a basic algorithm leads to worse results.

Another problem related to the basic alpha-beta search is the treatment of sekis. There are seki positions that are correctly assessed by Abstract Proof Search in a natural way and incorrectly assessed by basic alpha-beta search. To prevent basic search from failing in these situations, some special code has to be added or pass moves may be considered. These possible solutions may be search time and/or programming time consuming.

Table 4. Results for ggv2 with more time and nodes

Algorithm	Total time	Number of nodes	% of problems
Preventip3-10s-100000N	635.20	4607171	79.17%
ip3-10s-100000N	63.57	81302	90.28%

Table 5. Results for ggv3 with more time and nodes

Algorithm	Total time	Number of nodes	% of problems
Preventip3-10s-100000N	726.40	4319840	70.83%
ip3-10s-100000N	23.97	33936	73.61%

In order to check whether the algorithm scales well, we also did some experiments with relaxed controls, which are unrealistic for today's technology, but that show the evolution of the problem solving when more computation is available. The results are summarized in tables 4 and 5. With stopping criteria of 10 seconds and 100,000 nodes, the interest of Abstract Proof Search increases. It solves much more problems in the tenth of the time of basic alpha-beta search for ggv2, and for even more complex problems such as in ggv3, it still solves more problem in $1/30^{th}$ of the time of basic alpha-beta search. We can note that giving more time and more nodes to Abstract Proof Search for ggv3 does not change much the results, because for complex problems, Abstract Proof Search stops searching early as it does not find forced moves, whereas basic alpha-beta search, that relies on the number of liberties of strings is inaccurate in establishing the complexity of some problems and spends much time searching complex and useless sub-trees.

The average speed of basic alpha-beta search is approximately 7,000 nodes per second. It is much faster than Abstract Proof Search that only develops approximately 1,200 nodes per second. However, Abstract Proof Search finds the solutions to the problem in much less nodes than basic alpha-beta search. The verification of the game definitions functions at each nodes explains the relatively small speed of Abstract Proof Search. Each game definition function is equivalent to a small tree search. Transpositions Tables are not currently used in the game definition functions, their proper use may well speed-up Abstract Proof Search.

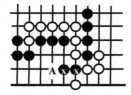

Fig. 6. A problem solved by Abstract Proof Search, not by basic alpha-beta search

Figure 6 gives an example of a problem that is solved in 443 nodes and 0.33 seconds with Abstract Proof Search and which is not solved in 8,844 nodes and 1.21 seconds with basic alpha-beta search. The search is stopped as soon as it exceeds one second or 10,000 nodes. When more time and nodes are given to basic alpha-beta search, it solves the problem in 46,227 nodes and 6.59 seconds.

Table 6. Results with null move forward pruning

Algorithm	Book	Total time	Nodes	%
Preventip3-1s-10000N-NM	ggv1	13.34	69582	98.25%
Preventip3-1s-10000N-NM	ggv2	66.55	518398	77.08%
Preventip3-1s-10000N-NM	ggv3	30.50	230724	65.28%
Ip3-1s-10000N-NM	ggv1	10.58	9401	99.12%
Ip3-1s-10000N-NM	ggv2	31.57	39220	88.89%
Ip3-1s-10000N-NM	ggv3	16.93	20902	73.61%

Null move forward pruning has been tried with a reduction factor of four. The results shows that it is beneficial to the search algorithm, despite that forward pruning may alter the result of the search in some cases.

7   Conclusion

The theorem proving approach to the capture game in Go gives excellent results. It solves more problems than the basic alpha-beta approach, the answers it finds are always correct on the contrary of heuristics, it solve problems faster and with less nodes, and it is more simple to program than other approaches. The only small drawback is the requirement for an abstract analysis of the game. It may be overcome by a dynamic selection of forced moves based on the intersections accessed during smaller proofs. Note that the abstract analysis is not sufficient by itself, it has to be used with game definition functions for selecting moves.

The principles underlying the method are not specific to the capture game. They can also be used with different search algorithms. In the near future, we will try this method in other sub-games of Go and in other games. The games and sub-games that are concerned with this method are the games where a simple definition of the goal to reach can be given. Such games are for example the connection sub-game of Go, the virtual connections at Hex, or the five in a row game. Some sub-games of other diffi-cult games such as mate search in Chess or Shogi may also benefit of our method. Generalizing the method to make it work with integer numbers could benefit to other search programs such as Chess programs. Improvements can be made by using trans-position tables in the game definition functions, and by being even more accurate on the complete sets of moves to prevent gi games (for example taking into account the number of shared liberties between strings). Other important improvements to our current approach are the development of tools in order to facilitate the abstract analy-sis of games and the comparison between a dynamic selection of forced moves by analyzing the set of intersections tested during a search, and a selection based on abstract analysis. A combination of the two may well be the best alternative.

References

1. Cazenave T.: *Metaprogramming Forced Moves*. Proceedings ECAI98 (ed. H. Prade), pp. 645-649. John Wiley & Sons Ltd., Chichester, England. ISBN 0-471-98431-0. 1998.
2. Cazenave T.: *Generating Search Knowledge in a Class of Games*. submitted. http://www.ai.univ-paris8.fr/~cazenave/papers.html. 2000.
3. Frank I.: *Search and Planning under Incomplete Information: a Study using Bridge Card Play*. PhD Thesis, Department of Artificial Intelligence, University of Edinburgh. 1996.
4. Kano Y.: *Graded Go Problems For Beginners*. Volume One. The Nihon Ki-in. ISBN 4-8182-0228-2 C2376. 1985.
5. Kano Y.: *Graded Go Problems For Beginners*. Volume Two. The Nihon Ki-in. ISBN 4-906574-47-5. 1985.
6. Kano Y.: *Graded Go Problems For Beginners*. Volume Three. The Nihon Ki-in. ISBN 4-8182-0230-4. 1987.
7. Marsland T. A., Björnsson Y.: *From Minimax to Manhattan*. Games in AI Research, pp. 5-17. Edited by H.J. van den Herik and H. Iida, Universiteit Maastricht. ISBN 90-621-6416-1. 2000.
8. Thomsen T.: *Lambda-search in game trees – with application to go*. CG 2000. This vol-ume.
9. Willmott S., Richardson J. D. C., Bundy A., Levine J. M.: *Applying Adversarial Techniques to Go*. Journal of Theoretical Computer Science. 2000.

Solving Kriegspiel-Like Problems: Examining Efficient Search Methods

Makoto Sakuta and Hiroyuki Iida

Department of Computer Science, Shizuoka University
3-5-1 Johoku, Hamamatsu, 432-8011 Japan

{sakuta,iida}@cs.inf.shizuoka.ac.jp

Abstract. We recently proposed a deterministic approach for solving problems with uncertainty, called the Uncertainty Paradigm. Under this paradigm, deterministic solving of such a problem is resolved into a plain AND/OR-tree search. The search under this paradigm is denoted by the Uncertainty Paradigm Search (UPS). As an application, we have chosen the domain of Tsuitate-Tsume-Shogi, which is the mating problem of Kriegspiel-like variant of Shogi. The early implementation of UPS was based on a simple depth-first full-width search with iterative deepening (ID), which was unable to solve several hard problems. This paper explores an efficient search method using UPDS (Uncertainty Paradigm PDS) algorithm, which is a specialized version of PDS (Proof-number and Disproof-number Search) for UPS. UPDS generally performs better than ID or PDS, but fails to solve some easy problems. In addition, several variations of UPDS and ID have also been examined to tackle some hardest problems. All problems in the test set have been solved by a particular variation of UPDS, which shows superiority of the depth-first variants of the proof-number search in UPS.

Keywords: Problem solving, Uncertainty Paradigm, Metaposition, Metamove, UPS, Tsuitate-Tsume-Shogi, PDS, Kriegspiel

1   Introduction

In the domain of computer-game research, most efforts have been devoted to the programming of games with complete information such as chess. There is less literature on computer-game research using imperfect-information games.

We know that developing a program of an incomplete-information game is usually much more difficult than developing the program of a perfect-information game. For searching the game tree with incomplete information, a program has to combine a tree search with information sets and find the most reasonable strategy, according to the conventional approach of a game in extensive form. The information sets are somewhat apart from the tree structure, so the search cannot be resolved into a plain tree search.

However, there are some kinds of problems with uncertainty or some kinds of endgames of the imperfect-information games that can be solvable deterministically. Nevertheless, there are a few studies on the deterministic solving of problems with uncertainty. We have tackled deterministic solving of the problems with uncertainty and

T.A. Marsland and I. Frank (Eds.): CG 2000, LNCS 2063, pp. 55–73, 2001.
© Springer-Verlag Berlin Heidelberg 2001

succeeded in two domains by introducing the novel paradigm, the Uncertainty Paradigm [19]. Moreover, we have enhanced the search method based on a technique of a transposition table [17].

In two-agent problems with complete information such as Tsume-Shogi, very efficient search algorithms with high solving ability have been developed. In this paper, we examine one of these algorithms together with our approach, and show the results of applying the search algorithm to the problems of Tsuitate-Tsume-Shogi. In addition, we also examine some variations of search algorithms to tackle hardest problems.

2 Uncertainty Paradigm for Problems with Incomplete Information

2.1 Overview

The Uncertainty Paradigm is a way of thinking that recognizes an uncertain situation as if it is a definite situation. Under this paradigm, an uncertain situation is a hybrid of several certain situations. Here we give a brief sketch of this paradigm. We use the term *solver* to mean the player or agent to solve the problem, and *opponent* as the players or agents to hinder the problem being solved.

A *metaposition* is a hybrid of possible positions that are not distinguishable for the solver. A metaposition corresponds to a super node that represents the positions in an information set of the solver, in a game in extensive form [13]. The count of possible positions of a metaposition can be recognized as an *uncertainty index*.

A *metamove* is simply a move for a metaposition. When a metamove for a metaposition is a definite move, making the metamove corresponds to making the move for each position of the metaposition. When a metamove for a metaposition is a hybrid of several moves, making the metamove corresponds to making all the moves for each position of the metaposition. Consequently, the number of positions of the metaposition increases. This is a *diffusion* of the metaposition. If the solver cannot get any clue in such metamoves, the uncertainty of the metaposition soon results in combinatorial explosion. However, there are *observations* or clues that help the solver. By these observations, a metaposition is split into several metapositions with less uncertainty, i.e., metapositions that have fewer positions. Since the solver has to accept this split in a passive manner, he has to solve all the split metapositions. Therefore, this split is an *AND-split* of a metaposition even in case of the single-agent puzzles.

For all solving problems with uncertainty, the search graph (tree) under the Uncertainty Paradigm is a plain AND/OR graph (tree), which we call the Uncertainty AND/OR graph (tree). In such a graph (tree), a node is a metaposition and an edge is a metamove. In addition, the search under the Uncertainty Paradigm is a plain AND/OR-graph (-tree) search, which we call the *Uncertainty Paradigm Search (UPS)*.

2.2 Related Works

The Uncertainty Paradigm is a new approach for the problems with incomplete information. However, since our approach omits the probabilities and focuses on deterministic solving, the applicable domains are restricted.

Our approach recognizes an information set as a node of a search tree. In the area of computer games with incomplete information, a similar approach is known. The work by Frank and Basin [4] uses "vectors" at nodes to conduct a game tree search. In the same way as our approach, these vectors represent collections of nodes that come from the same information sets. The differences from our approach are that a search is not for an AND/OR tree but for a minimax tree with probability, and that in Bridge the uncertainty is greatest at the start point of a game and decreases as the game proceeds. Consequently, a vector with a fixed length can reasonably used in their methods. In our approach, the uncertainty changes dynamically in searching. That is to say, the uncertainty sometimes increases by the diffusion of a metaposition and sometimes decreases by the split of a metaposition. Though representation of fix-length vectors can be applied by using very large vectors in our problems, it is quite inefficient because the maximum uncertainty in a search is unknown at the beginning of a search.

There is little study in computing chess-like games with incomplete information. In the domain of Kriegspiel [14], Ferguson studied an endgame problem whether the king, bishop and knight can win against the king alone and showed the result that the player with bishop and knight can win with probability one in general (with king initially guarding both bishop and knight) [3]. And Ciancarini *et al.* reported on their works on some simple endgames of Kriegspiel and showed the performance of their program that can play KPK endgames by the knowledge-based approach [2]. However, there had been no study on solving Kriegspiel problems deterministically like ours. In addition, there had been no study about computing Tsuitate-Shogi.

3   Tsuitate-Tsume-Shogi

As the application of the Uncertainty Paradigm to the two-agent problem with uncertainty, we have selected the domain of *Tsuitate-Tsume-Shogi*. We first refer to Tsuitate-Shogi, then explain Tsuitate-Tsume-Shogi.

3.1   Tsuitate-Shogi

Tsuitate-Shogi is a Shogi variant that is one of the best known and most popular among all Shogi variants, as Kriegspiel is in chess. We read 'Tsuitate' (in Japanese) as a screen, therefore we call Tsuitate-Shogi (screen Shogi).

The rules of Shogi are available on the Internet, for example, at:
    http://www.jwindow.net/LWT/SHOGI/INTRO/shogi_intro.html.
The most important difference between Shogi and chess is that the players can re-use the captured pieces in Shogi. A captured piece becomes a *piece in hand* for the side that captured it. There are two types of moves in Shogi. One type of move is moving a piece on the board to another square on the board. Another type of move is putting one of the pieces in hand onto a vacant square on the board, which is called *dropping* a piece.

Let us give a short summary of the rules of Tsuitate-Shogi. Note that the rules of Tsuitate-Shogi seem mostly similar to Kriegspiel. Tsuitate-Shogi requires two players, a referee, two Shogi boards with a set of pieces, and a screen. Both players are unable to see the other side because of the screen. The principle of a game is that each player moves normally but he is not told the opponent's moves, which he has to guess through judicious play.

After one plays a move on his own board, the referee approves the move to provide information to both players as required by the rules. Information is given to both players by the referee saying 'Illegal, ... times', 'Black/White has played', 'Check', and 'Mate'. The referee must not tell other information aloud. The referee performs capturing on the board instead of the players.

Each player is allowed to make illegal moves within a certain number of times, typically eight times. If one has made illegal moves more than the allowed number of times, he loses the game on fouls. A game also ends when the king on either side is in mate.

3.2   Rules of Tsuitate-Tsume-Shogi

Tsuitate-Tsume-Shogi (TTS) is a mating problem of Tsuitate-Shogi [7], which is also a variant of Tsume-Shogi, a mating problem of Shogi [6]. Here we give a summary of the rules of TTS. Note that the rules of TTS seem similar to Tsume-Shogi except several points. Since TTS is a two-agent problem, there are two agents: the *attacker* as the *solver* and the *defender* as the *opponent*.

Because the attacker's king often has no meaning in TTS, most TTS problems only have the defender's king but do not have the attacker's king. Nevertheless, there are some problems that have both kings, which we call *problems with double kings*. In TTS, the attacker is unable to see the opponent's pieces and responses except for the initial position of a mating problem. The goal of solving a TTS problem is to mate the opponent's king after the sequence of check moves and their responses. However, the attacker is unable to see the definite information on the defender's responses, while the defender has the perfect information on both sides. The attacker must lead to a checkmate whatever moves the defender may play.

In Tsuitate-Shogi, an illegal move is checked and announced by the referee. Hence, a player can even make an explicit nonsense move as a feint merely in order to confuse the opponent. In TTS, trying an illegal move is also allowed. However, there is no meaning of trying an explicit nonsense move, because it is illegal at all positions of the metaposition and the attacker cannot get any information by trying it. Here we have chosen a strict definition. We have defined a *foul move* as a move that belongs to a subset of possible illegal moves. A foul move is a move that seems to be legal with only the attacker's pieces on the board, but is illegal when all pieces are considered. Thus, the foul moves are:[1]

1. A sliding piece (rook, bishop, lance) jumping over an opponent piece.
2. Dropping a piece onto a square where there is an opponent piece.
3. A pawn-drop move that causes the dropped pawn mate.
4. A move after which the king of the attacker remains in check or comes to be in check, for problems with double kings.

---

[1]   Moreover, there may be a foul move in violation of the rule that forbids a fourfold repetition of a certain position with the successive check moves and their responses. However, since there is at present no problem that needs this type of foul moves, we have not yet implemented it.

Fig. 1. A sample problem

The attacker is allowed to try the foul moves up to a certain number of times, typically eight times or less. If the attacker has tried a foul move, the count of fouls increases by one and he only knows that it is a foul move. He does not know a type of the foul. The position does not change at all and the attacker has to choose another move that is possibly a check move. The attacker cannot make a foul move if the count of fouls exceeds the allowed limit. Notice that the attacker cannot cancel a move that has been tried once, and it has to be a check move if it has turned out to be a legal move. If the attacker has made a legal move that does not check the king, it turns out that solving is unsuccessful.

Here, we define the concept *steps* of the sequence as the sum of both the number of the check moves of the attacker and the number of the responses of the opponent from the initial (meta)position to the mated (meta)position in a certain solution tree. In TTS as well as in Tsume-Shogi, the steps are equal to the number of plies of a sequence. The longest steps of a sequence among all sequences in the solution tree are defined as the steps of the solution.

3.3   A Sample Problem of TTS and Its Solution

Here, let us show a sample TTS problem (Figure 1), which is a definite initial position, and its solution. In this figure, 玉, 飛, 角, 金, 銀, 桂, 香, and 歩 denote a piece of king, rook, bishop, gold, silver, knight, lance, and pawn of the attacker, respectively. Similarly, 王, 龍, 馬, 金, 銀, 桂, 香, and 歩 denote a piece of king, rook, bishop, gold, silver, knight, lance, and pawn of the defender, respectively. The pieces beside the board on the right side are the pieces in hand of the attacker, and those on the left side are the pieces in hand of the defender.

The solution is given below. The attacker first makes the silver-dropping move **1. 銀*2c**. There are two moves as its response, **1. …玉1c** and **1. …玉3a**. Next, the attacker tries to make the move **2. 角*1c**. If this move is not allowed (foul type 2), the attacker knows that the king is at **1c**. Then the attacker can mate the king by dropping a gold onto **1d**.

If **2. 角*1c** is allowed, the attacker knows that the king is at **3a**. Then the defender has eleven possible moves as the response. Two moves are **2. …桂x1c** and **2. …香x1c**. The attacker knows these moves because the bishop at **1c** disappears by capturing and can mate the king by dropping a gold onto **3b**.

There still remain nine possible moves as the second response by the defender. Two moves are **2. . . . 玉4a** and **2. . . . 玉4b**. The other seven moves are the interposing moves by dropping a piece onto **2b**. Next, the attacker tries to make the move **3. 角3a+** (with promotion). If this move is not allowed (foul type 1), the attacker knows that the defender has just made an interposing move. Then the attacker can mate the king by dropping a gold onto **3b**.

If **3. 角3a+** is allowed, the attacker knows that the king is at **4a** or **4b**. Then the defender has two possible moves as the response. One move is **3. . . . 玉x3a**. The attacker knows this move because the promoted bishop at **3a** disappears by capturing and can mate the king by dropping a gold onto **3b**. If no capturing has occurred, the attacker knows that the defender has made another move **3. . . . 玉5a** and can mate the king by dropping a gold onto **5b**.

The longest solution sequence with seven steps is either of the below:
> **1. 銀*2c (–) 2. 角*1c (–) 3. 角3a+ (3a) 4. 金*3b mate.**
> **1. 銀*2c (–) 2. 角*1c (–) 3. 角3a+ (–) 4. 金*5b mate.**

Here, '(–)' denotes a response of the defender after which no capturing has occurred, and '(xy)' denotes a response after which the piece at the square xy has been captured by some defender's piece.

3.4   TTS Problem under the Uncertainty Paradigm

Since TTS has uncertainty, the information sets of TTS often consist of many positions. We have dealt with the TTS problems under the Uncertainty Paradigm. In TTS, every metamove at the solver's turn is one of the definite moves and every metamove at the opponent's turn is the uncertain hybrid of moves.

At a certain metaposition where the solver is to move, possible moves are the moves that are either check or foul moves for all positions of the metaposition. One of the possible moves is the metamove to the metaposition. However, it is hopeless to make the metamove that is illegal for all positions of the metaposition, because the solver cannot get any information by making it. No split of the metaposition occurs and the situation is the same as before making that metamove except for increasing the count of fouls. Therefore, the metamove that is a foul move for all positions of the metaposition should be ignored. This is the correct and game-theoretical cut-off.

At a certain metaposition where the opponent is to move, possible moves are all the legal moves of all positions of the metaposition. A metamove is the hybrid of all the possible moves.

The solver can guess the current position and the opponent moves by some clues (observations). The possible observations are:

1. Whether the move of the solver is a check move or a foul move
2. Whether the position is checkmated
3. Whether and where capturing occurs after the opponent move
4. Whether capturing occurs and what kind of piece is captured after the solver's move
5. Whether the countercheck occurs after the opponent move, for problems with double kings

By these observations, a metaposition is split into some metapositions with less uncertainty. Finally, TTS problem is resolved into the search of a generic AND/OR tree under the Uncertainty Paradigm.

4 Implementation of a Problem-Solver

We have implemented the computer program that can solve the TTS problem under the Uncertainty Paradigm.

4.1 Implementation of a Metaposition

Implementation of a metaposition can be categorized into several types. One method is to represent it as itself directly. This method is only applicable to the domains in which a metaposition is very simple or the difference of possible positions of a metaposition is small. However, since a metaposition in TTS problems sometimes has more than hundred positions, we can hardly represent it directly.

Another method is to represent a metaposition as an array of possible positions. Though this method needs much greater memory resources, this is considered the simplest, and probably fastest, method for the domains in which we cannot represent a metaposition directly. Moreover, it is easy to implement and less likely to cause bugs.

A third method is to represent a metaposition as a base position and several sequences of moves from the base position to each possible position. This method can be used for implementing a metaposition in TTS problems. This method reduces the memory requirement drastically compared to the above method. However, computing every position of a metaposition may be time-consuming as the level of the search gets deeper. Moreover, it is difficult to deal with transpositions of metapositions.

Another representation of a metaposition is possible. Partition search [5] is a method to reduce the number of search nodes drastically by introducing the super-nodes that represent several nodes. If we use this method for positions of a metaposition, it seems that the number of positions in every metaposition can be much smaller. Then, it might be better that each metaposition is implemented using more descriptive notation. However, it is not clear that this method is effectively applicable to the domain of chess-like problems.

Consequently, our implementation in TTS represents a metaposition as an array of possible positions.

Compaction of a Metaposition. After a metamove has been made for a metaposition, several positions of the metaposition may happen to be the same. It is necessary to check the existence of same positions and remove duplicate positions of a metaposition after every metamove, assuming our implementation of a metaposition. We call this operation the *compaction of a metaposition*. The compaction of a metaposition is indispensable before encoding it.

Encoding of a Metaposition. Suppose that every position is encoded by some method, we have to manage to combine all codes of positions into a single code. Generally speaking, this is a problem of selecting an appropriate hashing function. Various methods are possible depending on the applying domain and the encoding of a position. We studied how to encode a metaposition in TTS problems [17]. In TTS, every position is encoded as an exclusive-OR summation of pseudo-random numbers by using the Zobrist method. We examined four encoding methods: simple arithmetic summation, exclusive-OR summation, CRC (cyclic redundancy code), and MD5 for all codes of possible positions. We have found that exclusive-OR summation is liable to cause type-1 errors[2] of codes and is inappropriate, and that no type-1 errors are found in arithmetic summation, CRC and MD5, and that arithmetic summation is slightly faster than the others are. Consequently, our implementation has adopted encoding by arithmetic summation.

5   UPS Exploiting a Transposition Table

At the beginning, our developing program had performed the depth-first full-width iterative deepening, which did not use a transposition table [16]. Therefore, it searched the same positions repeatedly during iteration. Now that we have succeeded to encode a metaposition, we are ready to exploit a transposition table for registration, update and retrieval. In this section, we discuss several methods of UPS exploiting transposition tables.

5.1   ID (Iterative Deepening of Depth-First Full-Width Search)

This is a depth-first, full-width iterative deepening using a transposition table, which we abbreviate as ID. Iteration is doubly nested to find a solution of minimum steps and minimum fouls. Outer iteration is for steps and inner iteration is for fouls. This doubly nested iteration guarantees that an obtained solution has minimum steps and fouls.

After deciding the solvable steps and fouls, the re-search is performed to settle an optimal solution sequence. The re-search performs internal iterative deepening at every OR node (at the solver's turn) to find a solution with the shortest steps and fouls. The sequence is recognized as better when:

> The steps of the sequence are fewer than those of another sequence.
> If the above is the same, the fouls of the sequence are fewer than those of another sequence
> If the above is the same, the set of pieces in hand at the mated position is a subset of that of another sequence. (In our implementation, the number of pieces in hand is used for comparison for simplicity instead of evaluating the relation of two sets.)[3]

The program selects the metaposition that has the best sequence at every OR node, and the metaposition that has the worst sequence at every AND-split.

---

[2] A type-1 error occurs when more than one metaposition happens to have the same code. This error is fatal because the error is hard to be detected and may lead to an incorrect result.

[3] However, we recognize the problems in which the solver has no piece in hand at the mated position as 'good' problems.

5.2 PDS (Proof-Number and Disproof-Number Search)

AO* [12] is a well-known and representative algorithm of AND/OR-tree search. However, recently, several novel and highly efficient algorithms have been developed and studied. In the domain of Tsume-Shogi, these algorithms have shown the marvelous solving ability and solved most hard problems with over hundred or thousand steps.

Allis developed the proof-number search introducing the concepts of proof number and disproof number [1]. Seo developed the C* algorithm [20] (which was recently renamed to PN* [21] because there had already existed another C* algorithm), which uses a transposition table and performs iterative deepening increasing the threshold of proof number. PN* was inspired by IDA* [8], which is an iterative deepening version of A* increasing the threshold of estimated cost. However, PN* has a new feature that it performs iterative deepening at every OR node. Seo named this the *multiple iterative deepening*. Moreover, Nagai developed the PDS algorithm [9,10], which performs iterative deepening increasing the thresholds of proof number or disproof number and performs multiple iterative deepening at every AND node as well as at every OR node. Nagai also developed df-pn [11] as a modification of PDS, which is basically a depth-first search but behaves the same as the proof-number search, and df-pn+, which uses the cost estimating function for proving and disproving.

We have compared these algorithms for solving endgames of 6×6 Othello and Tsume-Shogi, both of which are two-agent problems with complete information. Consequently, we have obtained the experimental data showing that the PDS algorithm is superior to others in view of both solving ability and solving speed in Tsume-Shogi [15,18]. Because UPS is a special kind of AND/OR-tree search, we can apply these algorithms in UPS. We have chosen the PDS algorithm, which seems best among them, in UPS of Tsuitate-Tsume-Shogi.[4]

However, the search tree is not a simple AND/OR tree as in Tsume-Shogi but is a generic AND/OR tree. The node at the solver's turn is basically OR node and the edges that stem from it have a relation of OR-condition. However, because some edges AND-split into plural edges, AND nodes and OR nodes come to intermingle. This problem can be settled by introducing an auxiliary node to separate an AND node and an OR node. See Figure 2. Consequently, AND nodes and OR nodes do not necessarily appear by turns in the obtained AND/OR tree.

Because the search algorithms such as PDS, which behave roughly in a best-first manner, are not full-width searches, they do not suit well finding an optimal solution sequence. We have managed to have the program find an approximately optimal solution sequence using the Algorithm 1 indicated below.

Algorithm 1 (Finding an approximately optimal solution in PDS).

1. The first search is performed not imposing any restriction on steps and fouls.
   If the result of this search is proved, it is assured that an optimal solution sequence has less or equal steps and fouls.

---

[4] However, a test set (30 problems) we have used in Tsume-Shogi is not large. Moreover, we have not sufficiently examined the df-pn algorithm. The df-pn algorithm should probably be examined in UPS later.

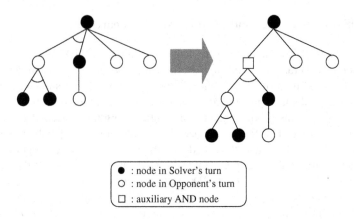

Fig. 2. Transformation of a general AND/OR tree

2. The second search is performed after setting the limit of search steps to the steps obtained minus two.
   If the result of the search is proved, continue the re-search decreasing the limit of search steps.
   If the result of the search is disproved, it is confirmed that no solution does exist with less than a certain steps.
3. Thirdly, the search is performed after setting the limit of allowed fouls to the fouls obtained minus one.
   If the result of the search is proved, continue the re-search decreasing the limit of allowed fouls.
   If the result of the search is disproved, it is turned out that an approximately optimal solution sequence with minimum steps and fouls have been found.

5.3   UPDS (Uncertainty Paradigm PDS)

The searches using the proof number and disproof number have the intention to minimize the cost of search by minimizing the number of nodes during search. This is based on the assumption that the cost of proving or disproving an unexpanded node is averagely equal over the search tree. This assumption is roughly valid in many complete-information two-agent problems including Tsume-Shogi. Therefore, these searches work very well for problems in which the solution tree has a large part of forcing move sequences.

However, in UPS, every node has a property of the uncertainty index (actually, the number of positions). The search cost generally increases as the uncertainty of a node gets larger. This may cause a problem for the variants of the proof-number search in UPS.

First, let us show a schematic example in Figure 3. In this figure, the uncertainty of a node is indicated as the size of a circle. The branches with an arc indicate the AND-split of a metaposition. Suppose that the search has already proceeded partially and the current search node is the root node of this figure after several iterations. The root node has two child nodes, which are denoted by L and R. The node L has more

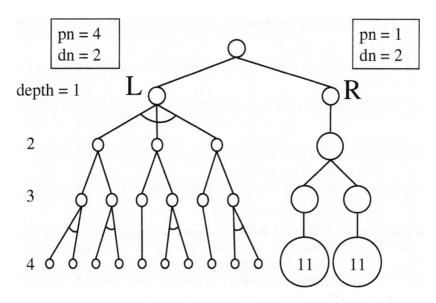

Fig. 3. Schematic example showing a problem of the proof-number search in UPS

descendant nodes, but the uncertainties of those nodes are small. On the other hand, the node R has fewer descendant nodes, but the uncertainties of those nodes get larger as the depth gets deeper. Suppose that all the nodes at the depth 4 are unexpanded. Then, every proof number and disproof number of those nodes is 1. As a result, the proof number and disproof number of the node L are 4 and 2, while the proof number and disproof number of the node R are 1 and 2. Therefore, the proof numbers indicate that the node R is easier to prove than the node L. This conclusion is generally correct in a normal AND/OR tree. However, is it also correct in an AND/OR tree with uncertainty?

To discuss a little quantitatively, let us compare the expenses of several typical parts of computation for a metaposition of which uncertainty index is N (that is, a metaposition that has N positions) with the expenses of computation for a definite position. (Here we restricted the case that a metaposition is represented as an array of positions.) The list of comparison with the case of a definite position is shown in Table 1. The cost of search seems to be at least $N \log N$ times (or N times, depending on the most critical parts) as much as that of the definite case.

In addition, it is almost impossible to unmake the previous metamove for a certain metaposition because every node has to be represented much dynamically in UPS. Consequently, it is also impossible to update the code of a metaposition linked with making/unmaking a metamove. Therefore, the cost of search and the memory requirement become still larger.

Other than the computational expenses, there may exist some relation between the uncertainty index and the easiness of proving/disproving. In TTS, because a metamove at the solver's turn is one element of an intersection of all possible moves for all positions of a metaposition, it is likely to fall into a situation having no metamove as the number of positions of a metaposition gets larger. That is to say, a metaposition that has a large

Table 1. Degree of computational expense in a metaposition compared to the cases in a position

	Solver's turn					
	Move generation	Move ordering	Intersection	Making metamove	Encoding	Compaction
position	M	no need	–	–	1	–
metaposition	NM	$NM \log M$	NM	N	N	$N \log N$
	Opponent's turn					
	Move generation	Move ordering	Intersection	Making metamove	Encoding	Compaction
position	M	no need	–	–	1	–
metaposition	NM	no need	–	NM	N	$N \log N$

N : Uncertainty index of a metaposition,
M : Average number of moves at all positions of a metaposition

uncertainty index is likely to be disproved. On the other hand, because a metaposition is mated at the opponent's turn only if all positions of the metaposition are mated, a metaposition that has a small uncertainty index is likely to be proved.

Considering all the above discussions, we come up with a search that sets the initial value of proof/disproof number of each unexpanded node to the value according to the formulae listed below, rather than 1 in the normal PDS.

$$\text{proof}_{\text{init}} = N^{\gamma_p}/C_p = (N/D_p)^{\gamma_p}$$

$$\text{disproof}_{\text{init}} = N^{\gamma_d}/C_d = (N/D_d)^{\gamma_d}$$

Here γ_p, γ_d, C_p, C_d, D_p, and D_d are the constant values. γ_p or γ_d indicates a degree trying to suppress the combinatorial explosion of uncertainty. As γ_p or γ_d is set to a larger value, the search is more likely to proceed in the way to select a metaposition with less uncertainty. C_p, C_d, D_p, and D_d are the damping factors of positive values. As they are set to larger values, the effect of γ_p or γ_d is suppressed.

This search can be recognized as a kind of PDS*, which is a variation of PDS that uses the cost estimating functions. However, since it also is a specialized version of PDS for the Uncertainty Paradigm, we have named it Uncertainty Paradigm PDS, which is abbreviated to UPDS. In TTS, it is expected that γ_p and γ_d such that

$$\gamma_p \geq 1, \quad \gamma_d \leq 1$$

are appropriate, considering the computational costs and the relation of easiness of proving/disproving. Nevertheless, if we set γ_p to a much larger value than 1, we can suppress the explosion of uncertainty at every metaposition, while we have to be under the apprehension that the search proceeds to a direction in which the number of nodes to be searched increases more and more. Incidentally, if we set $\gamma_p = \gamma_d = 0$, UPDS behaves exactly the same as PDS.

In our actual implementation of UPDS, we adopt the following formulae in order to set the initial values to the natural numbers within a limited range:

$$\text{proof}_{\text{init}} = \frac{\lceil (N/D)^{\gamma_p} \rceil}{\text{IPD}_{\text{max}}} \qquad \text{if } \lceil (N/D)^{\gamma_p} \rceil > \text{IPD}_{\text{max}}$$

$$\text{disproof}_{\text{init}} = \frac{\lceil (N/D)^{\gamma_d} \rceil}{\text{IPD}_{\text{max}}} \qquad \text{if } \lceil (N/D)^{\gamma_d} \rceil > \text{IPD}_{\text{max}}$$

Here IPD_{max} is the upper bound of initial values. We set D_p and D_d to the same value that is denoted by D.

Here, let us back to the example in Figure 3. Suppose that all the uncertainty indexes of ten unexpanded nodes in the left subtree are 1, and the uncertainty indexes of two unexpanded nodes in the right subtree are 11. For example, if we set the parameters γ_p, γ_d, $D(=D_p=D_d)$ to $1.5, 0.5, 4$, respectively, the initial proof numbers and disproof numbers on the left side are 1 and 1, while those on the right side are 5 and 2. As a result, the proof number and disproof number of the node L are 4 and 2, while the proof number and disproof number of the node R are 5 and 4. In this case, the proof numbers indicate that the node L is easier to prove than the node R. This is an opposite conclusion from the case of the normal PDS. In this way, the behavior of search can be controlled by changing these parameters.

5.4   Some Variations of UPDS and ID

dpUPDS (UPDS with Dynamic Parameters). At first, we thought that there must be certain parameters most suitable for all problems of TTS in the search using UPDS algorithm. However, after the experiments of solving all problems using various parameters, we have come to realize that the suitable parameters may vary depending on the characteristics of problems. Some problems have a feature that is nearly like that of complete-information problems. For such problems, it seems that $\gamma_p \sim 0$, $\gamma_d \sim 0$ is appropriate. Other problems need suppressing the uncertainty as much as possible to find a solution. For such problems, it seems that $\gamma_p > 1$, $\gamma_d \sim 1$ is appropriate. Then, we have added to the program a feature that performs the iterations changing these parameters. We have named it dpUPDS, which means UPDS with dynamic parameters. Unfortunately, since we have not known any guideline to find the suitable parameters, this iteration is something like area bombing and is liable to be time-consuming.

In our actual implementation, D is started from 1, then set to 4, and finally set to 16. (γ_p, γ_d) is set successively to the following values: (1.5, 1), (1.5, 0.5), (1.5, 0), (1.2, 0.9), (1.2, 0.4), (1.2, 0), (0.5, 0.3), (0.5, 0), (0.3, 0), (0, 0). Iteration is continued until the problem is solved.

uidUPDS (UPDS with Iterative Deepening of Uncertainty Index). A metaposition that has a large uncertainty index is likely to be disproved in general, though the computation to such a metaposition is much harder. Therefore, we have hit upon the alternative strategy that makes the program perform the iteration increasing the limit of uncertainty index. In a search, a metaposition that has a larger uncertainty index than the limit value is recognized as disproved and not searched further. We have named it uidUPDS, which means UPDS with iterative deepening of uncertainty index. The uidUPDS program seems to be a good variation in solving a problem that has a solution with a little uncertainty.

In our actual implementation, the limit of uncertainty index is increased according to the sequence: 16, 24, 32, 40, ..., 128, 512, 4096, 10000. Iteration is continued until the problem is solved.

uidID (ID with Iterative Deepening of Uncertainty Index). The above iteration of uncertainty index can be also applied to ID (iterative deepening of depth-first full-width search). We have named it uidID. We expect that uidID can find an optimal solution with a little uncertainty in a short time. However, since the level of nesting of iteration is two in ID, the level of nesting of iteration becomes three. Therefore, this search may be inefficient for problems with much uncertainty.

In our actual implementation, the limit of uncertainty index is increased according to the sequence: 16, 32, 48, ..., 128, 512, 4096, 10000. Iteration is continued until the problem is solved.

6   Experimental Results and Discussions

Experiments have been done under the following environment:
> Gateway2000, G6/GP6 Series (TB298-0109)   Pentium II 450MHz,
> RAM: 384MB   Windows 98.

Since both ID and PDS-variants are basically depth-first searches with transposition tables, the memory requirement is small except for transposition tables. The size of a transposition table is 100MB for ID and 134MB for PDS-variants. In addition, because metapositions change dynamically during the search, every metaposition is stored in the heap memory. Therefore, for problems with large branching factors, long steps and much uncertainty, much heap memory is required.

At first, we have examined PDS and UPDS with several sets of parameters. Both γ_p and γ_d are changed in an increasing order from (0,0), which corresponds to the PDS algorithm, to (1.8,0.6). D is fixed to 4. IPD_{max} is set to a fixed value 50 throughout all experiments.[5] The results of solving time are shown in Table 2. For references, the results using the ID search are also shown. Here, time was measured by seconds.

Solving of a TTS problem has multi-stages. The first stage is only for finding one solution. The first stage finds a solvable steps and fouls of the solution. The solution obtained in the stage is not necessarily optimal. The remaining stages are for finding an optimal solution with fewest steps and fouls. In the experiments using the PDS or UPDS search, we have assigned two hours to each stage. We have recognized the problems that have not been solved at the first stage in two hours as unsolved. Even if we give more time, the possibility to solve them is quite small, because the program has almost used up the memory resources or the number of searched nodes gets too large. On the other hand, we have not set a certain time limit for the ID program but continued running at most for 16 days and stopped manually if unsolved.

There are 39 problems in the source [7], which contains some easy problems with seven or nine steps and several hard problems with over twenty steps. Most problems have the solutions with less than four fouls except two problems. One problem (#35) has a solution with 33 steps and 8 fouls and the longest-step problem (#19) has a solution with 43 steps and 4 fouls.

---

[5] Using a few set of parameters, we examined the effect of IPD_{max} changing it to 10, 30, 50, 100, 1000, and 10000. The results when IPD_{max} was 1000 or 10000 seemed worse than the other results. However, we could not get the clear difference among the values from 10 to 100. The value 50 was selected somewhat arbitrarily.

The ID program solved 28 problems but it took much time to solve the relatively long-step problems. The PDS program failed to solve 16 problems. This is worst among the compared methods. Since PDS is a search that aims to minimize the number of search nodes as much as possible, it is likely to fall into the explosion of uncertainty in each metaposition and is unstable. Here, the unstability means that the algorithm happens to solve some hard problems with long steps, but is unable to solve some easier problems with short steps.

As for the parameters of UPDS, among six sets of parameters including PDS, the results of the parameters corresponding to the normal PDS seem to be worst. However, among the remaining five sets, it is not clear to distinguish which set is most suitable. A certain set of parameters is good for some problems, while another set is good for other problems. These results have led us to a new search method dpUPDS described above.

Let us check the speed of solving problems. When we see the results only for solved problems, the average value of solving time of PDS is smallest. (indicated in the 'time ave' row in Table 2) However, the PDS program has not solved several hard problems that the UPDS programs have solved. When we see the results for all problems, the average value of UPDS(1.1,0.4,4) is smallest. (indicated in the 'time all ave' row) Generally speaking, UPDS solves problems fastest among ID, PDS, and UPDS. However, UPDS shows some unstability. For instance, all the UPDS programs could not solve the problem #15 with eleven steps, which was solved in a short time by the ID program.

In view of settling an optimal solution, the iterative deepening that performs the full-width search is superior to the others of course. The Algorithm 1 for PDS or UPDS generally works well. However, for some problems, the search cannot find an optimal solution probably due to the effect of the transpositions.

We further experimented on solving all problems by the programs using uidID, dpUPDS, and uidUPDS. In uidUPDS, γ_p, γ_d, D was set to 0.8, 0.3, 4 respectively, because this set of parameters achieved the slightly superior results in Table 2. Thirty-five problems were solved using the above parameters. For the remaining four problems, the parameters were changed manually with trial and error. The results are listed in Table 3 and the logarithmic plot of solving time against the sum of the steps and fouls of each problem is shown in Figure 4. For references, the results using the ID algorithm are included. All problems were solved at least by one program. Especially, the uidUPDS program solved all problems. Moreover, it generally solved fastest among all programs. However, PDS variants including uidUPDS do not generally guarantee to find an optimal solution. Actually, the UPDS program did not give the optimal solutions in two problems. The uidID program solved 36 problems and mostly gave the optimal solutions. However, in two problems, uidID did not give an optimal solution. This was caused by the fact that the maximum uncertainty of the optimal solution is greater than that of the found solution with longer steps.

The solutions range from 7 steps with no foul to 43 steps and 4 fouls. The maximum value of the maximum number of positions in a metaposition of the solution tree is 66. Here, we show some statistical data of TTS problems based on the results by the ID program that performs a full-width search. The average and the maximum of the number of generated metapositions are 2921297 and 23382373. The average and the maximum of the effective branching factors are 2.20 and 2.89. The average and the maximum of

Table 2. Time of solving TTS problems

#	ST	F	ID	PDS	UPDS				
		γ_p	-	0	0.5	0.8	1.1	1.5	1.8
		γ_d	-	0	0.2	0.3	0.4	0.5	0.6
		D	-	-	4	4	4	4	4
1	7	0	2.5	2.1	0.7	1.2	1.3	1.4	1.3
2	13	1	48.1	12.1	7.7	8.0	7.6	7.9	7.7
3	9	2	37.5	7.7	4.8	6.8	5.0	6.6	6.5
4	9	0	1.0	2.8	1.1	1.0	1.0	1.1	1.0
5	11	0	3.4	1.2	1.1	1.1	1.1	1.2	1.2
6	11	0	14.5	3.7	2.2	1.8	3.0	2.9	2.7
7	11	1	16.8	63.3	34.8	35.6	50.0	35.9	162.9
8	13	3	862.6	398.1	274.8	253.1	296.5	214.0	191.0
9	11	1	27.3	×	32.6	16.2	17.7	16.8	16.8
10	13	1	59.4	20.2	4.9	9.7	92.1	11.5	24.9
11	15	1	640.8	×	×	×	×	52.8	76.7
12	19	2	192.2	×	×	6458.6	3595.7	1788.3	×
13	19	3	14124.8	×	×	×	×	×	×
14	19	2	2476.9	703.0	3793.6	2103.7	×	×	×
15	11	3	103.3	×	×	×	×	×	×
16	9	2	15.4	8.0	8.8	8.4	7.4	10.7	1081.2
17	13	1	323.9	53.7	73.3	18.6	20.6	29.1	32.0
18	15	2	1364240	×	×	×	×	×	×
19	43	4	2977.2	21.9	29.8	27.5	23.5	24.6	18.1
20	9	1	2.5	1.5	0.8	0.8	1.0	0.9	0.9
21	11	0	4.4	9.9	1.5	1.7	4.2	8.8	×
22	19	1	208696	×	×	×	×	×	1862.0
23	29	2	×	×	×	×	×	×	×
24	37	1	×	×	×	×	×	×	×
25	39	1	×	×	9.3	12.1	17.7	×	101.8
26	13	2	42.3	16.6	16.3	15.9	15.8	15.8	16.9
27	15	2	×	×	×	×	×	×	×
28	19	1	×	×	×	4106.3	2795.1	1982.2	3161.1
29	17	2	1249.8	×	1139.3	22.1	21.1	20.1	38.6
30	29	1	×	1822.7	595.9	485.0	562.2	7926.2	1145.0
31	15	2	2787.9	564.7	607.6	×	769.7	460.3	605.5
32	39	1	×	106.9	211.8	99.2	88.6	115.1	109.0
33	7	1	2.8	3.0	2.8	2.7	2.8	2.7	2.7
34	21	2	×	833.7	909.5	5825.3	1064.6	1413.9	×
35	33	8	×	1740.7	1101.2	1747.7	3485.8	14415	14416
36	13	2	784.9	×	×	×	×	×	×
37	27	3	×	×	×	×	×	×	×
38	9	1	1.1	0.6	0.8	0.8	0.8	0.8	0.9
39	21	1	×	×	8597.6	8608.7	×	×	×
solved			28	23	27	28	27	27	26
time total			1599739	6398	17465	29880	12952	28566	23084
time ave			57133.5	278.2	646.8	1067.1	479.7	1058.0	887.8
time all total			-	122395	105991	109632	99253	115246	109784
time all ave			-	3138.3	2717.7	2811.1	2544.9	2955.0	2889.0
depth ave				29.92	29.40	29.55	29.35	29.40	31.39
uncertainty ave			4.15	11.09	3.96	3.61	3.60	3.16	3.02
nodes ave			2921297	150256	415857	446240	440545	502017	532583

#:problem number, ST,F:steps and fouls of the solution (not necessarily optimal), γ_p, γ_d, D:parameters of UPDS, ×:unsolved,
solved:the number of solved problems, time total(ave):total(average) time for only solved problems, time all total(ave):total(average) time for all problems, depth ave:average depth of the search, uncertainty ave:average uncertainty of the search, nodes ave:average number of generated nodes in the search.

Table 3. Results of solving TTS problems by search variations

	ID	uidID	dpUPDS	uidUPDS
solved	28	36	32	39
unsolved	11	3	7	0
total time	1599739	355784	164647	165088
average time	57133.5	9882.9	5145.2	4233.0

the maximum number of possible positions among the searched metapositions are 987.7 and 10572. The average and the maximum of the average number of possible positions are 4.15 and 10.85.

7 Conclusions and Future Works

We have introduced the Uncertainty Paradigm for deterministic solving of problems with uncertainty. Then the search space of the problem is resolved into a plain AND/OR tree, which has a metaposition as a node and a metamove as an edge. Since solving such a problem is the same as solving an AND/OR tree, we can use the methods that have been developed and studied for years in the domain of artificial intelligence. The efficient search algorithms, which have been developed for the search of an AND/OR tree with complete information, are also applicable to our approach. However, there is a trade-off between reducing the search nodes and suppressing the explosion of uncertainty of each node. Therefore, it was necessary to find the algorithm that could proceed a search keeping the balance whether the search nodes should be minimized or whether the uncertainty should be minimized. We have selected the PDS algorithm and developed a modified algorithm UPDS specialized for UPS. Then, the search has come to be controlled by changing the parameters of UPDS. Moreover, we have developed some variations of these searches. Finally, we have succeeded to solve all problems in the domain of TTS. The problems that is unable to be solved by the full-width search, due to the explosion of uncertainty or due to the explosion of the number of searched nodes, have come to be solved smartly. The disproving ability of PDS, UPDS or its variants is also high, for example, as shown by the fact that the Algorithm 1 works well for not a few problems. On the other hand, the disproving ability of ID or its variants is quite poor. Therefore, the search algorithms using proof/disproof numbers are very promising in UPS, as well as in the normal AND/OR-tree search.

However, UPDS that we have tested in this paper is not yet sophisticated. We should examine the other formulae giving the initial value of proof/disproof number, or enable to change the parameters at run-time according to the property of the problem. Though the current UPDS is domain-independent, the domain-dependent UPDS might perform better. This is a remaining subject in the future.

We could take the next target of our approach to apply to the endgame problems of Kriegspiel. Our approach is directly applicable to Kriegspiel endgames. However, while the solver's moves are restricted to the check moves in TTS problems, they are not restricted in Kriegspiel endgames. Therefore, the search space might be larger than TTS, though the position of a problem is generally much simpler than that in TTS.

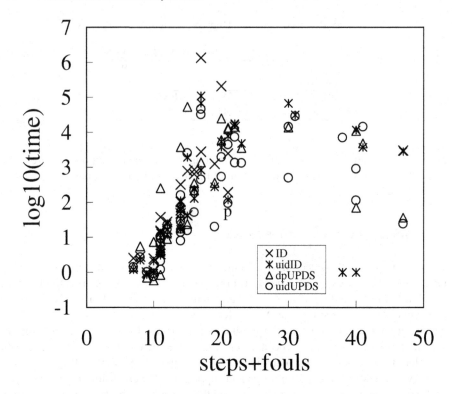

Fig. 4. Logarithmic plot of solving time against the sum of steps and fouls

In addition, the solutions of Kriegspiel problems often include a randomized mixed strategy of the multistage recursive game [2,3]. In order to solve this type of problem, we should relax the constraint of deterministic solving (that is to say, semi-deterministic or quasi-deterministic) and make the program exploit the randomized mixed strategy if necessary.

References

1. Louis Victor Allis. *Searching for Solutions in Games and Artificial Intelligence.* PhD thesis, University of Limburg, Maastricht, The Netherlands, 1994.
2. P. Ciancarini, F. Dalla Libera, and F. Maran. Decision making under uncertainty: A rational approach to Kriegspiel. In H. Jaap van den Herik and Jos W. H. M. Uiterwijk, editors, *Advances in Computer Chess 8*, pages 277–298. Drukkerij Van Spijk B.V., Venlo, The Netherlands, 1997.
3. Thomas S. Ferguson. Mate with bishop and knight in kriegspiel. *Theoretical Computer Science*, 96:389–403, 1992.
4. Ian Frank and David Basin. Search in games with incomplete information: a case study using Bridge card play. *Artificial Intelligence*, 100:87–123, 1998.
5. Matthew L. Ginsberg. Partition Search. In *Proceedings of Thirteenth National Conference on Artificial Intelligence*, pages 228–233, Portland, Oregon, 1996.

6. Reijer Grimbergen. A survey of Tsume-Shogi programs using variable-depth search. In H. Jaap van den Herik and Hiroyuki Iida, editors, *Proceedings of First International Conference on Computers and Games CG'98*, volume 1558 of *Lecture Notes in Computer Science*, pages 300–317, Heidelberg, 1999. Springer.

7. Tetsu Kato. Collection of Kapitein documents No.1. Explanations and problems of Tsuitate-Tsume-Shogi (in Japanese), 1995.

8. Richard E. Korf. Depth-First Iterative-Deepening: An Optimal Admissible Tree Search. *Artificial Intelligence*, 27:97–109, 1985.

9. Ayumu Nagai. A new AND/OR tree search algorithm using proof number and disproof number. In *Proceedings of Complex Games Lab Workshop*, pages 40–45, ETL, Tsukuba, November 1998.

10. Ayumu Nagai. A new depth-first-search algorithm for AND/OR trees. M.Sc. thesis, Department of Information Science, University of Tokyo, Japan, 1999.

11. Ayumu Nagai and Hiroshi Imai. Application of df-pn+ to Othello endgames. In *Proceedings of Game Programming Workshop in Japan '99*, pages 16–23, Hakone, Japan, October 1999.

12. Nils J. Nilsson. *Principles of Artificial Intelligence*. Tioga Publishing Company, Palo Alto, CA, 1980.

13. Guillermo Owen. *Game Theory*. Academic Press, New York, third edition, 1995.

14. D. B. Pritchard. *The Encyclopedia of Chess Variants*. Games & Puzzles Publications, Godalming, UK, 1994.

15. Makoto Sakuta and Hiroyuki Iida. An empirical comparison of innovative AND/OR-tree search algorithms on Tsume-Shogi (in Japanese, with English abstract). *Studies in Information, Shizuoka University*, 5:15–22, 1999.

16. Makoto Sakuta and Hiroyuki Iida. Solving problems with uncertainty: A case study using Tsuitate-Tsume-Shogi. In *Proceedings of Game Programming Workshop in Japan '99*, pages 145–152, Hakone, Japan, October 1999.

17. Makoto Sakuta and Hiroyuki Iida. Solving Kriegspiel-like problems: Exploiting a transposition table. *ICGA Journal*, 23(4):218–229, 2000.

18. Makoto Sakuta and Hiroyuki Iida. The performance of PN*, PDS, and PN Search on 6x6 Othello and Tsume-Shogi. In H. Jaap van den Herik and B. Monien, editors, *Advances in Computer Games 9*, pages 203–222. Universiteit Maastricht, The Netherlands, 2001.

19. Makoto Sakuta, Hiroyuki Iida, and Jin Yoshimura. Solving problems under Uncertainty Paradigm. In *Proceedings of International Conference on Advances in Infrastructure for Electronic Business, Science and Education on the Internet (SSGRR 2000)*, L'Aquila, Italy, August 2000. (in CD-ROM).

20. Masahiro Seo. The C* algorithm for AND/OR tree search and its application to a Tsume-Shogi program. M.Sc. thesis, Department of Information Science, University of Tokyo, Japan, 1995.

21. Masahiro Seo, Hiroyuki Iida, and Jos W.H.M. Uiterwijk. The PN*-search algorithm: Application to tsume-shogi. *Artificial Intelligence*, 129(1-2):253–277, 2001.

Strategies for the Automatic Construction of Opening Books

Thomas R. Lincke

Institut für Theoretische Informatik, ETH Zürich,
CH-8092 Zürich, Switzerland.
thomas.lincke@inf.ethz.ch

Abstract. An opening book is an important feature of any game-playing computer program. These books used to be constructed manually by an expert, by storing good moves suggested by theory, or simply by listing all games ever played by strong players [2,5,8]. Interest has recently shifted to automatic opening book construction where positions are selected by a best-first strategy, evaluated using a brute force search and then added to the opening book [3].

This paper presents the new "drop-out expansion" strategy for automatic opening book construction. It generalizes the previously used best-first strategy and reduces the opportunities for the opponent to force the player out of the book. The algorithm was used to calculate opening books for several games, including Awari and Othello, and helped to win the Awari tournament of the Computer Olympiad [10].

Keywords: opening book construction, expansion strategy, best-first, Awari, Othello

1 Introduction

Games are usually divided into three phases: opening, middlegame and endgame. In any of the three phases, the default action of a computer program is to start a brute force search for a "best" move. Since such a search has to be performed within a limited time, it can only examine nodes down to a certain depth, which means that the calculated value is only a heuristic approximation of the game-theoretic value.

However, in endgames a different approach is possible if the game has the convergence property [1], i.e. the number of pieces on the board decreases monotonically. In this case we are able to construct an endgame database of precalculated game-theoretic values for positions with a sufficiently small number of stones [9]. A computer program can benefit from such a database in two ways: first, the values of the endgame positions can be retrieved in one operation, and second, the retrieved values are exact game-theoretic values of the positions instead of approximations. The disadvantage is that we need additional space to store the database.

For openings an analogous approach is possible. Since the state space up to a few plies into a game is small, we can precalculate a database (called "opening book", or just "book") of values for positions that are likely to occur at the beginning of a tournament game. If the opening book is stored as a directed graph, with positions as nodes and

T.A. Marsland and I. Frank (Eds.): CG 2000, LNCS 2063, pp. 74–86, 2001.

moves as arcs, then the values can be propagated within the book. This way the computer program does not only save time compared to brute force search, but also obtains better values, assuming the search depth used for precalculation is greater than the one used during play. Again there is a tradeoff: on the one hand we save time (required for brute force search) and improve value accuracy, on the other hand we need additional space to store the opening book.

Although the benefits of using endgame databases and opening books are similar, algorithms for the construction of endgame databases have received more attention up to the present. The reason for this seems to be twofold: values in endgame databases are exact game-theoretic values, whereas values in the opening book are (mostly) heuristic values. This makes it hard to judge the usefulness of an opening book. Another reason is that efficient indexing functions for complete enumeration can be constructed for positions in endgame databases, whereas for positions in the opening book no efficient indexing functions do exist, due to the fact that the relevance of a position for the opening book stems from the position value and not from the configuration of stones on the board.

There are two approaches to automatic opening book construction. In the first approach, which may be called *passive book construction*, games are collected and, after analysis, used to update or extend the opening book. This approach is discussed in [5]. The basic idea is that if a move has been played before, then it is likely to be good and therefore worth to be added to the book. The main advantage of this approach is that it avoids losing twice using the same opening line: whenever a game is lost, that game is added to the book, which is then modified to play a different line the next time.

In the second approach, which may be called *active book construction*, the current opening book is analysed, and an expansion priority is assigned to all opening lines. The line with the highest priority is then expanded and the whole process is repeated [3]. The main advantage of this approach is that it does not depend on the availability of previously played games.

My main motivation for this work was the construction of an opening book for Awari. Because only a few dozen published Awari games exist, only active book construction could be used. However, for best results both approaches should be combined whenever possible.

In this paper I present a new expansion strategy for active book construction, called "drop-out expansion", which generalizes the previously used best-first strategy [3]. Section 2 sets the basic framework. Section 3 introduces the new expansion strategy and discusses its benefits compared to best-first strategy. Section 4 is a short description of OPLIB, an opening book construction tool, and section 5 summarizes the results.

2 Book Construction Basics

2.1 Book Representation

An opening book is represented as a directed graph. The nodes of the graph represent positions, and an arc between two nodes represents a legal move. One node, called the start node, represents the start position, and all other nodes must be reachable from it. If a node has an edge for each of its moves, then it is called an *interior node*, otherwise it is called a *leaf node*.

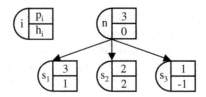

Fig. 1. Node representation

Every node i has two attributes: the heuristic value h_i and the propagated value p_i. The value h_i is computed by the search engine, using a brute force search. For interior nodes, p_i is the negamax value of p_{s_j} of all successor nodes s_j. For leaf nodes, p_i is equal to h_i.

$$p_i = \begin{array}{ll} \max_{s_j}(-p_{s_j}) & \text{for interior nodes} \\ h_i & \text{for leaf nodes} \end{array}$$

For book construction it is not actually necessary to keep h_i once i has become an interior node. However, for testing the expansion strategies, it is useful to be able to compare h_i and p_i during the calculation. Unless explicitly stated otherwise, let *value* mean *propagated value*.

Figure 1 shows an expample of a node with three successors. To improve readability, all figures in this paper use max-propagation only.

2.2 Book Expansion

When an opening book is created, it contains only the start node. Expansion is done in three steps:

1. Choose a leaf node and add all successors to the book
2. Calculate the values of the new successors
3. Propagate the values

This paper deals with the first step: how to choose the next node for expansion. Step two, the calculation of a (heuristic) value, is a topic of ongoing research in the field of computer games. I simply assume that a state-of-the-art search engine is available. Once the value is calculated, it will be negamax propagated through the graph (step three).

2.3 Goals

When is an opening book good? Of course, the ultimate goal in every tournament game is to win. Since it will rarely be possible to win a game straight from the book, its main benefit is that during the opening phase of the game search-time is saved, which may be used later in the game to outsearch the opponent. Therefore, the primary goal is to maximize the expected number of moves one can play within the book.

Which nodes should we expand to achieve this goal? A naive approach to guarantee a minimal number of moves from the book is to enumerate all positions at depth 1 from the start position, calculate their values and store them. Next, the same is done for all positions at depth 2, and so on. This will, however, waste a lot of time and space on positions which are very unlikely to occur in a tournament game, because to reach them some player would have to make an obviously bad move. Therefore the secondary goal must be to achieve the primary goal with as few expansions as possible.

3   Expansion Strategies

The goals in Section 2.3 suggest that an expansion strategy should choose nodes for expansion according to the likelihood of their occurrence in a game. Since we can assume that good moves are more likely to occur than bad moves, the most straightforward strategy is best-first. This strategy was implemented in [3], it is used here as a reference against which I want to compare the new strategy.

For the following discussion of expansion strategies I will assume that the game engine using the book will only play optimal book moves. This is a reasonable assumption because there is no point in constructing an opening book and then to ignore it. However there are also good reasons to choose suboptimal moves now and then, for example to reduce predictability of the game engine.

3.1   Best-First Expansion

The rule for best-first expansion is: Expand the leaf node that is reached by following a path of best moves from the start node. If an interior node has more than one best move, choose one of them at random. See Figure 2.

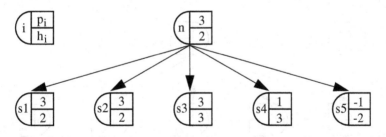

Fig. 2. Example for best-first expansion. The best value among all successors of n is 3, thus the candidate set for expansion is $\{s_1, s_2, s_3\}$

This strategy is simple and ignores bad moves, but it has a major flaw: suppose we just expanded the start position of chess, and the engine returned 0.1 (measured in pawns) for e2–e4, and 0 for all other moves. Expansion will now continue along e2–e4, and it is easily possible that the value for that move will always be > 0, so all other moves from the start node will be ignored forever. This violates the goal of maximizing the expected number of book moves, because the common move d2–d4 by the opponent leads to a position that is not in the book, i.e. we "drop out" of the book.

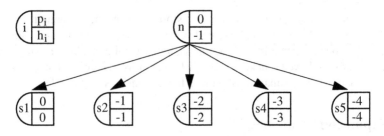

Fig. 3. Successor s_1 is the only best move

Of course the assumption that the value of the best move will only increase is weak. But even if we assume that the propagated value of a node will oscillate in an interval around its original heuristic value, the problem remains.

Take, for example, the situation in Figure 3, where successor s_1 currently is the only best move. Assume that during further expansion the propagated values remain in the range $[h_i - 2, h_i + 2]$. Then s_2 and s_3 may eventually have values larger than that of s_1, the favorite candidate, and thus become eligible for expansion. However if their values ever become less than -2 they will never be selected again, because the value of s_1 will never go below -2. At some point, a situation similar to Figure 4 will be reached, where the values of the alternative moves get stuck just below the lowest value which the optimal move has ever reached. The resulting successor values are misleading, because they are biased to look worse than what they are. This again leads to early drop-out, because the depth of the book after move s_2 will remain shallow, but it is still likely to be played by an opponent.

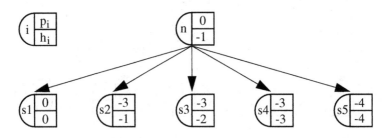

Fig. 4. The successors s_2 and s_3 are expanded until their values fall below -2

The heart of the problem with early drop-out is the fact that we use the same engine to search for values for both the first and the second player. We implicitly assume that the opponent uses the same evaluation function as we do, which is not necessarily true. It would be more reasonable to assume that the evaluation function of the opponent is only similar, and to expand some suboptimal move alternatives in a controlled way.

There is another minor flaw in best-first expansion. Because we make a random choice if more than one best move is available, the probability of choosing a specific leaf node on a best path depends on the number of best moves at any node along the path. This means that the book grows faster along some paths. We could solve this if we first made a list of all leaf nodes reachable with best moves, and then chose one at random, but that would be both time and space consuming. Instead we prefer to have a strategy that solves this by making a series of local decisions.

3.2   Drop-Out Diagrams

A drop-out diagram shows the depth and value of all leaf nodes that can be reached, under the assumption that the book player only makes optimal moves and that the opponent is allowed to make any move. Figure 5 shows the possible drop-out nodes of a small Othello opening book calculated with best-first expansion. The root value of the book is 0.9, a leaf node with this value is only reached if the opponent only plays optimal moves too. If the opponent plays any suboptimal move, then leaf nodes with higher values may be reached.

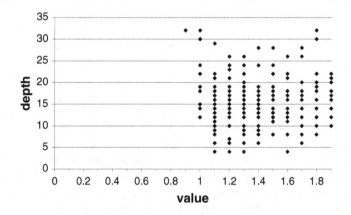

Fig. 5. Drop-out diagram of a small Othello book calculated with best-first expansion

Each dot in the diagram represents one or several candidate leaf nodes for expansion: Which one should we choose? Or, to reformulate the question, if you were the opponent of the book player: Which leaf node would you try to reach? Obviously, if the opponent only plays optimal moves, the book player will be able to play at least 32 plies from the book. However, as the plot shows, there is a line in the book which ends at ply 4, with a value that is only 0.2 points worse than the optimal value. If any opponent of this book player ever finds out about this line, he will start playing this line, and therefore this is probably the best candidate for expansion.

In a drop-out diagram like Figure 5, best-first strategy expands nodes from left to right. On the other hand, breadth-first strategy expands nodes from bottom to top. But what we are looking for is a strategy that expands nodes from bottom-left to top-right.

3.3 Drop-Out Expansion

The problem with best-first expansion was, that in situations as in Figure 4 only the best move is considered for expansion, whereas it is not unlikely that an opponent will play the second best move. To solve this we now consider all moves for expansion, and give each move a priority depending on the depth of the book following the move and the difference between the best value and the value of the move. A successor has high expansion priority if it is a good move and/or it has a shallow subbook, and a successor has low expansion priority if it is a bad move and/or it has a deep subbook. To calculate the expansion priorities we add two new attributes, epb_i and epo_i, to the nodes.

epb_i is the expansion priority for when it is the book player's move (1). It is initialized to zero in leaf nodes, and depends only on the expansion priority of the optimal successors. The $+1$ is the depth penalty, it guarantees that shallow nodes have higher priorities.

epo_i is the expansion priority for when it is the opponent's move (2). It is initialized to zero in leaf nodes, and depends on the expansion priority of all successors. Besides the depth penalty ($+1$), suboptimal moves get an additional penalty which depends on the value difference to the optimal move.

$$epb_i = \begin{array}{ll} 1 + \min_{optimal\ s_j}(epo_{s_j}) & \text{for interior nodes} \\ 0 & \text{for leaf nodes} \end{array} \tag{1}$$

$$epo_i = \begin{array}{ll} 1 + \min_{s_j}(epb_{s_j} + \omega(p_i - p_{s_j})) & \text{for interior nodes} \\ 0 & \text{for leaf nodes} \end{array} \tag{2}$$

ω is the weight for the difference $p_i - p_{s_j}$ between the optimal value and the value of the suboptimal successor s_j. It should be ≥ 0, the right choice of ω is game specific and depends on the heuristic value resolution, i.e. $+1$ may mean "one piece ahead" or "0.01 pieces ahead". A low value for ω means higher priority for suboptimal moves, if $\omega = 0$ then all successors will be expanded to the same depth, regardless of their values. On the other hand, if $\omega \to \infty$ then drop-out expansion degenerates into best-first expansion because only optimal moves will be expanded.

What do we gain if we use drop-out expansion? Obviously a move that is only slightly worse than the best move will not be ignored forever, even if the best value never decreases. With increasing depth of the best move, the priority for the expansion of suboptimal moves will increase too and this will eventually lead to their expansion. For the same reason it will not happen that a suboptimal move gets stuck with a bad value, as was shown in Figure 4 for best-first strategy. Thus all the problems observed with best-first expansion have been solved.

An additional benefit from drop-out expansion is that the parameter ω can be used to control the shape of the opening book. The user can choose any shape between full expansion of every line and best-first expansion.

The benefit from drop-out expansion may also be understood as a kind of insurance: if the opponent wants to force a drop-out, he has to pay with a move that is so bad that it has not been considered for expansion yet. If he keeps playing good moves, we will not drop out of the book early. So, at the end of the opening, we have either a good position, or we have saved lots of time, or a combination of both.

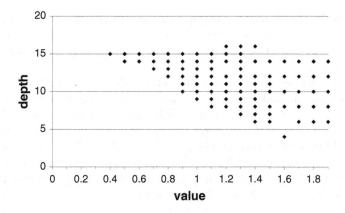

Fig. 6. Drop-out diagram of a small Othello book calculated with $\omega = 1.0$

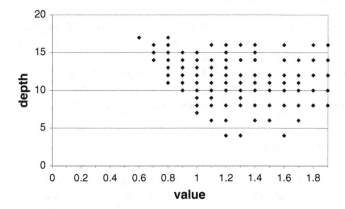

Fig. 7. Drop-out diagram of a small Othello book calculated with $\omega = 2.0$

3.4   Further Enhancements

So far our reasoning has led us to consider not only the values of successor nodes, but also the depths to which they have already been expanded. Thereby the subset of leaf nodes considered for expansion was changed to avoid the problems found with best-first expansion. However, the candidate set of leaf nodes changes most radically when the value of the start node changes. It would be of advantage if we could settle that value first.

This is exactly what conspiracy number search [6] [7] can do: Expand a leaf node that is reached by following a path of best moves from the start node and which is most likely to change the value of the start node. Two new attributes need to be added to each node, one to count the number of leaf nodes that have to change to increase the value of the node, and one to count the number of leaf nodes that have to change to decrease the value.

While conspiracy number expansion is useful to try to settle the value of a node with the smallest number of expansions possible, the expanded nodes may be way off in a deep remote line, hardly useful with respect to the primary goal, the maximization of the number of expected moves in the book. Therefore conspiracy number expansion should be used only in conjunction with drop-out expansion, and only if the value of the start node is unstable.

Another enhancement is the use of fractional depth. The depth of the successor nodes is not incremented by 1 (as in (1) and (2)), but instead a value within the range $0.01 \leq fractd_i \leq 1$ is used, which depends on the difference between the best and the second best value. If the difference is large (or if there is only one move), then the fractional depth is small, otherwise large.

This favors the expansion of lines with unique or almost unique moves. The increased expansion of these moves is justified by the fact that they do not contribute to the exponential growth of the book as the other moves do.

3.5 Other Considerations

The task of opening book construction would be simplified if we had a good model of the opponent, i.e. if we could predict the opponent's moves with high accuracy. For drop-out expansion I proposed a linear function, $\omega(p_i - p_{s_j})$, for the value-dependant penalty. This is a very simple opponent model, where ω is the estimated similarity of the book player's engine and the opponent's engine. I also considered some non-linear functions, but abandoned the idea because of a lack of efficient implementations.

4 Implementation

All the strategies described in this paper, plus a few experimental ones, were implemented in OPLIB, a game independent software tool for two player games on cyclic graphs. OPLIB is based on an implementation of DCGs with attributed nodes and arcs. On top of the internal DCG, it provides a variety of features such as adding and editing of nodes and arcs, retrieval of lists of successors and predecessors, value propagation, transposition and cycle detection, and statistics. A well defined interface exists through which OPLIB can access game specific functions. So far, the game specific functions have been implemented for Awari, checkers, chess and Othello. Large opening books with about 500000 nodes have been constructed for Awari and Othello, and smaller ones for chess and checkers.

Storing an opening book as a DCG induces some overhead with respect to disk space. For instance, to store one Othello position in the book requires approximately 120 bytes on average, compared to a straightforward encoding of an Othello position requiring only 16 bytes. About 50% of the 120 bytes is used to store the graph structure (nodes and arcs), another 25% to store attribute information and the rest for a hash table for position lookup.

However, it does not make sense to optimize disk usage anyway. With 120 bytes per position we could store more than 80 million positions in 10GBytes, which is well within the limits of current disk storage technology.

The real bottleneck in opening book construction is the preprocessing time. In tournament games, the time available for one move is usually limited to about 3 minutes.

To ensure a competitive quality of the values in the leaf nodes, the search engine of the opening book expander must search for a comparable amount of time per position. Calculating 20 positions per hour, or 500 positions per day, it would take about 440 years to fill an opening book with 80 million positions.

To speed up the construction process, OPLIB can also be run in a distributed mode. Search engine clients running on a network of workstations can then fetch a position from the server, calculate a value and send it back to the server, which then updates the opening book and finds a new position for the client. Up to 50 workstations at a time have been used for book construction.

5   Results

The calculation of large opening books is an ongoing project. Currently I have books with about 300000 nodes for Othello and about 800000 nodes for Awari. Both were constructed using a mix of the strategies described in this paper, as well as some experimental ones, with drop-out expansion responsible for more than 80% of all expansions.

Figure 8 shows the depth distribution and the drop-out diagramm of the nodes in the Othello book. The majority of the nodes is concentrated in the depth range from 10 to 20 plies. The deepest lines were expanded to a depth of 41 plies. The fraction of the number of leaf nodes from depth 15 to 41 is almost constant between 90% and 95%, meaning that about one out of ten successors was expanded. With average successor counts between 10 and 15 moves, this suggests that the deep lines were expanded by best-first expansion. Note that state-of-the-art Othello engines usually can solve positions at depths between 35 and 40 plies. To avoid the influence of solved positions on the shape of the book, the engine was configured to solve positions at depth 43. Thus the book does not yet contain any solved positions.

In the Awari book (Fig.9), the nodes are spread out over a much larger range, the deepest lines go down to 91 plies. This node distribution can partially be attributed to the low average successor count of about 5 moves. But the Awari engine also makes use of endgame databases, which help to solve a substantial number of positions from the 5th ply on, and also help to narrow the book by easily refuting bad moves. This seems to prevent exponential growth of the number of nodes with increasing depth. It is conjectured that the shape of the book was mostly influenced by the endgame databases, and not by the choice of expansion strategy.

An open question is how to measure the performance of the expansion strategies against each other. For instance, a self-play test will always favor the best-first strategy, because it works best against similar opponents. On the other hand, if tested against a different engine, then the outcome might be more influenced by the relative strengths of the search engines than by the chosen expansion strategy.

6   Conclusions

I have shown that, for opening book construction, best-first expansion has certain deficiencies. For instance, it may completely ignore alternative moves with values only slightly inferior to the best value, and it has a tendency to stop expansion of suboptimal

Depth	Nodes	Leafs	%Leafs
0	1	0	0%
1	1	0	0%
2	3	0	0%
3	14	0	0%
4	60	23	38%
5	180	58	32%
6	631	282	45%
7	1709	1003	59%
8	3786	2456	65%
9	7240	5148	71%
10	13053	10149	78%
11	18851	15239	81%
12	25699	21867	85%
13	29230	25549	87%
14	30774	27479	89%
15	28681	25763	90%
16	26758	24317	91%
17	22911	20876	91%
18	18489	16927	92%
19	15091	13783	91%
20	12259	11091	90%

Depth	Nodes	Leafs	%Leafs
21	11234	10434	93%
22	7652	7001	91%
23	6341	5841	92%
24	4419	4021	91%
25	3963	3662	92%
26	2623	2374	91%
27	2139	1941	91%
28	1674	1487	89%
29	1556	1464	94%
30	861	798	93%
31	681	639	94%
32	402	368	92%
33	370	337	91%
34	272	245	90%
35	283	259	92%
36	191	175	92%
37	186	168	90%
38	136	128	94%
39	61	59	97%
40	24	23	96%
41	10	10	100%

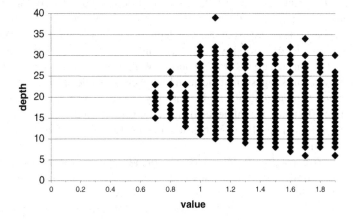

Fig. 8. Depth and drop-out statistics for the Othello book

moves with a misleadingly low value. In both cases, a lucky or an informed opponent can force us to drop out of the book with only a small penalty for him. Both problems are related to the implicit assumption that the opponent uses the same evaluation function.

I propose a new strategy, drop-out expansion, which, in a user controlled way, also considers suboptimal moves for expansion. This not only avoids the problems with best-first strategy, but also gives the user the flexibility to control the growth of the opening book between full-breadth expansion and best-first expansion. For best results, drop-out expansion can be combined into a mixed strategy with conspiracy number expansion.

The flexibility of drop-out expansion can also be used to tune a book to an opponent: if the opponent is known to play similar moves, then parameters can be chosen to construct a best-first like opening book. If the opponent is known to play differing moves often, then parameters can be chosen to expand more alternative moves.

Depth	Nodes	Leafs	%Leafs	Solved	%Solved
0	1	0	0%	0	0%
1	6	0	0%	0	0%
2	36	21	58%	0	0%
3	77	40	52%	0	0%
4	190	128	67%	0	0%
5	319	185	58%	1	0%
6	698	462	66%	1	0%
7	1161	844	73%	17	1%
8	1554	1122	72%	15	1%
9	2103	1560	74%	52	2%
10	2578	1934	75%	102	4%
11	3045	2310	76%	126	4%
12	3452	2643	77%	233	7%
13	3853	3018	78%	233	6%
14	3903	3055	78%	367	9%
15	3874	3021	78%	356	9%
16	3899	3008	77%	474	12%
17	4034	3118	77%	551	14%
18	4062	3183	78%	691	17%
19	3934	3089	79%	650	17%
20	3710	2865	77%	787	21%

Depth	Nodes	Leafs	%Leafs	Solved	%Solved
21	3734	2848	76%	830	22%
22	3948	3126	79%	1029	26%
23	3642	2809	77%	990	27%
24	3681	2918	79%	1134	31%
25	3400	2725	80%	1152	34%
26	3000	2372	79%	1114	37%
27	2857	2240	78%	1032	36%
28	2726	2101	77%	1161	43%
29	2839	2248	79%	1092	38%
30	2493	1867	75%	1271	51%
31	2810	2225	79%	1123	40%
32	2484	1845	74%	1580	64%
33	2850	2243	79%	1052	37%
34	2547	1922	75%	1726	68%
35	2863	2303	80%	1009	35%
36	2416	1787	74%	1769	73%
37	2927	2311	79%	1048	36%
38	2625	1960	75%	2017	77%
39	3129	2522	81%	1206	39%
40	2617	2031	78%	2029	78%
41	2721	2208	81%	1086	40%

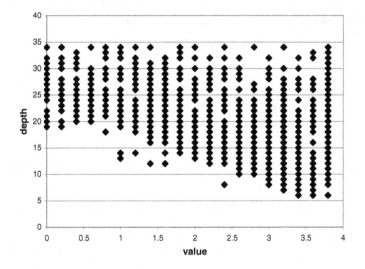

Fig. 9. Depth and drop-out statistics for the Awari book

All expansion strategies mentioned in this paper were implemented on OPLIB, an opening book tool. The results for Othello and Awari show the feasibility of automatic construction of large opening books. A tool like OPLIB may also play a major role in solving games in the future, because it supports the user in managing the solution tree (or solution graph) manually, as was the case with previously solved games [1][4].

Acknowledgments

My thanks go to Alvaro Fussen for letting me use his Othello engine, and to Nora Sleumer for her many helpful comments on earlier versions of this paper.

References

1. L. V. Allis. *Searching for Solutions in Games and Artificial Intelligence*. PhD thesis, University of Limburg, Maastricht, The Netherlands, 1994.
2. M. G. Brockington. KEYANO Unplugged - The Construction of an Othello Program. Technical Report 97-05, Department of Computing Science, University of Alberta.
3. M. Buro. Toward Opening Book Learning. *ICCA Journal*, 22(2):98–102, 1999.
4. R. U. Gasser. *Harnessing Computational Resources for Efficient Exhaustive Search*. PhD thesis, ETH Zürich, 1995.
5. R. M. Hyatt. Book Learning - A Methodology to Tune an Opening Book Automatically. *ICCA Journal*, 22(1):3–12, 1999.
6. D. A. McAllester. Conspiracy Numbers for Min-Max Search. *Artificial Intelligence*, 35:287–310, 1988.
7. J. Schaeffer. Conspiracy Numbers. *Artificial Intelligence*, 43:67–84, 1989.
8. J. Schaeffer, R. Lake, P. Lu, and M. Bryant. Chinook: The World Man-Machine Checkers Champion. *AI Magazine*, 17(1):21–29, 1996.
9. K. Thompson. Retrograde Analysis of Certain Endgames. *ICCA Journal*, 9(3):131–139, 1986.
10. R. van der Goot and T. R. Lincke. Marvin Wins Awari Tournament. *ICGA Journal*, 23(3):173–174, 2000.

Awari Retrograde Analysis

Roel van der Goot

Department of Computing Science,
University of Alberta, Edmonton AB, Canada,
roel@cs.ualberta.ca

Abstract. This paper introduces an alternative algorithm for Retrograde Analysis in Awari. A parallel version of this algorithm created Awari endgame databases of up to 35 pebbles. Both sequential and parallel versions of the algorithm are compared to sequential and parallel versions of the traditional Retrograde Analysis algorithm.

Keywords: Retrograde Analysis, Awari, endgame database.

1 Introduction

Board games can more or less be divided into two categories. The first category consist of games where the pieces on the board increases, because a player's move consists of putting a piece on the board. Examples of these games are Tick Tack Toe, Connect Four, Go Muko, Go, Othello, Hex, and Amazons. The second category consists of games where the number of pieces on the board decreases because a player can do a move that captures pieces. Games that belong to this category are Chess, Checkers, Draughts, Backgammon, Awari, and Lines of Action.

For both categories search algorithms can proof the game result for positions near the end of a game. However, for games in the first category the number of endgame positions is so big that enumerating all of them is sheer impossible (except for trivial games like Tick Tack Toe). For games of the second category, the number of positions near the end of the game is small. Usually small enough to traverse them all, and collect their game values in a database, a so called *endgame database*.

2 Retrograde Analysis

Retrograde Analysis (RA) calculates endgame databases by pushing the values of final positions towards the initial position. In contrast, *alpha-beta* searches from an initial position towards the final positions.

First, RA identifies all final positions in which the game value (usually win, draw or loss) is known. By making *reverse moves* from these final positions the game value of some non-final positions can be deduced. And by making reverse moves from these newly proven non-final positions, the game value of other non-final positions can be deduced, *ad infinitum*.

Ströhlein was the first researcher who came up with the idea to create endgame databases. He applied his idea to chess [1]. Ken Thompson computed many four and

T.A. Marsland and I. Frank (Eds.): CG 2000, LNCS 2063, pp. 87–95, 2001.

five piece chess databases [2], which he later extended with many of the six piece databases. Lewis Stiller calculated five and six piece chess databases, but only recorded the positions with the most moves to win [3]. In Checkers, Jonathan Schaeffer and others built eight piece Checker databases [4], which enabled his program Chinook to win the world man-machine checkers championship. Schaeffer also created endgame databases for single agent search (the 15 puzzle) [5]. Ralph Gasser proved that the game of Nine Men's Morris is a draw by creating a database and then searching from the initial position towards the database [6].

3   Awari

Awari is one of the oldest game known to mankind and is still being played today. The game originates from Africa and many rules variations exist. We will focus at the rules as they have been used in the Computer Olympiads. Worth noting is that Awari is also the only remaining game on Victor Allis's list of games to be solved by the year 2000 [7].

3.1   Rules of Awari

Awari is a board game for two players (called North and South). The Awari board (see Figure 1) consists of twelve pits, six on North's side labeled 'a' through 'f,' and six on South's side labeled 'A' through 'F.' Two mancala pits that are not part of the playing area are used to collect captured pebbles. North collects its pebbles in the mancala labeled 'm,' South collects its pebbles in the mancala labeled 'M.' The initial position contains four pebbles in each pit, making 48 pebbles total (see Figure 2). South moves first.

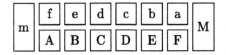   

Fig. 1. The Awari Board   **Fig. 2.** The Initial Position

A player's move consists of choosing a non-empty pit on his/her side of the board, grabbing all the pebbles in it, and sowing them counter-clockwise around the board starting with the right hand neighboring pit while skipping the original pit. If the last pebble ends on the opponent's side of the board in a pit that contains two or three pebbles (including the ones sown by the move), the player who just moved captures all pebbles in that pit. He/she continues capturing pebbles in preceding (clockwise) pits as long as they are on the opponents side of the board *and* contain two or three pebbles.

If the player to move has a choice, he must leave his opponent with a move. If however the player to move runs out of moves, his opponent captures all remaining pebbles on the board. If during a game, a position repeats itself, the game is over. The player who captures most pebbles wins. In case both players capture the same number of pebbles the game is drawn.

4 Retrograde Analysis for Awari

Retrograde Analysis for Awari is different from the RA for the games discussed in Section 4. Where all other games only store the result as win, loss or draw, Awari stores the number of pebbles that can be won. This means that for the n pebble endgame database the results range from $-n$ to $+n$. This is actually an optimization to reduce the number of positions in the database that can be achieved by excluding the mancalas from the positions.

4.1 Previous Sequential Retrograde Analysis for Awari

Victor Allis and others introduced a RA algorithm for Awari [8] based on the above observation. Unfortunately the description of his algorithm is at a very high level, so my description of it may not be completely accurate. The algorithm uses the same technique used for chess databases. However, Awari has $2p + 1$ values (from $-p$ to $+p$) and Chess has only three values (win, draw or loss). Hence their algorithm for Awari iterates p times. In the nth iteration ($n < p$) all positions with values $p - n$ and $n - p$ are deduced. After p iterations, all remaining positions are assigned the value 0 (similar to chess where after one iteration all remaining positions are assigned the value draw).

Allis noticed that the calculation of the index in the database for a position was very time consuming. As a result he wanted to refrain as much as possible from computing them. He considered creating a file that contained for each position the indices of its children. But he found that that would make the program I/O bound. He settled for a solution where every position is extended with a counter and a state. The position's counter indicates how many of its children still have to send their final game values. Once this counter reaches zero, the position sends its final game value to its predecessor and updates its state to final.

4.2 Alternative Sequential Retrograde Analysis for Awari

The alternative RA algorithm introduced in this paper tries to push the game values from the final positions towards the initial value just like Allis's RA algorithm. Where their algorithm uses *reverse moves* and propagates *fixed* game values, the alternative algorithm uses *moves* and propagates *evolving* game values.

The alternative RA algorithm is explained in a simplified version first. Later we will see that an optimization can be performed that saves memory space as well as computation time. The simple RA algorithm consists of two phases: an *initialization phase* and an *iteration phase*. The initialization phase initializes the database. The iteration phase repeatedly iterates over all positions in the new endgame database. Each position looks up the evolving game value of the positions that are reachable from it in one move. The best evolving game value found is stored in the database for the current position. The iteration phase iterates until the endgame database converges on a stable value assignment.

One improvement to the above algorithm is to introduce a third phase that is executed before the other two phases. This *capture phase* calculates an intermediate capture database that reduces the work that has to be done in the iteration phase and reduces the memory usage of the algorithm. The capture database has the same size as the eventual endgame database, however it only looks at capture moves from the positions in the database. Hence, the capture database contains minimum values for the endgame database. Figure 3 shows the pseudo code for this enhanced algorithm.

What are the advantages of this algorithm over the traditional algorithm?

- The programming of Awari moves instead of reverse moves is much easier. Awari's reverse moves are unnatural and contain many exceptions, and are a source for programming errors.
- Since moves are easier to implement than reverse moves, it is easier to optimize them, which means that it is much faster to do moves than it is to do reverse moves.
- The proposed algorithm uses less memory than the previous one because a position does not have to remember its state nor the number of children that still have to send their game value. This allows the alternative algorithm to calculate larger databases.
- The proposed algorithm does not need to have the smaller endgame databases in memory. Again resulting in the possibility to create bigger databases.

For a speed comparison between both RA algorithms we use a time that is documented in [9]. Creating the 17 pebble endgame database takes 39 hours and 25 minutes on a 50MHz Sparc Classic for the traditional algorithm. The alternative algorithm uses 1 hour and 13 minutes on a 300MHz Mips R12000 processor.

4.3   Previous Parallel Retrograde Analysis for Awari

Henri Bal and Victor Allis calculated Awari endgame databases on a cluster of workstations [9]. They parallelized the sequential algorithm described in Section 4.1. Every workstation gets a disjoint piece of the endgame database. The RA algorithm pushes the value of every position to its preceding positions one reverse move away. On a cluster this usually means a communication between two workstations. Since communication is inherently slow (several thousands of cpu cycles), it is becoming easily a bottleneck in the distributed RA algorithm.

Remember from Section 4.1 that Allis already reduced the number of updates (and hence communications) by maintaining extra data for each position. The number of communications is reduced further by combining communications between workstations.

4.4   Alternative Parallel Retrograde Analysis for Awari

The algorithm described in Section 4.2 has been parallelized for a shared memory machine with 16 Gbyte of memory. The parallel algorithm uses Posix threads. Every thread is responsible for a part of the capture database and the endgame database. In the first iteration of the iteration phase, every thread gets the same amount of work. Because some positions have more non-capture moves than others the amounts of work that the threads get is updated after the completion of each iteration. This allows for better parallelism in the later iterations.

The parallel RA algorithm calculated optimal play for all positions with 35 or less pebbles on the board. Note that the 35 pebble endgame database is the biggest database

```
procedure create_capture_db(number_of_pieces)
begin
    for every position with number_of_pieces do
        max_value = -INFINITY;
        for every capture move in position do
            do_move(position, move);
            max_value = max(max_value,
                    - get_endgame_value(position));
            undo_move(position, move);
        od
        set_capture_db_value(position, max_value);
    od
end

procedure initialize_endgame_db(number_of_pieces)
begin
    for every position with number_of_pieces do
        set_endgame_db_value(position, 0);
    od
end

procedure create_endgame_db(number_of_pieces)
begin
    create_capture_db(number_of_pieces);
    initialize_endgame_db(number_of_pieces);
    do
        changed = false;
        for every position with number_of_pieces do
            max_value = get_capture_db_value(position);
            for every non-capture move in position do
                do_move(position, move);
                max_value = max(max_value,
                        - get_endgame_value(position));
                undo_move(position, move);
            od
            if (max_value ≠ get_endgame_db_value(position))
                set_endgame_db_value(position, max_value);
                changed = true;
            fi
        od
    while changed
end
```

Fig. 3. The alternative Retrograde Analysis algorithm.

Table 1. Endgame databases: their sizes, number of iterations it took to stabilize and the time (in seconds) of the first iteration.

Pebbles	Positions	Cumulative	Iterations	Time/Iteration
0	1	1	1	0
1	12	13	2	0
2	78	91	5	0
3	364	455	15	0
4	1,365	1,820	19	0
5	4,368	6,188	33	0
6	12,376	18,564	36	0
7	31,824	50,388	34	0
8	75,582	125,970	34	0
9	167,960	293,930	45	0
10	352,716	646,646	37	0
11	705,432	1,352,078	60	0
12	1,352,078	2,704,156	84	0
13	2,496,144	5,200,300	60	1
14	4,457,400	9,657,700	44	1
15	7,726,160	17,383,860	45	1
16	13,037,895	30,421,755	50	1
17	21,474,180	51,895,935	46	2
18	34,597,290	86,493,225	45	3
19	54,627,300	141,120,525	42	7
20	84,672,315	225,792,840	50	11
21	129,024,480	354,817,320	51	11
22	193,536,720	548,354,040	57	19
23	286,097,760	834,451,800	49	37
24	417,225,900	1,251,677,700	46	51
25	600,805,296	1,852,482,996	63	61
26	854,992,152	2,707,475,148	65	89
27	1,203,322,288	3,910,797,436	58	128
28	1,676,056,044	5,586,853,480	63	168
29	2,311,801,440	7,898,654,920	77	238
30	3,159,461,968	11,058,116,888	56	343
31	4,280,561,376	15,338,678,264	64	461
32	5,752,004,349	21,090,682,613	60	639
33	7,669,339,132	28,760,021,745	68	937
34	10,150,595,910	38,910,617,655	66	1345
35	13,340,783,196	52,251,400,851	71	1820
36	17,417,133,617	69,668,534,468		
37	22,595,200,368	92,263,734,836		
38	29,135,916,264	121,399,651,100		
39	37,353,738,800	158,753,389,900		
40	47,626,016,970	206,379,406,870		
41	60,403,728,840	266,783,135,710		
42	76,223,753,060	343,006,888,770		
43	95,722,852,680	438,729,741,450		
44	119,653,565,850	558,383,307,300		
45	148,902,215,280	707,285,522,580		
46	184,509,266,760	891,794,789,340		
47	0	891,794,789,340		
48	279,871,768,995	1,171,666,558,325		

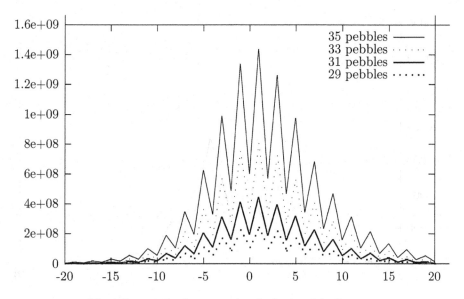

Fig. 4. Frequencies for game values in the Awari databases.

that we can calculate with the current algorithm without the machine starting to swap. Table 1 shows how many iterations the parallel iteration phase took. This is an indication of how many moves it takes before a capture occurs. The table shows also how long (in seconds) the first iteration lasted on 64 processors. The table does not contain total calculation times because of run-time restriction on the Origin (it is shared among researchers). The redistribution of the work over the threads results on average in an 8% speed improvement in the later iterations.

The parallel RA algorithm of Bal and Allis uses 49 minutes and 36 seconds on 64 Sparc Classics running at 50MHz and connected by a 10 BaseT Ethernet to calculate the 17 pebble endgame database. The alternative algorithm uses 1 minute and 23 seconds on a 64 300MHz processor Origin 2400 to calculate the same database. Actually the 17 pebble endgame database is too small for this machine.

5   Endgame Database Statistics

Figure 4 shows the frequency of game values for several endgame databases. The figure shows a big odd/even effect in the database. This has two reasons, (*i*) captures usually remove two pebbles from the board, and (*ii*) optimal play only results in repeating positions in the endgame.

An interesting question is: "What is the game value of the initial position?" Figure 4 gives the impression that North has the edge, because it is a big advantage to have the move. The mean of all graphs is positive (+1.83 pebbles for the 35 pebble database). However this is for an arbitrary position in the database. Many of the positions in the database for example allow the initial player to capture immediately.

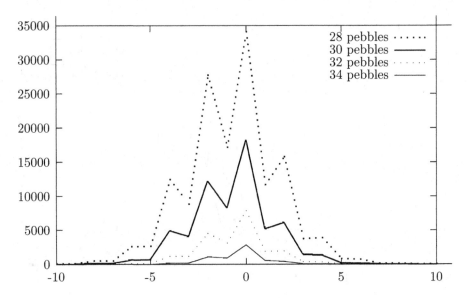

Fig. 5. Frequencies for game values of symmetric positions with at most four pebbles in a pit and no immediate captures.

To give a better answer to the aforementioned question, the database is queried for positions that resemble the initial position. Figure 5 shows the frequency of game values for symmetric positions with at most four pebbles in a pit and no immediate captures. The mean for these positions in the 34 pebble database is -0.43 pebbles, indicating that North has a small edge to win the game.

However this advantage is higher for the smaller databases and lower for the bigger databases, which may indicates that the initial position in Awari is a draw. This is according to me the most likely scenario.

6   Solving Awari

Will Awari survive the year 2000? [7]

- Yes, it will. The 35 pebble endgame database only contains 4.4% of all Awari positions ($52 * 10^9 / 1.2 * 10^{12}$ from Table 1).
- No, it will not. The 35 pebble endgame database covers 73% of an Awari game (35/48).

In short, it all depends on how you look at it. According to me Awari will be solved soon. This prediction is based partially on the knowledge that Thomas Lincke from ETH Zürich, who has an excellent web page on Awari endgame databases [10], and the author of this article are competing to solve the game. And as always competition accelerates progress and sparks innovation. The author is considering another RA algorithm that uses less memory (one bit instead of one byte) but more time to calculate. This algorithm would allow to build the 44 pebble endgame database.

References

1. Ströhlein, T.: Untersuchungen über Kombinatorische Spiele. PhD thesis, TH München, Germany (1970)
2. Thompson, K.: Retrograde Analysis of Certain Endgames. Journal of the International Computer Chess Association **9** (1986) 131–139
3. Stiller, L.: Group Graphs and Computational Symmetry on Massively Parallel Architecture. Journal of Supercomputing **5** (1991) 99–117
4. Lake, R., Schaeffer, J., Lu, P.: Solving Large Retrograde Analysis Problems Using a Network of Workstations. Advances in Computer Chess VII (1994) 135–162
5. Culberson, J., Schaeffer, J.: Efficiently Searching the 15-Puzzle. Technical Report TR 94-08, Department of Computing Science, University of Alberta (1994)
6. Gasser, R.: Efficiently Harnessing Computational Resources for Exhaustive Search. PhD thesis, ETH Zürich, Switzerland (1995)
7. Allis, L., van den Herik, H., Herschberg, I.: Which Games Will Survive? Heuristic Programming in Artificial Intelligence 2 (1991) 232–243
8. Allis, L., van der Meulen, M., van den Herik, H.: Databases in Awari. Heuristic Programming in Artificial Intelligence 2 (1991) 73–86
9. Bal, H., Allis, L.: Parallel Retrograde Analysis on a Distributed System. In: Supercomputing'95, IEEE (1995)
10. Lincke, T.: Awari Endgame Database. http://wwwjn.inf.ethz.ch/games/awari/ (2000)

Construction of Chinese Chess Endgame Databases by Retrograde Analysis[*]

Haw-ren Fang[1], Tsan-sheng Hsu[2], and Shun-chin Hsu[3]

[1] Department of Computer Science
University of Maryland
A.V. Williams Building, College Park, Maryland 20742, USA
hrfang@cs.umd.edu
[2] Institute of Information Science
Academia Sinica
No 128, Section 2, Academia Road, Nankang, Taipei 115, Taiwan
tshsu@iis.sinica.edu.tw
[3] Department of Computer Science and Information Engineering
National Taiwan University
No 1, Section 4, Roosevelt Road, Taipei 106, Taiwan
schsu@csie.ntu.edu.tw

Abstract. *Retrograde analysis* is well-known and has been successfully developed in the design of Western chess[1] endgame databases. However, there is little achievement using this technique in the construction of Chinese chess endgame databases. Although the two types of chess have the same number of pieces, similar individual characteristics for pieces, and comparable scales of the size of the boards, the fundamental differences in their playing rules lead to different construction schemes and results of endgame databases.

In this paper, we describe our approach to the construction of Chinese Chess Endgame Databases when only one of the players possesses attacking piece(s). We show the results we have in constructing and analyzing a set of 151 endgame databases with a total of at most two attacking pieces, four defending pieces and two Kings. Our databases can be used by Chinese chess computer playing systems and computer aided Chinese chess training systems.

Keywords: computer Chinese chess, endgame databases, retrograde analysis, index scheme, algorithm.

1 Introduction

The study and playing of games and puzzles benefitted from knowledge databases abstracted from them. In Western chess, retrograde analysis has been successfully applied to construct endgame databases. For example, a constrained 6-piece endgame,

---

[*] This work was done when the first author was with Institute of Information Science, Academia Sinica, Taiwan.

[1] To avoid the ambiguous usage of the term chess, we use Western chess and Chinese chess in this paper.

T.A. Marsland and I. Frank (Eds.): CG 2000, LNCS 2063, pp. 96–114, 2001.
© Springer-Verlag Berlin Heidelberg 2001

KRP(a2)KbBP(a3), is constructed using *heuristic* retrograde analysis [7], and the complete 5-piece databases to solve KPPKP endgame are also available [3]. On the other hand, little is known of Chinese chess endgame databases in comparable scale until recently [9,10]. It is prerequisite to construct them for better endgame studies and improve the Chinese chess playing programs.

The structure of this paper is as follows. Section 2 introduces the notations and rules of Chinese chess, fundamental definitions and models for our problem, and the indexing scheme used by our databases. Section 3 shows our algorithm for constructing the databases. Section 4 describes rules of Chinese chess that have no similar counterparts in Western chess and how it might affect the databases. Section 5 gives conclusions. Finally, in Appendix A, we show the statistics of the set of databases that we have constructed. Appendix B gives some interesting information obtained from mining those databases.

2   Preliminaries

2.1   Notations and Rules

In Chinese chess, two sides are called *Red* and *Black*. Each side has one King, two Guards, two Ministers, two Rooks, two Knights, two Cannons and five Pawns, whose naming convention follows that of ICCS[2] and is employed through this paper[3]. These pieces are abbreviated as K, G, M, R, N, C and P, respectively. As shown in Fig. 1(a), the board of Chinese chess consists of nine vertical and ten horizontal lines. Pieces are located at the intersection of a vertical and a horizontal line, which is called an *address* of the board. For convenience, we use the addressing method proposed by ICCS, which is similar to what is commonly used in Western chess. When the game starts, the pieces of each side occupy half of the board separated by the *river*, the imaginary stream between the two central horizontal lines, as represented in Fig. 1(b). For example, the Red King is at e0 at the beginning of a game. We term the pieces that can move across the river, i.e., Rook, Knight, Cannon and Pawn, to be *attacking pieces*. In contrast, King, Guard and Minister are termed as *defending pieces* because they are confined in the domestic region[4]. A side is *armless* if it has no attacking pieces. This paper studies the endgames with exactly one armless side.

A *position* in Chinese chess is an assignment of a subset of pieces to distinct addresses on the board with a given player-to-move. Two positions are *conjugate* if they have the same player-to-move and each position is a mirror image to the other, i.e., every piece of one position is in a contra-position with respect to the central vertical line to the same piece of the other.

---

[2] ICCS with one server located at `iccs.ixa.org` stands for *Internet Chinese Chess Server*, which is widely used by Chinese chess players all over the world, including many masters.

[3] Another version of naming convention of pieces is proposed by *WXF, World Xiangqi Federation*. They are King, Adviser, Elephant, Chariot, Horse, Cannon and Pawn, respectively.

[4] More detailed information of Chinese chess such as notations and rules in English can be found in FAQ of the Internet news group `rec.games.chinese-chess`, which is available on `http://txa.ipoline.com/`.

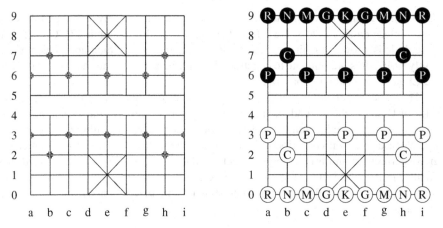

Fig. 1. An empty board and the starting position of Chinese chess.

In Chinese chess, if the opponent King is in check, denoted as *OKIC*, then the player who makes the move to **OKIC** loses. In addition, if the opponent makes a move to a position where the two Kings facing each other, denoted as *KFK*, which means the two Kings are in the same vertical line and there are no other pieces in-between, then the player whose move leads to **KFK** position loses.

The moves toward **OKIC** or **KFK** positions are *illegal* because they make the moving player immediately lose the game. A position is called *checkmate* if the current player has no legal moves and his King is in check. A position is called *stalemate* if the current player has no legal moves and his King is not in check. The current player of a checkmate position loses in both Western and Chinese chess. On the other hand, the current player of a stalemate position loses in Chinese chess, while it is considered a draw in Western chess.

2.2   Graph Representation and Properties

Given a subset of pieces S, we use a *state graph* to denote all possible positions with pieces $S' \subseteq S$ and the moves of them. This state graph is a finite, directed, bipartite and cyclic graph. The vertices are the positions, or *states*, and each directed edge indicates the corresponding move from one state to another, with the relationship of the *parent* state moving to the *child* state.

In a directed graph, the *out-degree* of a vertex v is the number of edges that are directed out of v, i.e., the number of children of v. We denote the specific stalemate state which has no child by *STALEMATE*. Note that all the children of a stalemate state are either **OKIC** or **KFK**, if any. An *end state* is defined as either **OKIC**, **KFK**, or **STALEMATE**. Note that the out-degree of an end state is 0.

Our databases ignore the *60-move-rule*[5]. A state, or position, is called a *win state* if the current player always has a way to (1) reach an **OKIC** or **KFK** state, (2) force

---

[5] According to [1], if there are no pieces being captured in sixty moves, the game ends in a draw. It is called the 60-move-rule, which is similar to the 50-move-rule in Western chess.

the opponent to reach a **STALEMATE** state, or (3) force the opponent to violate some special rules defined and discussed in Section 4 no matter how the opponent reacts. It is a *loss state* if the opponent always has a way to reach a win state no matter how the current player reacts. It is a *draw state* if the current player always has a way to prevent the opponent from reaching a win state, and vice versa.

For any finite, two-player, zero-sum and perfect information game, such as Western or Chinese chess, the following properties are observed for the child states of a non-end state in a state graph: (1) a win state must have at least one loss child state, (2) a loss state has nothing but win child states, and (3) a draw state has at least a draw child state and no loss child states. A *playing strategy* of a game for a player is a set of rules that the player follows to make the next move using the information available in the state graph. An *infallible* playing strategy is one that can be used to win the game if starting from a win state, and to not lose the game if starting from a draw state. A *perfect* playing strategy is not only infallible but also able to play optimally in the strong sense of fastest win and slowest loss.

2.3 Structure of Chinese Chess Endgames

We associate each position of the game with a *position value* that is either one of the following types: *win-draw-loss*, *distance-to-mate* or *distance-to-conversion*. The details of the above three types of position values are discussed in Section 2.4.

Similar to the notations widely used in Western chess endgame databases, a string of capital characters indicates all positions with the corresponding remaining pieces. For example, KCCKGGM endgame refers to all positions with one King and two Cannons on the Red side and one King, two Guards and one Minister on the Black side. Since an endgame with no attacking pieces on either side is trivial, this kind of endgame is not considered in this paper.

KPK is called a *supporting database* of KPKG since one state in KPKG may convert to another in KPK after the capture of the Black Guard by the Red side. The relationship between databases is represented by a directed and acyclic graph in which the vertices are the databases and each directed edge from P to Q indicates that Q is a supporting database of P. Given a string of characters S representing the pieces on the board, we call the endgame S and all its supporting databases together a *complete hierarchy of endgame databases* of S. In such a hierarchy, the endgame S is called a *source* database. The endgames without any supporting databases are called *sink* databases. An example for the string KRKGGMM is illustrated in Fig. 2. A complete hierarchy of endgame databases of S corresponds to the state graph of the positions with pieces $S' \subseteq S$.

2.4 Position Values

The three most popular types of position values, win-draw-loss, distance-to-mate and distance-to-conversion in Western Chess, are introduced here and applied to our Chinese chess endgame databases. In the first type, each position is associated with a flag indicating that whether the current player can win/draw/lose the game starting from this position, respectively. Using this type of position values in the endgame database, one

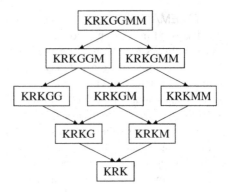

Fig. 2. The hierarchy of the KRKGGMM endgame database and all of its supporting databases.

cannot derive an infallible playing strategy as a player might wander in the win states without reaching an end one.

Starting from a state q and no matter how the opponent reacts, the *distance-to-mate* value of q is the maximum number of plies the current player can win the game if q is a win state; it is the minimum number of plies that the current player can defer his loss if q is a loss state. Hence, the distance-to-mate value of a non-end win state is exactly one greater than the minimum distance-to-mate value of all its loss child states. Note that a non-end win state has at least one loss child state. The distance-to-mate value of a non-end loss state is exactly one greater than the maximum distance-to-mate value of all its child states, which are win states by definition. With the databases in the distance-to-mate metric, one can play perfectly as follows: for a non-end state, move it to the child state whose distance-to-mate value is one less than it has if it is not a draw state; otherwise, move it to a draw child state. Thus, we can easily design a perfect playing strategy using the distance-to-mate type of position values.

Although the distance-to-mate encoding can be used to design a perfect playing strategy, its values are considerably larger than those of distance-to-conversion. Given a non-end win position q in a database, its distance-to-conversion value equals that of an end win state if it has a loss child state in supporting databases; otherwise, it is one plus the minimum distance-to-conversion value of all its loss child states. For a non-end lose state q, its distance-to-conversion value equals that of an end loss state if all of its child states are in supporting databases; otherwise, it is one plus the maximum distance-to-conversion value of its all win child states.

A player can reach an end state with the optimum number of plies by the databases in distance-to-mate metric. Similarly, a player can reach an end state or a state in some supporting databases with the optimum number of plies using the distance-to-conversion type of position values. Moreover, a player from a win state can always reach an end state and win by the databases in the distance-to-conversion metric by a playing strategy similar to that using the distance-to-mate metric. Thus, distance-to-conversion is infallible.

In summary, storing win-draw-loss position values is most compact, but moving among the win positions endlessly may lead to a draw. Distance-to-conversion is infallible but imperfect because of the possibility of detour to win. Finally, distance-to-mate is perfect; however, it has the disadvantage of having larger position values. All our databases listed in Appendix A are in distance-to-mate metric. Distance-to-conversion is applied only when overflow occurs in some position values, which are currently encoded in one byte.

The encoding of positions in our databases is slightly different from that described in [8]. In [8], one major meaning of the position value is the extreme number of plies to the checkmate position, the distance-to-mate, whereas in our databases, it means the extreme number of plies to the capture of King, which is exactly two greater than that in [8]. The detailed explanations are listed in Table 1. Note that all position values of win states are odd and those of loss states are even.

Table 1. The meanings of position values in our databases.

Position Value	Meaning of Position Value	
	distance-to-mate	distance-to-conversion
0	draw	draw
1	OKIC or KFK	the win position has a loss child in the supporting databases
2	checkmate or stalemate	the loss position has all its children in the supporting databases
$n=3, 5,..., 253$	current player can capture opponent's King in n plies	current player can capture some piece in n plies and then win
$n=4, 6,..., 254$	opponent can capture current player's King in n plies	opponent can capture some piece in $n-2$ plies and then win
$n=255$	temporary flag *Unknown* used during the construction	temporary flag *Unknown* used during the construction

By using this convention, the positions are ordered according to their values to achieve the goal of fastest win and slowest loss. In general, win positions are favored over draw positions, which are favored over the loss positions. Moreover, for the win positions, the smaller the value the better. For the loss positions on the contrary, the larger the value the better.

2.5 The Indexing Scheme

We follow more or less the same fundamental principles of Western chess [5,6] in designing our indexing scheme to compactly map all the possible positions into the databases. Here we focus our discussion on how our indexing scheme of Chinese chess differs from that of Western chess because of the subtle differences between these two games.

Given a state, we first partition the pieces of each side into the following subsets: Rook(s), Knight(s), Cannon(s), Pawn(s), and the *rest*, which includes the King and the defending piece(s). For each subset S, we first define an ordering on all possible placements of the piece(s) in S. The number of all possible arrangements is called the *unit-weight* of S. For convenience, the unit-weight of S is 1 if $S = \emptyset$. Then we find the *local index*, which is the index of the current arrangement of piece(s) in S within the above order. Thus each subset S is first locally optimally indexed. Then we order the subsets by placing the subsets of the Red piece(s) before the subsets of the Black piece(s). Within subsets of piece(s) of each side, we first place the subset of the rest pieces, following by the subsets of Rook(s), Knight(s), Cannon(s), and Pawn(s), if any. The *weight* of a subset is the multiplication of the unit-weights of the subsets proceeding it. The final index value of a given state is the sum of products of the local index and the weight of each subset.

In both Western chess and Chinese chess, two conjugate positions have the same position value, which results in space redundancy called *conjugate redundancy*.

In Chinese chess, with the board having nine columns, the size of index space after removing conjugate redundancy for each subset can be a bit more than half of the original one by our modified indexing scheme described as follows. The subset of four or five Pawns are excluded here because we feel they should not be appearing in an endgame by intuition. To index a state, at most one of its subsets can be further compactly indexed to remove conjugate redundancy; otherwise, the endgame databases are unreliable because two positions with different assignment of pieces may have the same index number but different position values. To remove the conjugate redundancy within a given subset of pieces S, we generate a new order on all possible placements of S on the board such that there is no conjugate positions in the new order. Hence this new unit-weight of S is reduced. The amount of reduction is shown in Table 2. In our scheme for removing conjugate redundancy, we pick one subset of attacking pieces for removing the most conjugate redundancy[6].

3   A Sequential Algorithm of Retrograde Analysis

Here we show a sequential algorithm to construct the endgame databases using retrograde analysis which is similar to the one in [2].

In the following, we first state a general algorithm to compute the distance-to-mate position values for the states in a hierarchy of databases all together. Then we describe the modification needed if we want to construct a database at a time using a a bottom-up fashion, i.e., a database is constructed after all of its supporting databases are done. With the fact that a non-**KFK** position is draw if both sides have no attacking pieces, endgame databases consisted of the two Kings and one attacking piece has no supporting databases. Thus those sink databases can be built first. We then show the modification when we need to use the distance-to-conversion type of position values. Finally, we describe an algorithm that independently verifies the correctness of the constructed database.

3.1   Constructing a Perfect Endgame Database

Given a set of pieces S, we first associate each vertex i.e., state or position, in the state graph for S with two variables. One, called $UnknownChildren$ is the current number

---

[6] For some databases, picking the set of the King and the defending piece(s) to be indexed to remove the conjugate redundancy can result in more efficient index scheme. However, we have not implemented it yet.

Table 2. The number of all possible assignments for each subset of pieces on the board.

Subset of Pieces	Number of all possible assignments of piece(s) (unit-weight)		Ratio (a)/(b)
	(a) after removing conjugate redundancy	(b) before removing conjugate redundancy	
PPP	13135	26235	50.068%
RR, NN or CC	2045	4005	51.061%
PP	765	1485	51.515%
R, N or C	50	90	55.556%
P	31	55	56.364%
KGM	138	275	50.182%
KGMM	408	810	50.370%
KGGM	242	480	50.417%
KGGMM	714	1410	50.638%
KM	32	62	51.613%
KMM	96	183	52.459%
KG	21	40	52.500%
KGG	38	70	54.286%
K	6	9	66.667%

of child states whose final position values are still unknown. This variable is set to the special value $Propagated$ if the position value of the current state is finalized. The other, called $BestValue$, is the current best position value for this state derived from the child states whose position values are finalized. Because a state may have no child states with finalized position values, we initialize the special value $Unknown$ to a state if needed. For technical reasons, we make $Propagated$ to be a value that does not equal to any possible numbers of unknown child states, and $Unknown$ to be a special value that will be updated by any propagated values. Note that the edges of the state graph is not stored and is computed whenever needed. Note also we apply the redundancy removing scheme described in Section 2.5 in indexing the vertices.

The constructing process consists of three phases: initialization, propagation, and final. In the initialization phase, each end state, i.e. **OKIC**, **KFK** or **STALEMATE**, sets its $UnknownChildren$ to be 0 and its $BestValue$ the position value according to Table 1. For each non-end state, we set its $BestValue$ to be $Unknown$ and its $UnknownChildren$ to be the number of its child states. After initialization, the $UnknownChildren$ and $BestValue$ variables of all the states with position value 1 have been properly set.

In the propagation phase, we iteratively traverse the vertices until no $BestValue$ variable changes during a complete round of traversal. We keep a traversal index initialized as 1 and increased by 1 after each round of traversal. During each traversal, we maintain the invariant that after the n-th traversal, all non-draw states with position value less than or equal to n have propagated their values to their parent states. Note that this invariant is achieved in the first traversal since all states with position value 1 are properly set in the initialization phase. In the state graph it is easy to observe that if there is no state with the position value n, there is no state with the position value $n+1$.

Therefore, the invariant is also maintained in the subsequent round of traversals and the propagation phase ends when no states are propagated in a complete round of traversal.

During each round of traversal, we use the following propagation principle. The position value of a win or loss state s represents the number of plies to the end states. Each parent state of s needs one more ply to the end states if it moves to s. Therefore, s propagates its position value plus one to all its parent states. For a draw state, the propagated position value is still draw. The $BestValue$ of each state is the current best value of those propagated to it. A state updates this value whenever it receives a better value from a child state and sets $UnknownChildren$ to be $Propagated$ when it has propagated its value to its parent states. It can be verified that the better the position value is, the worse the value it propagates. This coincides with the zero-sum property.

According to the above discussion, all states with position value n can be propagated during the nth round of traversal with the assumption that all win or loss states with position values less than n are propagated as shown below. For an even n, $Unknown - Children$ of each loss state with position value n must be 0, because its all child states are win states with position value less than n. And $BestValue$ of such states is n, since it has at least one child state with position $n - 1$ and no other values propagated from its child states can be better than n. For an odd n, each state with position value n has $BestValue = n$ because it has at least one child state with position value $n - 1$ propagated in the previous traversal. When $BestValue = n$, this value will not be updated in the future since it cannot be updated by any loss higher win position values in the later traversals. Thus, all these states with $BestValue = n$ are propagated in this traversal, even though they do not necessarily have $UnknownChildren = 0$.

The final phase sets all un-propagated states to draw. After the propagation phase, each un-propagated state has at least one un-propagated child and no loss children. Therefore, a player should move to a un-propagated child rather than a win child to get loss. Then the game goes on indefinitely. Thus, all un-propagated states are draw.

3.2   Splitting of the State Graph

As described in Section 2.3, we split the state graph of a set of pieces into a hierarchy of databases, and then build one database at a time in a bottom-up fashion. Note that while constructing a database, the position values of the states in all of its supporting databases are known. In the initialization phase for each non-end state in the current database, $BestValue$ is set to be the best value derived from the child states in the supporting databases, and $UnknownChildren$ equals the number of child states in the current database. Moreover, the maximum $BestValue$ except $Unknown$ in this phase is recorded and stored in w. After the initialization phase, all states in the supporting databases are no longer used and can be released.

In the propagation phase, we first make w rounds of traversal even if there might have no changes in the $BestValue$ values during one round of traversal. The reason for this modification is that the states with $BestValue$ higher than the current traversal index still have a chance to be propagated in the future rounds. As a result they might possibly update the position values of their parents. After that, we ends the propagation phase when there is no changes in $BestValue$ during one round of traversal. There is no modification needed for the final phase.

3.3   Modifications for Distance-to-Conversion

To construct a database in the distance-to-conversion metric, we only need to retrieve the win-draw-loss type of position values for the states in its supporting databases. It implies a distance-to-conversion database can be constructed with the supporting databases using either the distance-to-mate or distance-to-conversion metric. In the initialization phase, each loss state in the supporting databases propagates the value 1, the smallest number indicating a win, to all its parents, and each win state in the supporting databases propagates 2, the smallest number indicating a lose, to all its parents. There is no need for modification in the propagation and final phases.

3.4   Verifying the Correctness

As described in [8], large databases may contain some errors during construction because of hardware or software failures. For reliable the endgame databases, one needs processes to verify the correctness independently. Our verification algorithm traverses the databases once and performs the following verification step for each state. For each end state, verify its correctness according to Table 1. For each non-end state, verify if its position value is the best value of those propagated from its children. This process guarantees the correctness of the databases. All of our databases when using the distance-to-mate metric have been independently verified using this approach.

4   Chinese Chess Special Rules

As mentioned in Section 2.2, the end states for Chinese chess are **KFK**, **OKIC** and **STALEMATE**. The game is draw if it exceeds a certain number of plies. Besides, there is a set of rules, where we can find no counterpart in Western chess, called *special rules*[7] to define the win, draw, or loss of a game. The algorithm described above might not correctly construct an endgame database since it neglects these special rules. We begin by briefly describing the special rules and then our strategy of dealing with them in this paper.

4.1   Special Rules in Chinese Chess

Among these special rules, the most influential one is *indefinitely checking*, which means that a player checks his opponent continuously without ending[8]. It is not allowed in Chinese chess unless the opponent also continuously checks him. Therefore, if a player cannot prevent his King from being captured without checking his opponent's King

---

[7] There were more than one versions of rules in Chinese chess including the once used Asia, China, and Taiwan rules of Chinese chess. The differences between them are all about detailed special rules. In 1998, three conferences were held in China to solve this problem. Finally, a new version of playing rules of Chinese chess was born. Our discussions in this chapter are all based on this new version of rules described in [1].

[8] In real games, checking indefinitely is determined by the appearance of the same position for three times in a sequence of moves in which a player checks his opponent continuously.

endlessly, then he loses. This implies the possibility that one position is recorded as a draw in the databases but is considered as a loss because of this special rule. Another special rule, which is also possible to stain the databases, is *indefinitely chasing*. The term *chase* is defined similarly as the term check. The difference is that the prospective piece to be captured is not the King but another opponent's piece. In Chinese chess, a player cannot chase his opponent continuously without ending, unless he uses the King or Pawns to do so. This special rule may also possibly cause the constructed databases in Section 3 to contain incorrect data. Nevertheless, the ruling of the outcome of a game by OKIC, KFK and STALEMATE is not affected by any special rule. Hence it guarantees the correctness of win and loss information stored in the constructed databases.

All the special rules are dealing with certain kinds of sequences of moves and can be classified into two groups. The first group is about which sequences are forbidden (e.g., checking indefinitely) for one of the two players (e.g., the side to check), and the second is about which sequences are permitted. If a player violates the rules in the first group, then he loses. All the rules in the first group deal with indefinitely checking and chasing. Notice that not all rules relevant to checking and chasing fall in the first group. For instance, the rule that Kings and Pawns have the right to chase other pieces endlessly belongs to the second group. The special rules that have no relation to checking and chasing are all in the second group.

4.2 Avoidance of Spoilage by the Special Rules

According to the above discussion, the win and loss positions computed using the algorithm in Section 3 are correct, but the recorded draw positions might be mistaken. The first group of rules may stain the draw positions because someone might lose because of violating these rules. In contrast, if someone has done what is permitted and both the two players stay on playing without change, then the game is treated as a draw. Therefore, the second group of the special rules does not stain our constructed databases. Only the first group of rules should be dealt with to correct the mistaken draw positions in the databases. It implies only special rules about indefinitely checking or chasing might spoil the databases.

In our databases, we assume that only one side possesses attacking piece(s), and thereby the other side is armless. Because the armless side has no power to checking or chasing his opponent endlessly without using his King, only the attacking side can do what is forbidden by the first group of the special rules. However, if the attacking side cannot win the game, he can move his King or defending pieces purposelessly to get a draw rather than loses the game. As a result, the databases are in concord with those special rules. Since the attacking side will never want to do what is forbidden by the special rules, the databases constructed by the algorithm described in Section 3 are correct.

5   Conclusions

We have reported here the preliminary achievement of applying retrograde analysis in constructing a set of 151 endgame databases in which there are at most two attacking

Table 3. Detailed Statistics for the KNP(9)KGM endgame.

Distance-to-mate	Number of Positions		Distance-to-mate	Number of Positions		Distance-to-mate	Number of Positions	
	Red Turn	Black Turn		Red Turn	Black Turn		Red Turn	Black Turn
KFK	219090	219090	33	8566	0	68	0	11766
OKIC	81504	0	34	0	5350	69	12103	0
mate	0	690	35	9274	0	70	0	17981
1	1155	0	36	0	4884	71	16640	0
2	0	492	37	8686	0	72	0	21110
3	1458	0	38	0	5628	73	17570	0
4	0	66	39	9000	0	74	0	20229
5	714	0	40	0	4633	75	16071	0
6	0	69	41	8399	0	76	0	17613
7	1095	0	42	0	5583	77	14817	0
8	0	153	43	10132	0	78	0	19323
9	2414	0	44	0	6854	79	16895	0
10	0	258	45	12513	0	80	0	27449
11	4779	0	46	0	9562	81	20938	0
12	0	714	47	14594	0	82	0	41697
13	7990	0	48	0	9449	83	25393	0
14	0	713	49	13844	0	84	0	53994
15	12574	0	50	0	9211	85	24221	0
16	0	1398	51	11563	0	86	0	49831
17	17532	0	52	0	7620	87	16175	0
18	0	2505	53	11064	0	88	0	31089
19	21532	0	54	0	8622	89	4763	0
20	0	2886	55	12073	0	90	0	9170
21	16057	0	56	0	10173	91	655	0
22	0	3043	57	12473	0	92	0	1423
23	13087	0	58	0	11019	93	131	0
24	0	2807	59	11208	0	94	0	300
25	13364	0	60	0	9888	95	22	0
26	0	3931	61	10575	0	96	0	67
27	16653	0	62	0	8942	97	6	0
28	0	5265	63	9033	0	98	0	30
29	15930	0	64	0	8629	draw	4316	127586
30	0	4722	65	8145	0	total	839205	839205
31	11205	0	66	0	9579	-	-	-
32	0	4119	67	9214	0	-	-	-

pieces and four defending pieces in addition to the two Kings, and only one side having attacking piece(s). The details are described in Appendix A.

Without loss of generality, we let the Red side be the attacking side in the databases. An endgame is called *Red to win* if the Red side of this endgame can generally win the game except in some positions where a Red piece is doomed to be captured in a few moves. On the contrary, an endgame is called *Black to draw* if the Black side of the endgame can generally draw the game except in some positions where a Black piece is doomed to be captured in a few moves. Note that since only the Red side possesses attacking piece(s) in our setting, the Red side can never lose the game.

Some of these 151 endgames are complicated enough to be studied mainly by masters, such as KCP(8)KGG[9], KCP(5)KGM, KNP(9)KGM and KCCMKMM. There are literatures of the studies of many endgames by human experts. For most endgames, there is consensus between human literatures and computer databases of the types the endgames belong to. For example, KNK endgame is Red to win whereas KCK is Black to draw. All our databases coincide with the results stated in the literatures. For some complicated ones, there are conflicting statements in the literatures, such as KNP(9)KGM is regarded Black to draw in some literatures, but it is considered Red to win in some others. From our statistics, there are only 4316 draw positions, and 834889 Red-to-win positions including 219090 **KFK** ones. Our database gives an explicit answer that KNP(9)KGM endgame is Red to win. The detailed statistics of this endgame is listed in Table 3. Moreover, we illustrate an example of this endgame in Appendix B.

In the future, we plan to tackle the problems caused by the special rules. Besides, the topic of knowledgeable encoding and querying of endgame databases [4] is an approach to condense the win-draw-loss information in physical memory to improve the nowadays Chinese chess programs. When these two topics are well-developed in the future, Chinese chess computer system may have a chance to beat human champions as in the case of Western chess.

Acknowledgment

We thank anonymous referees for very helpful comments.

References

1. China Xiangqi Association. *The Playing Rules of Chinese Chess*. Shanghai Lexicon Publishing Company, 1999. In Chinese.
2. H. Bal and V. Allis. Parallel retrograde analysis on a distributed system. In *Proceedings of the 1995 ACM/IEEE Supercomputing Conference*, 1995.
3. E.A. Heinz. Endgame databases and efficient index schemes for chess. *ICCA Journal*, 22(1):22–32, 1999.
4. E.A. Heinz. Knowledgeable encoding and querying of endgame databases. *ICCA Journal*, 22(2):81–97, 1999.
5. L. Stiller. Some results from a massively parallel retrograde analysis. *ICCA Journal*, 14(3):91–93, 1991.
6. K. Thompson. Retrograde analysis of certain endgames. *ICCA Journal*, 9(3):131–139, 1986.
7. H.J. van den Herik, I.S. Herschberg, and N. Nakad. A six-men-endgame database: KRP(a2)KbBP(a3). *ICCA Journal*, 10(4):163–180, 1987.
8. C. Wirth and J. Nievergelt. Exhaustive and heuristic retrograde analysis of KPPKP endgame. *ICCA Journal*, 22(2):67–80, 1999.

[9] In Chinese chess, the Pawns can neither move backward nor promote to other pieces as reaching the opposite boundary of the board. Therefore, the horizontal level of each Pawn is important for endgame studies. We indicate the level by adding a number within the parentheses right after the corresponding Pawn.

9. R. Wu and D. Beal. Solve chinese chess endgames. In *Proceedings of Joint Conference on Information Sciences*, pages 970–973, Atlantic City, USA, February 2000.

10. R. Wu and D. Beal. Computer analysis of some chinese chess endgames. In *Advances in Computer Chess*, volume 9. To appear.

Appendix A: Statistics of All 151 Endgame Databases

Here we list the statistics for a set of 151 databases in which there are at most two attacking pieces and four defending pieces in addition to the two Kings and only one side owns attacking piece(s). We call a database *useless* with the strong sense that a player or a program can easily design an infallible playing strategy using only its supporting databases. Otherwise, it is *useful*.

For instance, defending pieces are function-less for attacking in a position if there are no Cannons[10]. Therefore, if one can play KNKG endgame infalliblely, we expect he or she can play KNGKG endgame infalliblely. Thus, KNGKG Endgame is useless. All the source databases in our set of 151 endgame databases are useful. In addition, for any given database in our set, all of its supporting databases are also in it.

In the following statistics, two conjugate positions are always regarded as the same one. It helps to compare the amount of disk space used with the number of all possible valid positions. Tables 4–7 list the statistics of all the 151 endgame databases. The last column of the tables is the maximum distance-to-mate values among those positions that are neither **KFK** nor **OKIC** . Note that for the databases with no checkmate and stalemate positions, we put '**KFK/OKIC**' instead of the distance-to-mate maximum. Notice that the database size and valid position size are in terms of units (2-bytes). Both **OKIC** and **KFK** are regarded as valid positions. Each position which has the properties of not only **OKIC** but also **KFK** is regarded as a **KFK** and not counted as an **OKIC**. Because of page limitation, we list only the detailed statistics of KNP(9)KGM endgame in Table 3.

---

[10] In Chinese chess, a Cannon can capture an opponent's piece if there is exactly one piece in between. The piece being jumped over can belong to either side and possibly be a defending piece.

Table 4. Statistics of 151 endgame databases — Part I.

Database Name	Database Size	Valid Position Size	# of KFK positions	# of Draw Positions Red Turn	# of Draw Positions Black Turn	Distance-to-mate Max.
KRKGGMM	634500	533196	112653	64719	204192	64
KRKGGM	216000	183642	40626	0	7854	32
KRKGMM	364500	309888	77949	0	15411	36
KRKGG	31500	27162	6273	0	654	20
KRKGM	123750	106434	27870	0	4350	16
KRKMM	82350	70929	20547	0	2826	24
KRKG	18000	15681	4260	0	366	10
KRKM	27900	24294	7290	0	768	12
KRK	4050	3600	1116	0	54	4
KNNKGGMM	25951050	22120452	4347957	116547	2257626	62
KNNKGGM	8834400	7711560	1588662	32266	653002	48
KNNKGMM	14908050	13013190	3043869	57921	1272474	51
KNNKGG	1288350	1151928	248090	2209	52471	42
KNNKGM	5061375	4523145	1102295	9950	334366	46
KNNKMM	3368115	3010797	811260	5637	229932	31
KNNKG	736200	673563	170559	0	78	38
KNNKM	1141110	1043922	291660	317	49776	32
KNNK	165645	155340	45030	0	0	14
KNCKGGMM	57105000	44238852	8693862	212210	4524812	71
KNCKGGM	19440000	15422652	3177324	53963	1297660	56
KNCKGMM	32805000	26025678	6087387	107608	2577736	58
KNCKGG	2835000	2303154	495478	3202	107293	42
KNCKGM	11137500	9046170	2204590	25197	715816	46
KNCKMM	7411500	6020541	1621467	19920	495429	36
KNCKG	1620000	1346886	340998	948	33291	38
KNCKM	2511000	2087604	583320	2736	120776	32
KNCK	364500	310320	89700	0	5046	14
KNPKGGMM	34897500	26332884	5170518	14589937	19991293	129
KNPKGGM	11880000	9253620	1902519	820899	2258067	125
KNPKGMM	20047500	15615450	3655458	2229456	4774313	113
KNPKGG	1732500	1392678	298698	2091	84843	42
KNPKGM	6806250	5469780	1332672	16758	552771	98
KNPKMM	4529250	3640437	982440	282189	674142	56
KNPKG	990000	820530	207498	465	18054	38
KNPKM	1534500	1271772	355740	1050	87956	34
KNPK	222750	190458	55056	0	432	20
KNKGGMM	634500	533196	112653	397065	420030	1
KNKGGM	216000	183642	40626	134238	142809	43
KNKGMM	364500	309888	77949	218811	231936	35
KNKGG	31500	27162	6273	18300	20760	43
KNKGM	123750	106434	27870	67266	78078	43
KNKMM	82350	70929	20547	47067	50382	29
KNKG	18000	15681	4260	30	408	38
KNKM	27900	24294	7290	12081	15973	31

Table 5. Statistics of 151 endgame databases — Part II.

Database Name	Database Size	Valid Position Size	# of KFK positions	# of Draw Positions		Distance-to-mate Max.
				Red Turn	Black Turn	
KNK	222750	190458	55056	0	432	20
KCCGKGGM	39264000	33465816	6089640	44809	2695860	56
KCCGKGMM	66258000	56473524	11490426	138363	5732635	62
KCCMMKMM	68485005	58374000	14964372	94242	4895394	71
KCCMKGGM	60859200	51871896	10242321	96719	4318504	61
KCCMKGMM	102699900	87533784	19886811	46954552	64711518	129
KCCKGGMM	25951050	22120452	4347957	10576457	16392207	104
KCCGKGG	5726000	4998360	955950	0	174	48
KCCGKGM	22495000	19635050	4190786	13488	1446410	44
KCCGKMM	14969400	13066740	3043869	13200	1055234	42
KCCMMKM	23202570	20253150	5325120	20847	1385781	34
KCCMKGG	8875300	7747260	1588662	5153	374462	29
KCCMKGM	34867250	30434300	7162148	47935	2515273	46
KCCMKMM	23202570	20253150	5325120	9841018	13751960	83
KCCKGGM	8834400	7711560	1588662	14169	629388	66
KCCKGMM	14908050	13013190	3043869	7500728	9730354	71
KCCGKG	3272000	2924110	652320	0	0	28
KCCGKM	5071600	4532310	1102295	0	0	30
KCCMMK	3368115	3010797	811260	807	107811	16
KCCMKG	5071600	4532310	1102295	2628	221421	25
KCCMKM	7860980	7025020	1905882	6817	465890	33
KCCKGG	1288350	1151928	248090	1194	55341	27
KCCKGM	5061375	4523145	1102295	6828	365517	49
KCCKMM	3368115	3010797	811260	1555559	2077626	57
KCCGK	736200	673563	170559	0	0	18
KCCMK	1141110	1043922	291660	254	35485	16
KCCKG	736200	673563	170559	369	31917	25
KCCKM	1141110	1043922	291660	957	67125	34
KCCK	165645	155340	45030	33	5031	14
KCPGKGGM	52800000	40636900	7365405	4810083	10120605	76
KCPGKGMM	89100000	68574750	13939986	8321664	17410949	90
KCPMKGGM	81840000	62628580	12362256	15963421	23889033	90
KCPMKGMM	138105000	105685710	24069600	33574339	45414260	72
KCPKGGMM	34897500	26332884	5170518	18167910	21015935	131
KCPGKGG	7700000	6115416	1163774	2467	157117	50
KCPGKGM	30250000	24024016	5117056	1516346	4471125	77
KCPGKMM	20130000	15987204	3725760	1328397	3120218	59
KCPMKGG	11935000	9425856	1930434	990741	1752034	52
KCPMKGM	46887500	37029316	8726292	6680193	11196885	54
KCPMKMM	31201500	24641640	6502374	7126266	9904010	54
KCPKGGM	11880000	9253620	1902519	4973458	6845837	133
KCPKGMM	20047500	15615450	3655458	10338647	11908493	97
KCPGKG	4400000	3604050	801618	0	83124	30
KCPGKM	6820000	5586250	1357828	0	179511	36

Table 6. Statistics of 151 endgame databases — Part III.

Database Name	Database Size	Valid Position Size	# of KFK positions	# of Draw Positions Red Turn	# of Draw Positions Black Turn	Distance-to-mate Max.
KCPMKG	6820000	5555650	1351890	572116	900755	28
KCPMKM	10571000	8611230	2342436	1027769	1780531	37
KCPKGG	1732500	1392678	298698	153375	259158	45
KCPKGM	6806250	5469780	1332672	1849072	2545682	98
KCPKMM	4529250	3640437	982440	2259230	2633056	49
KCPGK	990000	836028	211344	0	2129	22
KCPMK	1534500	1288818	360780	133447	170990	20
KCPKG	990000	820530	207498	83346	130269	28
KCPKM	1534500	1271772	355740	299246	425298	42
KCPK	222750	190458	55056	19356	24690	20
KCGKGGM	960000	806424	157572	561718	643125	51
KCGKGMM	1620000	1360836	297723	1008484	1060191	35
KCMMKMM	1674450	1406862	388638	1002801	1018224	KFK/OKIC
KCMKGGM	1488000	1249944	265311	902507	979834	3
KCMKGMM	2511000	2109276	515844	1542008	1593342	1
KCKGGMM	634500	533196	112653	388248	418986	1
KCGKGG	140000	119042	24423	57	3156	48
KCGKGM	550000	467505	107198	300095	353912	43
KCGKMM	366000	311163	77949	215332	231691	37
KCMMKM	567300	482268	136512	342726	345756	KFK/OKIC
KCMKGG	217000	184492	40626	130379	142990	3
KCMKGM	852500	724630	183402	526204	541192	1
KCMKMM	567300	482268	136512	340724	345756	KFK/OKIC
KCKGGM	216000	183642	40626	131475	142398	1
KCKGMM	364500	309888	77949	224709	231939	KFK/OKIC
KCGKG	80000	68812	16476	0	1738	28
KCGKM	124000	106652	27870	0	3678	30
KCMMK	82350	70929	20547	50265	50382	KFK/OKIC
KCMKG	124000	106652	27870	76930	78776	1
KCMKM	192200	165304	48240	116149	117064	KFK/OKIC
KCKGG	31500	27162	6273	18984	20775	1
KCKGM	123750	106434	27870	76503	78564	KFK/OKIC
KCKMM	82350	70929	20547	49680	50382	KFK/OKIC
KCGK	18000	15681	4260	0	268	18
KCMK	27900	24294	7290	16983	17004	KFK/OKIC
KCKG	18000	15681	4260	11166	11421	KFK/OKIC
KCKM	27900	24294	7290	16872	17004	KFK/OKIC
KCK	4050	3600	1116	2484	2484	KFK/OKIC
KPGKGGM	595200	489620	95295	373736	394181	13
KPGKGMM	1004400	826230	180603	609498	645627	KFK/OKIC
KPMKGGM	922560	754580	160119	563026	594227	13
KPMKGMM	1556820	1273350	312183	906390	961167	KFK/OKIC
KPKGGMM	393390	317412	66996	237339	250326	1
KPGKGG	86800	72836	14868	54454	57898	17

Table 7. Statistics of 151 endgame databases — Part IV.

Database Name	Database Size	Valid Position Size	# of KFK positions	# of Draw Positions Red Turn	Black Turn	Distance-to-mate Max.
KPGKGM	341000	286004	65448	207626	220543	21
KPGKMM	226920	190374	47709	134115	142665	KFK/OKIC
KPMKGG	134540	112246	24684	82095	87448	17
KPMKGM	528550	440829	111732	309499	329079	21
KPMKMM	351726	293403	83349	197160	210054	KFK/OKIC
KPKGGM	133920	110190	24327	81318	85827	13
KPKGMM	225990	185940	46809	131220	139131	KFK/OKIC
KPGKG	49600	42410	10124	28065	31545	23
KPGKM	76880	65730	17166	44015	48538	13
KPMKG	76880	65370	17091	41682	47044	23
KPMKM	119164	101318	29646	64916	71636	13
KPKGG	19530	16434	3780	11856	12636	13
KPKGM	76725	64356	16848	44676	47505	9
KPKMM	51057	42903	12447	28584	30456	KFK/OKIC
KPGK	11160	9738	2640	1021	1299	22
KPMK	17298	15003	4509	1514	1924	20
KPKG	11160	9555	2592	6003	6780	13
KPKM	17298	14802	4446	9369	10350	13
KPK	2511	2214	684	216	273	20

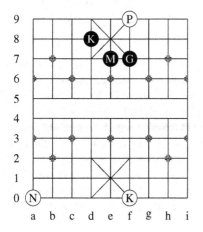
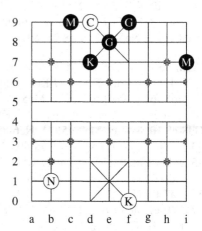

Fig. 3. (a) An example of the KNP(9)KGM endgame. (b) An example of the KNCKGGMM endgame

Appendix B: Mining the Endgame Databases

As mentioned in Section 2.1, the state space of Chinese chess is a finite, directed, bipartite and cyclic graph. If the propagated value of a child is the same as the position value of its parent, the edge between them indicates a optimal move. If we keep only these optimal moves and remove the rest of the edges, the resulting state graph represents the perfect playing strategies. We call the states without parent states in this trimmed state graph *sources*. In general, if one knows how to play optimally starting from all the sources of a certain endgame, we would expect that he or she can play this endgame starting from any position in the endgame. In this regard, the sources with the maximum position value are usually most interesting to the human.

Below we provide two endgame studies of Chinese chess. They are both Red to move and win with the maximum odd position value in their databases. We list the optimal move series for each study. Some branches are also recorded within parentheses after their brothers in the graph representation. A move is described by the abbreviation of the moving piece and the addresses of starting point and destination separated by a dash sign '-'. Some notations used in [8] are also employed here. An equal sign '=' means the game ends in a draw and the sign '+—' means the Red side is going to win. An exclamation mark '!' after a move by Red indicates the unique winning move; by Black indicates the unique optimal defense move in the distance-to-mate metric or the only way to draw. A question mark '?' indicates a seemingly good but actually stupid move.

Figure 3(a): 1. Pf9-e9! (1. Na0-b2? Kd8-d9! 3. Nb2-c4 Gf7-e8! =) Kd8-e8 2. Pe9-d9 Ke8-d8 3. Pd9-c9 Gf7-e8 4. Na0-b2 Kd8-d7 5. Pc9-b9 Kd7-d8 6. Nb2-d3 Kd8-d9 7. Nd3-f4 Kd9-d8 8. Nf4-e6 Ge8-f7 9. Pb9-c9 Gf7-e8 10. Kf0-e0 Kd8-d7 11. Ke0-e1 Me7-g9! (Me7-c9? 12. Ne6-c5! Kd7-d8 13. Nc5-b7! Kd8-d9 14. Nb7-c9 +—) 12. Ne6-g7 Ge8-d7! 13. Pc9-d9 Mg9-i7 14. Pd9-e9 Mi7-g5 15. Ng7-d5 Gf7-e8 16. Nf5-d4 Kd7-d8 17. Nd4-e6 Mg5-e7 18. Ne6-g7 Ge8-f7 19. Ke1-d1 Me7-g5 20. Ng7-e6 Mg5-e7 21. Ne6-f4 Kd8-e8 22. Nf4-g6 Me7-g5 23. Ng6-e5 Mg5-i9 (Ke8-e9 24. Ne5-f7 Ke9-e8 25. Nf7-g5 +—)24. Ne5-d7 +—

Figure 3(b): 1. Cd9-e9! Mi7-g5! 2. Nb1-c3! Mg5-e7! 3. Nc3-d5! (3. Nc3-e4? Ge8-f7! 4. Ne4-d6 Kd7-d8! 5. Nd6-f7 Kd8-d9! 6. Nf7-h8 Kd9-e9 =) Me7-c5! 4. Nd5-f4! (4. Nd5-f6? Kd7-d8 5. Nf6-d5 Mc5-a7! =) Mc5-a7! 5. Nf4-e6! Ge8-d9! (Ge8-f7? 6. Ce9-d9! Gf7-e8 7. Cd9-d8 +—) 7. Kf0-f1 Gd9-e8 8. Kf1-e1 Ge8-d9 9. Ne6-d4 Gd8-e8 10. Nd4-c6 Ma7-c5! 11. Nc6-a5 Mc5-e7 (Mc5-a7 12. Na5-b7 Ge8-d9 13. Ce9-e4 +—) 12. Na5-b7 Ge8-d9 13. Nb7-d6 Gd9-e8 14. Ke1-e0 Ge8-f7 (Ge8-d9 15. Nd6-c4 Mc9-a7 16. Nc4-e5 Kd7-d8 17. Ne5-c6 Kd8-e8 18. Ce9-e7 +—) 15. Nd6-f7 Kd7-d8 16. Nf7-h8 Kd8-e8 (Kd8-d9 17. Ce9-c9 Me7-c9 18. Ng7-f9 Mc9-a7 19. Nf9-g7 Ma7-c5 20. Ng7-f5 Mc5-a7 21. Nf5-d4 Ma7-c9 22. Nd4-e6 +—) 17. Ce9-e7 Mc9-e7 18. Nh8-g6 Ke8-d8 19. Ng6-e7 Kd8-e8 20. Ke0-e1! Ke8-e9 21. Ke1-d1 Ke9-e8 22. Ne7-c8 +—

Learning from Perfection

A Data Mining Approach to Evaluation Function Learning in Awari

Jack van Rijswijck

Department of Computing Science,
University of Alberta,
Edmonton, Alberta, Canada T6G 2H1
javhar@cs.ualberta.ca

Abstract. Automatic tuning of evaluation function parameters for game playing programs, and automatic discovery of the very features that these parameters refer to, are challenging but potentially very powerful tools. While some advances have been made in parameter tuning, the field of feature discovery is still in its infancy. The game of Awari offers the possibility to achieve both goals. This paper describes the efforts to design an evaluation function without any human expertise as part of the Awari playing program *Bambam*, as being developed by the Awari team[1] at the University of Alberta.

Keywords: Machine learning, heuristic search, game playing, alpha-beta, Awari.

1 Introduction

An evaluation in a game playing program typically takes as input a vector of *feature values* that have been extracted from the game position. These features, to be called the *base features*, are typically not those that form the actual description of the game position. The features that describe the game state are the *atomic features*. For instance, the atomic features in chess are the states of the 64 squares; examples of states are "empty", "white pawn", and "black rook". The base features may include concepts such as mobility, passed pawns, and king safety.

Since the choice of base features impacts both the speed of the evaluation function and its ability to approximate the game theoretic value of a position, the automatic discovery of base features, if it is possible, is a potentially very powerful tool. The notion of automatically discovering features "from scratch" is very attractive, but unfortunately seems to be too far fetched for most games. Humans have typically already identified key features in a given game. A language that allows these features to be expressed is usually either so high level that it is specialized and biased towards these features, or it is low level and the feature descriptions are discouragingly complex. In neither case is there much hope of discovering genuinely new high level features.

The ideological objection against adding human defined base features often leads to machine learning systems that are indeed able to make progress from their initial random state, but whose performance falls far short of that of existing world class programs. For

---

[1] See http://www.cs.ualberta.ca/~awari.

T.A. Marsland and I. Frank (Eds.): CG 2000, LNCS 2063, pp. 115–132, 2001.

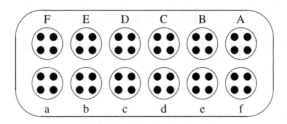

Fig. 1. The starting position in Awari

our purposes, the main concern is to come up with a program that plays as strongly as possible. Whether or not it includes human knowledge, the performance of the program should in this context be compared to that of existing strong programs.

One situation in which feature discovery from scratch is relevant is when there is little or no human strategic knowledge available for the game at hand. This is the case in the game of Awari. While developing the Awari playing program *Bambam*, the Awari team at the University of Alberta has been able to come up with only a handful of high level features that may or may not be important in Awari. Perhaps better features can be constructed.

The outline of this paper is as follows. Section 2 introduces the game of Awari and its rules. Section 3 describes the problem of evaluation function learning and lists some previous work that has been done in this area. Base features and evaluation functions were learned automatically via methods explained in Section 4, as well as hand-developed by humans as reported in Section 5 to enable a comparison. Results are presented in Section 6, followed by conclusions and a discussion.

2   Awari

Awari is one of a family of ancient African games, played with seeds or pebbles that are distributed over rows of pits. The variant described here is the one used in the Computer Olympiad. The Awari board consists of two rows, each containing six pits. Figure 1 shows the starting position of Awari, where each pit contains four pebbles. The two players are commonly called *North* and *South*. The six pits on the bottom row, named '*a*' through '*f*', form South's *home area*. Pits '*A*' through '*F*' are North's home area. The pits '*a*' and '*A*' are the *first pit* of the respective players.

A move consists of taking all of the pebbles from one of the pits in the home area, and "sowing" them counterclockwise around the other eleven pits of the board by dropping one pebble in each consecutive pit. If more than eleven pebbles are sown, then the starting pit is skipped. There is no limit on how many pebbles a single pit can contain. Pebbles can be captured, and the object of the game is to capture more pebbles than the opponent.

Capturing occurs in *vulnerable pits*. A pit is vulnerable if it contains two or three pebbles after the sowing is completed. If the last pebble lands in a vulnerable pit, then the pebbles in that pit are captured by the sowing player. A multiple capture occurs when the preceding pit is also vulnerable; capturing then continues until a non-vulnerable pit is reached. Captures only occur in the opponent's zone; a multiple capture cannot continue beyond the opponent's first pit.[2]

Since the number of pebbles on the board decreases throughout the game, it is possible to build endgame databases that contain the result with perfect play for each possible configuration up to a given number of pebbles. The Awari team at the University of Alberta has built databases that contain all positions containing up to 35 pebbles. This is only 13 pebbles away from the starting position, which is seemingly quite close. However, it can take many moves before 13 pebbles are captured; experiments suggest that this may take on the order of 50 ply with perfect play. Before the stage of the game is reached where the endgame databases can be used, there still is the need for a good evaluation function.

3 Evaluation Function Learning

The biggest problem that evaluation function learners face is getting feedback on how good any given instance of the function is. This is usually solved by playing many games and using reinforcement learning techniques. However, in Awari we have a radically different situation. The quality of an evaluation function can be measured directly against the perfect information in the endgame databases.

3.1 Previous Work

Previous attempts to mine endgame databases include research done by Terry van Belle [9] and Paul Utgoff et al. [7,8]. Van Belle attempted to apply genetic algorithms to checkers endgame databases, which proved to be unsuccessful. Utgoff developed the ELF learning algorithm, which was able to make progress in identifying features in checkers by using reinforcement learning techniques rather than learning from endgame databases. However, the program did not reach a high level of play when compared to other checkers playing programs.

Human efforts to gain understanding from endgame databases include work by Michie, Roycroft and Bratko [2], as well as John Nunn [4], all using chess databases. Most attempts have had limited success, with Nunn's impressive work [3,5,6] being a favourable exception. All of these results have only come with a considerable investment of time.

Automatic feature discovery and evaluation function construction without using perfect information was done by Buro, with experimental results for the game of Othello [1]. Buro introduced GLEM, a Generalized Linear Evaluation Model, and was able to extract useful features that led to an improvement in playing strength. Each feature corresponds to a set of board squares, and weights are assigned to every possible instantiation of the squares question. This leads to a table containing on the order of 1.5 million weights, which were tuned using a hill climbing approach.

The Awari application described in this paper sidesteps several issues that are relevant in Buro's approach, since in Buro's case learning is driven by game play, not by databases.

---

[2] There are several additional rules that deal with specific situations; see
http://www.cs.ualberta.ca/~awari/rules.html for a detailed explanation.

The Awari training data covers all possible board positions up to a given number of pebbles, and contains perfect evaluation information. This eliminates the problems of acquiring representative training data and possible overfitting, and enables the accuracy of any evaluation function to be measured directly. Unlike in GLEM, the Awari evaluation functions described in this paper are not linear combinations of features; instead, each board position is mapped to one unique class whose evaluation values are not learned but extracted directly from the databases. The class mapping is done by calculating a very small number of Boolean comparisons, which allows the evaluation value to be computed quickly.

When using a linear evaluation function, each evaluation feature is to apply in every possible position. One can however imagine that some particular feature f_0 may only be relevant in certain types of positions, perhaps only when another feature f_1 is present. The weight of feature f_0 then depends on the value of f_1, which makes the function nonlinear. To cope with this and still keep the function linear, one can introduce the new feature $f_0 \wedge f_1$. GLEM uses this approach on a large scale, generating many such features. Another way of coping with nonlinearity is to have two evaluation functions, where the value of f_1 decides which of the two functions to use. The two evaluation functions can have different weights, and possibly even different features; feature f_0 could appear in one but not in the other. The method in this paper takes this approach to the extreme; all of the work is done by the feature decisions, and the functions that are used after all the decisions are made are simple value lookups. The evaluation function is iteratively refined by targeting its weakest spot and automatically generating the single feature that reinforces the accuracy the most. Thus, the number of generated features is kept small.

3.2   Data Mining in Awari

There are two reasons why an evaluation function is still needed when perfect information is available. In game playing programs that rely on search, there is a tradeoff between the quality of the information returned by the evaluation function and the speed with which it can be computed. A program with a "fast and dumb" evaluation can reach a larger search depth than a program with "slow and smart" evaluation, which may enable it to play stronger. For positions in the large databases that do not fit into memory it may therefore be better to approximate the value rather than going to disk to retrieve the correct value, since a disk lookup is relatively very slow. Experiments show that disk retrieval can be on the order of a thousand times slower than calculation. For positions outside the databases, an evaluation function is needed regardless.

This introduces two criteria for success for an evaluation function: its ability to *interpolate*, i.e. approximate an existing database while using less memory, and its ability to *extrapolate*, i.e. generalize to positions with larger numbers of pebbles. Evaluation functions can be tested according to these criteria, as well as in actual game play when combined with an Awari playing program.

Exploiting the symmetry of the Awari board, the database values are all computed from the viewpoint of South to move. In the remainder of the text, the player to move will always be South.

4   Framework

Consider a board position p whose *value*, defined as the difference in the number of captured pebbles when both players play perfectly from p onward, is $v(p)$. If $v(p)$ is positive, South will capture more pebbles; if it is negative then North will capture more. The *material difference* m indicates the difference between the number of pebbles already captured by South and North, respectively.[3] The material difference at the end of the game with perfect play will be $m + v(p)$, and South wins if $m + v(p) > 0$. South is therefore interested in knowing whether $v(p) > -m$.

The Awari endgame databases contain the values for all positions with at most 35 pebbles. From these databases, statistical information can be gathered that may be used to estimate the values of positions that are not in the database, or for situations in which retrieving the database value may be too costly. The databases also enable an accurate assessment of the quality of a given evaluation function. There is no danger of overfitting, since the information in each database covers the entire space of board positions with a certain number of pebbles, rather than just a sample of it.

The automatic construction of an evaluation function consists of two subtasks. The first subtask involves discovering base features, starting with a collection of atomic features. The second subtask involves synthesizing the base features into an evaluation function.

4.1   High-Level Features

The starting point for building high-level features is a set of *atomic features*. In this framework we use only Boolean features, of the following two types:

- pit i contains exactly j pebbles, denoted for example by $C_{=3}$ where 'C' refers to the pit name and '$= 3$' refers to the target value j;
- pit i contains more than j pebbles, denoted for example by $d_{>3}$ where 'd' is the pit name and '> 3' is the threshold value j.

A high level feature consists of a Boolean operator and two features, each of which can itself be an atomic or a high level feature. If the set of Boolean operators is complete, then *any* Boolean function can be expressed in this way. The set of functions {and, not, or} is complete, as is the set {nand, or}, but these have the disadvantage that some of the 16 possible Boolean functions of two variables can only be expressed by chaining together functions, which for the feature discovery would essentially mean one or more extra steps of inference. This complicates the task, and makes it potentially much less feasible. Consider for example the exclusive-or function '$\oplus$'. It can be built by combining the and-function '$\wedge$' and the or-function '$\vee$', for example by $x \oplus y = (x \wedge \neg y) \vee (\neg x \wedge y)$. However, the two terms $(x \wedge \neg y)$ and $(\neg x \wedge y)$ might themselves not be of any value. Unless the system keeps track of *all* previously generated features, this greatly reduces the probability of $x \oplus y$ being discovered.

---

[3] The board position p itself does *not* contain any information about the number of pebbles already captured by either player.

It would therefore be beneficial to have all 16 different Boolean functions of two variables available. These functions form eight pairs of mutually complementary functions. Since the expressive power of the system does not change when a function is added that is the complement of an existing function, eight of the functions can be omitted. Of the remaining eight functions, there are the superfluous functions $f(x, y) = \text{true}$, $f(x, y) = x$, and $f(x, y) = y$. The five remaining functions are the Boolean $\wedge$, $\vee$, $\oplus$, $\Rightarrow$, and $\Leftarrow$.

In practice, the set of possible features quickly becomes too large. The feature discovery system must therefore do a heuristic search of the space of features. Some metric must be defined by which the quality of a feature can be measured, according to the overall goal of building a good evaluation function. Consistent with the evaluation function fitness measure described in Section 4.3, the "fitness" of a feature can be measured by computing the statistical correlation between the feature value x and the database value y over all positions in the database. This correlation $r(x, y)$ is defined as

$$r(x, y) = \frac{\text{cov}(x, y)}{\text{var}(x)\ \text{var}(y)} = \frac{\mathcal{E}(xy) - \mathcal{E}(x)\mathcal{E}(y)}{\mathcal{E}(x^2) - \mathcal{E}^2(x)\ \ \mathcal{E}(y^2) - \mathcal{E}^2(y)}$$

where $\mathcal{E}$ is the expected-value operator. With this definition, the correlation r will be in the interval $[-1, +1]$, where -1 indicates strong negative correlation and $+1$ indicates strong positive correlation. Since these are both desirable, $r^2(x, y)$ can be used as a fitness measure for a Boolean feature.

4.2 Approximating Values

Let M_i be the set of all possible board positions containing exactly i pebbles, and let H_i be a partitioning of M_i into subsets. Suppose that the statistical average μ_S and deviation σ_S of the values in each subset $S \in H_i$ be known, and that the database values follow a Gaussian distribution.[4] If $v(p)$ is not accessible, but it is known that $p \in S(p) \in H_i$, then the probability that $v(p) > -m$ is equal to

$$\Phi(\frac{\mu_{S(p)} + m}{\sigma_{S(p)}}),$$

where Φ is the cumulative normal distribution

$$\Phi(x) = \frac{1}{\sqrt{2\pi}} \int_{-\infty}^{x} e^{-\frac{1}{2}\xi^2}\, d\xi.$$

This probability can be used as the heuristic evaluation of p. Since an evaluation function only needs to impose an ordering on the set of game positions and the function Φ is strictly monotonically increasing, we can equivalently use just

$$\text{eval} = \frac{\mu_{S(p)} + m}{\sigma_{S(p)}}$$

---

[4] Plots of the distributions indicate that this is indeed the case.

and omit the Φ function. If $\sigma_{S(p)} = 0$ then the value $v(p)$ is known, and the evaluation function will assume an infinite value corresponding to a proved win or proved loss. The potentially arbitrarily large evaluation value does not present a problem, since the observed values for $\sigma_{S(p)}$ do not get arbitrarily small in practice.

4.3 Measuring Fitness

It is clear that the quality of this evaluation depends on the value of $\sigma^2_{S(p)}$, which is the mean squared difference between $\mu_{S(p')}$ and $v(p')$ over all $p' \in S(p)$. If $\sigma_{S(p)} = 0$, then the evaluation function perfectly predicts the database value. The mean squared difference between $\mu_{S(p')}$ and $v(p')$ over the entire database M_i is equal to

$$\sigma^2(H_i) = \frac{1}{|M_i|} \sum_{S \in H_i} |S|\sigma^2_S.$$

The number $\sigma(H_i)$ may be called the *fitness* of the partitioning H_i, where $\frac{|S|}{|M_i|}\sigma^2_S$ is the *contribution* that a particular subset S has to $\sigma^2(H_i)$.

The fitness is a direct measure of the accuracy of H_i when used as an evaluation function. The fitness of the partition $H_i = M_i$ is equal to the standard deviation of the entire database itself, which corresponds to an evaluation function that only looks at material difference and ignores the position. This may be used as a baseline comparison for evaluation functions; the *relative fitness* $\sigma_{rel}(H_i)$ is thus defined as $\sigma(H_i)/\sigma(M_i)$. The relative fitness of the "material-only" evaluation function is 1; a lower relative fitness implies a better evaluation function.

4.4 Constructing a Partition

The partition H_i can be refined by targeting one of its subsets S and dividing it further according to some criterion. To do this, we consider a set of features and simply choose the feature that reduces the contribution of S the most. The maximum reduction is achieved by the feature with the highest fitness within S. This fitness, as mentioned before, is measured by the statistical correlation between the feature value and the database value over all positions $p \in S$.

For each of the subsets $S \in H_i$, we can define the *potential* as the difference between its contribution before and after splitting S according to its highest correlating feature. An evaluation function can now be built up by starting with the initial partition $H_i = M_i$ and iteratively refining it by splitting the subset with the highest potential. The resulting structure is a *decision tree*. Notice that this method will eventually construct a perfect evaluation function by partitioning M_i into subsets of size one, but that partition cannot feasibly be used as an evaluation function since the evaluation function is a lookup table that contains the μ and σ values for each subset.

5   Baseline Comparison

The main goal of this project is the automatic discovery of evaluation functions that *do better than hand-crafted ones*. It is therefore important to establish a baseline for comparison. For the features, the baseline is formed by the straightforward atomic features

Table 1. Highest scoring atomic features

10 pebbles		15 pebbles		20 pebbles	
atom	r^2	atom	r^2	atom	r^2
$a_{=0}$	0.09017	$b_{=0}$	0.04320	$b_{=0}$	0.04294
$b_{=0}$	0.07374	$a_{=0}$	0.03669	$a_{=0}$	0.04234
$B_{=0}$	0.05579	$c_{=0}$	0.02989	$c_{=0}$	0.03467
$A_{=3}$	0.04815	$A_{=3}$	0.02494	$A_{=3}$	0.02193
$C_{=0}$	0.03565	$B_{=3}$	0.02386	$d_{=0}$	0.02143
$A_{>2}$	0.09178	$A_{>2}$	0.07800	$A_{>2}$	0.09017
$a_{>1}$	0.06976	$B_{>2}$	0.06154	$B_{>2}$	0.07126
$a_{>2}$	0.05290	$A_{>3}$	0.04688	$a_{>2}$	0.06235
$B_{>1}$	0.05048	$a_{>2}$	0.04415	$A_{>3}$	0.05727
$A_{>1}$	0.04685	$a_{>1}$	0.04139	$a_{>1}$	0.05611

Table 2. Maximum possible feature scores

k	number of pebbles		
	10	15	20
0	0.61149	0.52256	0.56437
1	0.61222	0.55392	0.57530
2	0.61641	0.56141	0.57677
3	0.61175	0.54993	0.56821

of a position. For the evaluation functions, human-defined concepts such as mobility and safety are used.

5.1 Comparison for Features

The atomic features with the highest correlation r^2 are listed in Table 1. The collection of atomic features consists of the set of equality features and the set of inequality features, each of which is capable of uniquely encoding all board states. Using both sets, the p pebble database is described by $2 \cdot 12 \cdot p$ atomic features.

The correlations as listed in the Table 1 are relatively weak; perhaps better correlating features can be discovered automatically. It is possible to get an indication of the maximum possible correlation of a binary feature. This can actually be measured from the database, by setting a threshold k and defining the feature value to be 1 if and only if the database value exceeds k. This artificially produces a perfect separation of the database, which is the best that a binary feature could possibly do. Table 2 lists the r^2 scores of these maximum features; the best score is achieved in each case when $k = 2$.

These experiments establish a score range for useful high-level features. An useful feature should have a correlation r^2 exceeding that of the best atoms, as listed in Table 1. At the same time, it is known that its r^2 cannot exceed that of the theoretical maximum as listed in Table 2. It should be noted that the latter results do not impose a maximum attainable accuracy on evaluation functions, since an evaluation function can achieve better scores by combining multiple binary features.

Table 3. Performance of hand crafted features

feature	number of pebbles		
	10	15	20
mobility difference	0.85485	0.88392	0.86392
mobility South	0.86620	0.89240	0.89088
mobility North	0.92840	0.95363	0.94150
safety difference	0.97414	0.91441	0.85615
safety North	0.97186	0.91666	0.87136
safety South	0.99230	0.95167	0.91030
attack difference	0.99266	0.95284	0.93025
attack South	0.98705	0.95901	0.95385
attack North	0.99930	0.97697	0.96124
balance	0.90811	0.89057	0.86243
none	1.00000	1.00000	1.00000
none (absolute)	5.20733	4.41549	5.02372

5.2   Comparison for Evaluation Functions

To assess the quality of an evaluation function, it can be compared to the best evaluation based on human constructed features. The human members of the Bambam team have been able to come up with the following features:

- *mobility*: the number of available moves;
- *safety*: the number of "safe" pits, ie. the number of pits containing three or more pebbles;
- *attack*: the number of enemy pits under attack;
- *balance*: the difference between the number of pebbles in the two home areas.

The first three features can be computed either for South, for North, or the difference between the two. Table 3 lists the relative fitness scores $\sigma_{i,e}$ for these features on the 10, 15, and 20 pebble databases. Recall that lower relative scores correspond to better evaluation functions. The absolute fitness values of the entire databases are listed on the bottom row.

The evaluation functions from the feature discovery will be constructed by combining several features. To compare these with what can be gained by combining the human defined features, Table 4 lists the scores for the combinations of human defined features. The best score overall is achieved by a combination of safety difference and mobility difference.

The scores correspond to an evaluation function that contains a lookup table for μ and σ for each of the possible input values. Since both mobility difference E_{mob} and safety difference E_{saf} can assume thirteen different values, the function based on both E_{mob} and E_{saf} can take on 169 values.

The numbers indicate that the amount of "noise" in the evaluation function can be reduced by some 15% to 20% by using human-defined features, when using the databases to tune the parameters. Evaluations that were hand-tuned by the Bambam team members turned out to have a relative fitness of at best 0.95. Section 6.4 contains

Table 4. Combining two hand crafted features

feature 1	feature 2	number of pebbles		
		10	15	20
mobility South	mobility North	0.84209	0.87195	0.85521
safety South	safety North	0.97109	0.90897	0.85175
attack South	attack North	0.98680	0.94649	0.92906
mobility difference	safety difference	0.84843	0.85077	0.79654
mobility difference	balance	0.83790	0.85905	0.82903
mobility difference	attack difference	0.83073	0.87541	0.84633
safety difference	balance	0.89245	0.88615	0.84126
attack difference	balance	0.88296	0.88690	0.85534
safety difference	attack difference	0.96568	0.90757	0.85121

results that indicate how much increase in playing strength can be gained by improving the fitness.

6 Results

This section contains the results of three experiments. In the first experiment, high-level features were discovered by a genetic search algorithm. In the second experiment, an evaluation function corresponding to a decision tree was generated. The decisions in the decision tree are based on atomic features only; in a future experiment, the two approaches can be combined by using high-level evolved features in the decision tree. Both the features and the evaluation functions were subsequently tested for generalization to larger databases. In the third experiment, the playing strength of hand-crafted and of automatically constructed evaluation functions was tested by playing matches.

6.1 High-Level Features

Starting with the atomic features, higher level features were constructed by an evolutionary approach with two different recombination rules. The first rule takes two features at random and combines them with one of the five Boolean functions. The second method randomly replaces a subtree of one feature by a subtree of another. The lowest scoring features are removed from the population, where the score is equal to the correlation r^2 with the database value with a small penalty for size. This penalty, in combination with the second method, is necessary to prune out "evolutionary junk" that would otherwise accumulate in the features. Without it, features proved to grow in size very quickly.

It should be noted that the actual details of the evolutionary scheme are not of main interest in this case. The main question is whether high-level features exist and can be discovered relatively quickly. Several heuristic search approaches can be considered; the genetic algorithm used here is merely one example.

For the 10 pebble database, the average score r^2 of the population was 0.19 with an average feature size of 10.0 after 2000 generations.[5] The size of a feature is defined

---

[5] This evolutionary run is quite feasible computationally; it took about an hour on a 400 MHz PC.

Table 5. Best discovered features for 10 pebbles

size	r^2	feature
1	0.09178	$A_{>2}$
2	0.14977	$(b_{>0} \Leftarrow a_{=0})$
3	0.15118	$(a_{=0} \wedge (b_{>0} \Rightarrow f_{=5}))$
7	0.22012	$(A_{>2} \vee ((B_{>2} \vee ((a_{=0} \wedge (b_{>0} \Rightarrow f_{=5})) \vee a_{=8})) \oplus A_{>7}))$

as the number of atoms it contains. If a smaller feature has a higher score than another feature, then the former *dominates* the latter. The best non-dominated features that were discovered are listed in Table 5.

The feature $(b_{>0} \Leftarrow a_{=0})$ can be explained in human terms: if pit 'a' is empty, then 'b' should not be empty.[6] However, the larger features have no concept that is understandable to humans. But the feature "$(A_{>2} \vee ((B_{>2} \vee ((a_{=0} \wedge (b_{>0} \Rightarrow f_{=5})) \vee a_{=8})) \oplus A_{>7}))$" is as transparent to computers as the feature of a "passed pawn" is to humans, and conversely the former is as opaque to humans as the latter is to computers. The feature has good correlation and is easy to compute. If the goal is to produce a high level evaluation function, rather than to teach humans something about Awari, then the correlation is the important issue.

Note that the 7-atoms feature, though it seems big, can be computed very quickly as it consists entirely of Boolean operations. If a feature is kept as a tree structure, it can be updated incrementally during an alpha-beta search by only computing the value of a subtree if one of its two children changes its value.

For the 5 pebble database, features with an average score of 0.329 and average size of 12.9 were discovered within half an hour on a 400 MHz PC. But the 5 and 10 pebble databases are too small to be useful for our purposes, as the information contained in them likely does not generalize well to higher databases. For the 15 pebble database, it took about a day on the same PC to reach an average score of 0.12 and average size of 13.8. Some of the best non-dominated features are listed in Table 6. The 5 atoms feature is much smaller and yet reaches almost the same score as the 22 atoms feature, so it may be more useful under the time constraints of actual game play.

It appears that evolved features of relatively small size correlate about twice as well as the best atomic features. For the 10 pebble database, the best evolved feature mentioned achieved a correlation of about 36% that of the theoretical maximum, while the best atom scored about 15%. For the 15 pebble database, these numbers are 28% and 14%, respectively.

6.2   Decision Tree

An evaluation function corresponding to a decision tree can be generated automatically via the methods described in Section 4.4. In this case, only atomic features were used to split the subsets.

Figure 2 shows the resulting decision tree when the 15 pebble database is split into nine segments, numbered 0 through 8, according to the learning mechanism described

---

[6] This can be either good or bad; in this case, since r is greater than zero, it is apparently good not to have both 'a' and 'b' empty.

Table 6. Best discovered features for 15 pebbles

size	r^2	feature
1	0.07800	$A_{>2}$
2	0.13577	$(A_{>2} \vee B_{>2})$
5	0.15191	$(((B_{>2} \oplus e_{=10}) \oplus d_{=8}) \vee (A_{>2} \vee C_{>2}))$
22	0.15650	$(C_{=4} \Leftarrow (((D_{>8} \Leftarrow (e_{=13} \vee B_{>2})) \wedge (c_{=13} \oplus (e_{=12} \oplus (E_{>8} \Leftarrow (B_{=9}$ $\vee A_{>2})))))) \wedge (e_{=3} \Leftarrow (((((b_{=2} \Leftarrow A_{>2}) \Rightarrow D_{>6}) \vee (A_{=8} \oplus b_{=0})) \wedge$ $(C_{>4} \vee ((D_{=14} \Leftarrow (B_{>1} \vee (A_{>12} \Leftarrow (C_{>6} \oplus a_{>0))))) \Rightarrow D_{>8})))))$
34	0.16423	$(A_{>2} \Leftarrow (b_{=3} \Leftarrow (((((a_{>1} \Leftarrow a_{=0}) \Rightarrow (A_{>2} \vee ((c_{>13} \Leftarrow (D_{=12} \Leftarrow a_{>0}))$ $\Rightarrow (c_{=0} \vee B_{>4})))) \Rightarrow ((((D_{=10} \Leftarrow a_{>2}) \wedge (A_{=10} \oplus ((c_{>11} \vee ((B_{=7} \oplus$ $a_{>2}) \vee B_{=0})) \Rightarrow b_{=0}))) \Rightarrow (b_{>0} \wedge ((F_{=0} \vee f_{>1}) \vee (e_{=0} \Rightarrow B_{=9})))) \vee$ $b_{=8})) \Rightarrow (C_{>10} \oplus (((c_{>11} \vee ((B_{=7} \oplus a_{>2}) \vee B_{=0})) \Leftarrow A_{>3}) \Rightarrow (A_{=3}$ $\vee (e_{=13} \vee B_{>2))))) \vee c_{=9})))$

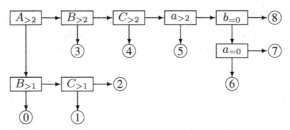

Fig. 2. A nine-valued decision tree

above. Vertical arrows correspond to "yes", horizontal arrows correspond to "no". Note that most of the decisions involve the leftmost pits for both players, and are based on "> 2", which corresponds to safety, and "$= 0$", which corresponds to mobility. When the tree is expanded further, other features will naturally appear; for example, segment 6 is fourth in line to be expanded and its split involves the feature $e_{>4}$.

Figure 3 lists the relative fitness score σ_{rel} of the decision tree evaluation as it acquires more and more leaves. By the time it contains nine leaves, the score is better than that of the human-defined evaluation function E_{saf}, as indicated by the upper dashed line. The safety difference score can take on thirteen different values, so the decision tree achieves a better accuracy with fewer nodes. The best human defined feature $E_{\mathrm{mob}}\&E_{\mathrm{saf}}$, which is the combination of mobility and safety, corresponds to the lower dashed line. Its score is better still, but it needs 169 values to achieve this. When compared to the best hand-tuned evaluation function[7] whose σ_{rel} is about 0.95, the decision tree already does better when it contains three leaves.

6.3 Generalization

An important issue is generalization. The high-level features and the decision tree were obtained by probing the 15 pebble database, but how well do they generalize to positions with more pebbles on the board? The features and decision tree were tested against the

---

[7] See Section 5.2.

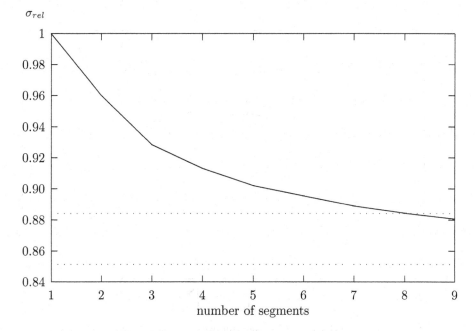

Fig. 3. Progressive scores for the decision tree

databases up to 24 pebbles. Table 4 shows the correlation scores r^2 for two features. The first feature, f_7, is the 7-atom feature discovered in the 10 pebble database, as listed in Table 5. The second feature, f_5, is the 5-atom feature evolved in the 15-pebble database and shown in Table 6.

As can be seen, the scores of the features are quite stable from 15 pebbles onward. The feature f_7 actually does better than f_5. These findings suggest that it may be possible to mine the smaller databases for high level features that can be used in the larger databases. An evaluation function based on f_5 alone would achieve $\sigma_{rel} = 0.91642$ on the 15 pebble database. According to Figure 3 this is comparable to the decision tree of Figure 2 when based on four atomic decisions.

Figure 5 plots the relative fitness score for two evaluation functions, E_M and E_D. The lower line corresponds to E_D, a data mined evaluation function containing five parameters. The upper line represents E_M, a simple hand-tuned evaluation function based only on mobility. The fitness score appears stable for positions with twenty or more pebbles. The numbers were obtained from the databases containing up to 35 pebbles, which are currently the largest databases built for Awari.

Finally, to test the stability of the evaluation parameters themselves, Figure 6 displays the μ values of E_D. The values also appear to be stable, and extrapolation of the values to positions with more than 35 pebbles appears straightforward.

6.4   Playing Strength

The fitness measure used to test evaluation features and functions indicates the correlation with perfect information. However, the goal of producing an evaluation function is

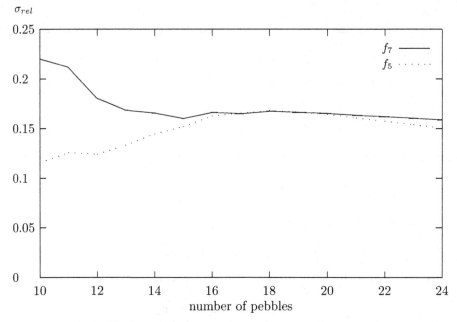

Fig. 4. Progressive scores for two high-level features

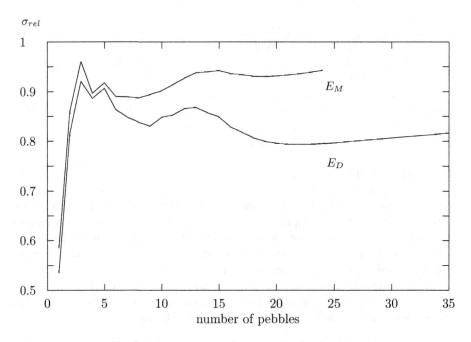

Fig. 5. Progressive scores for two evaluation functions

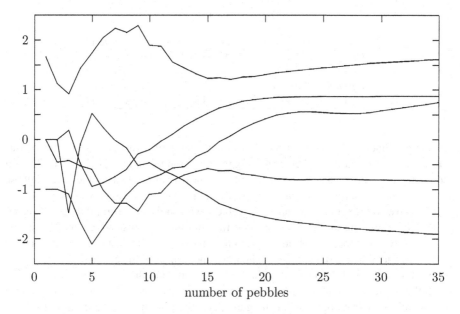

Fig. 6. Progressive parameter values for E_D

ultimately to maximize the playing strength. To test this, a round-robin match was played between four evaluation functions:

- E_P: "pebble count", material only;
- E_T: mobility used only as tie-breaker;
- E_M: mobility;
- E_D: data-mined evaluation.

The pebble count function corresponds to a program that only looks at the number of captured pebbles, and effectively ignores the position. The fitness score σ_{rel} of this evaluation function is by definition equal to 1. The functions E_T and E_M both consider mobility, but E_T uses it only as a tie-breaker between positions that have equal material advantage, while E_M counts each point of mobility as equal to one captured pebble.

The evaluation functions were tested with Bambam, a program containing state-of-the-art search techniques. During the tests, the databases themselves were not used. Each encounter between two evaluation functions consisted of 72 games, corresponding to all 36 possible two-move openings in Awari played once from the north side and once from the south side by each program. The results are listed in Table 7. Each box lists the match score in percentage of games won, and the average difference in number of pebbles captured. The data mined evaluation E_D did not exploit the two weaker evaluations better than E_M did, but it did win each of its matches convincingly, including the match against E_M itself.

These experiments test the playing strength of evaluation functions in the absence of databases. When one considers a hybrid program that uses large databases to look up values for all positions containing up to a certain number of pebbles, and a heuristic

Table 7. Match results

	E_D	E_M	E_T	E_P	final result
E_D		65% +2.7	60% +3.4	73% +4.9	66% +3.7
E_M	35% -2.7		71% +3.6	81% +7.4	63% +2.8
E_T	40% -3.4	29% -3.6		78% +4.4	49% -0.9
E_P	27% -4.9	19% -7.4	22% -4.4		23% -5.6

evaluation function for all other positions, then a game of Awari consists of two phases. During the first phase, the value of the game position cannot be proved yet, although some positions in the search tree may contain known values. During the second phase, the value of the root position can be proved, and the program can play perfectly.

The average length of the games in Table 7 was on the order of 50 to 60 moves per side. This allows the stronger evaluation sufficient time to gain an advantage. On the one hand one would expect a hybrid program to achieve a smaller gain in playing strength when enhanced with a data mined evaluation function, since the evaluation function can only make a difference in the first phase of the game. On the other hand, experiments with the 35 pebble database have indicated that database positions are reached during the search right from the start of the game, and the presence of these large databases tends to extend the length of the first phase to where it is typically half as long as an entire game without databases.

7   Conclusions

The evaluation features and functions obtained by data mining compare favourably with those defined by humans. The evaluation function is a decision tree, in which each node represents a binary decision based on a Boolean feature. The evaluation function contains a lookup table for the μ and σ values for each of the leaves. The values in the lookup table are specific to a given number of pebbles on the board; one table is needed for every number of pebbles whose database does not fit into memory. Building a high performance evaluation function would thus involve three steps:

1. identify features;
2. build a decision tree;
3. extrapolate to positions with more than 35 pebbles.

The first two steps have been demonstrated; the first step may even be skipped. The methods used were relatively simple genetic evolution and decision tree learning, respectively. The third step is a matter of regression, which is feasible since the values follow a stable trend from one database to the next.

The evolution and decision tree learning techniques are computationally feasible when applied to the 15 pebble database. On the 20 pebble database it would take about ten times as long; on the very large databases it is not feasible. However, the generalization results indicate that this is not strictly necessary, since the fitness of the discovered features and evaluations appear to remain stable for larger numbers of pebbles. Taking advantage of this, only the third step needs to involve the largest databases.

Thus, the conclusion is that it is possible in the game of Awari to build a high performance evaluation function entirely from scratch, without any human expertise. The strength of the evaluation functions has been measured both in terms of correlation with the perfect information from end game databases, and in terms of actual playing strength when used in a state-of-the-art game playing program.

8   Discussion

The performance of the evaluation functions discussed are of the order of a 15% reduction in the amount of noise relative to a material-only evaluation function. Compared to other games, this may be quite low. In our experience, Awari seems to be a game that exhibits a relatively high amount of "randomness", where clustering of good and bad positions hardly seems to occur, unless there are powerful strategic concepts that have as yet escaped us. Consulting various sources on Awari strategy has not convinced us that there are such concepts.

While this may explain why the noise in Awari is difficult to suppress, it may also be what enables automatic evaluation function design and learning to work well in Awari. In other games we may get better correlation, but those games may have more strategic "depth" enabling humans to do much better as well at analyzing the game positionally.

Any game of sufficient complexity allows for a very large set of possible features and evaluation functions, of which the ones that can be explained in human terms, and thus discovered by humans, are only a vanishingly small subset. There can be countless good features that do not mean anything to humans, and countless ways to combine them into good evaluation functions. It is generally unrealistic to expect automatically built evaluation functions to discover or rediscover human semantics in a game, but that does not mean "computer semantics" cannot do just as well or better, as mentioned in Section 6.1.

On the other hand, in many games there may be a favourable bias towards human-understandable features, for the reason that the game itself was designed by humans, and the rules have semantics that humans can relate to. Since the strategy of a game is more or less related to its rules, it is not surprising if powerful strategic concepts thus also have human semantics. This is perhaps where the game of Awari is relatively well suited to computers relative to humans. The strategy may be sufficiently chaotic and random that computers can both do better in search *and* in "understanding" the game.

References

1. Michael Buro. From simple features to sophisticated evaluation functions. In *Computers and Games '98*, pages 126–145. Springer, 1998.
2. Donald Michie and Ivan Bratko. Ideas on Knowledge Synthesis stemming from the KBBKN Endgame. *Journal of the International Computer Chess Association*, 10(1):3–13, 1987.
3. John Nunn. *Secrets of Rook Endings*. Batsford, 1992.
4. John Nunn. Extracting Information from Endgame Databases. *Journal of the International Computer Chess Association*, 16(4):191–200, 1993.
5. John Nunn. *Secrets of Pawnless Endings*. Batsford, 1994.

132   Jack van Rijswijck

6. John Nunn. *Secrets of Minor Piece Endings*. Batsford, 1995.
7. Paul E. Utgoff and Doina Precup. Constructive Function Approximation. Technical Report 97-04, Department of Computer Science, University of Massachusetts, Amherst, MA, 1997.
8. Paul E. Utgoff and Doina Precup. Constructive Function Approximation. In H. Liu and H. Motoda, editors, *Feature Extraction, Construction and Selection: A Data Mining Perspective*, volume 453 of *The Kluwer International Series in Engineering and Computer Science*, chapter 14. Kluwer Academic Publishers, 1998.
9. Terry van Belle. A New Approach to Genetic-Based Automatic Feature Discovery. Master's thesis, University of Alberta, 1995.

Chess Neighborhoods, Function Combination, and Reinforcement Learning

Robert Levinson and Ryan Weber

University of California Santa Cruz
Santa Cruz, CA 95064 U.S.A
{levinson,weber}@cse.ucsc.edu

Abstract. Over the years, various research projects have attempted to develop a chess program that learns to play well given little prior knowledge beyond the rules of the game. Early on it was recognized that the key would be to adequately represent the relationships between the pieces and to evaluate the strengths or weaknesses of such relationships. As such, representations have developed, including a graph-based model. In this paper we extend the work on graph representation to a precise type of graph that we call a piece or square neighborhood. Specifically, a chessboard is represented as 64 neighborhoods, one for each square. Each neighborhood has a center, and 16 satellites corresponding to the pieces that are immediately close on the 4 diagonals, 2 ranks, 2 files, and 8 knight moves related to the square. Games are played and training values for boards are developed using temporal difference learning, as in other reinforcement learning systems. We then use a 2-layer regression network to learn. At the lower level the values (expected probability of winning) of the neighborhoods are learned and at the top they are combined based on their product and entropy. We report on relevant experiments including a learning experience on the Internet Chess Club (ICC) from which we can estimate a rating for the new program. The level of chess play achieved in a few days of training is comparable to a few months of work on previous systems such as Morph which is described as "one of the best from-scratch game learning systems, perhaps the best" [22].

Keywords: linear regression, value function approximation, temporal difference learning, reinforcement learning, computer chess, exponentiated gradient, gradient descent, multi-layer neural nets.

1 Introduction

For years, researchers have sought a model of computer chess play that was similar to that used by humans. In particular, we are assuming a model of humans applying patterns learned from experience as opposed to employment of deep brute-force search as in the top chess systems. In addition, attempts have been made to create a chess program that is both autonomous and adaptive, while still being competitive. A wide range of machine learning techniques have also been tested, but most have met with limited success, especially when applied to playing complete games. Brute-force

T.A. Marsland and I. Frank (Eds.): CG 2000, LNCS 2063, pp. 133–150, 2001.
© Springer-Verlag Berlin Heidelberg 2001

search and human supplied static knowledge bases still dominate the domain of chess as well as other more complex domains. Therefore we have attempted to create such a system, which conforms more to cognitive models of chess. Our research will not be completed until we can compete effectively against the best programs. We believe the research presented here represents a significant step in that direction compared to other previous attempts.

Our approach can be divided in three main parts that will be discussed in detail throughout this paper. First, since the representation is the foundation of the learning system, we have spent a lot of time developing a representation which can be processed efficiently and provide an even balance between detail and generalization. Secondly, we focus on assigning appropriate evaluations in the range [0,1] to board positions. Our system steps back through the temporal sequence of moves in a training episode to assign credit to each of the board states reached, using Temporal Difference learning [24]. This relies only on the given knowledge that a win is worth 1 and a loss is worth 0. This credit assignment can be very difficult, given little a priori knowledge. Finally, once the system has assigned evaluations to all the positions in a game, the next step is to update the internally represented weights of the global optimization function, to predict more accurately in the future based on the loss incurred between the prediction $\hat{Y}_t$ and the result Y_t. This is achieved with the use of internal interconnected nodes with associated weights in a 2-layer regression network. Multiple layers of representation are very important for representing complex domains, but they are often difficult to deal with in practice. In many of our experiments, a multiplicative update on the upper level and non-linear combinations of input vector products at the lower level perform best. However there are still many additional modifications that must be made to the standard feed forward neural network to achieve adequate performance within reasonable time constraints. For the first time, our system has achieved a level of play, which is competitive against skilled amateur opponents after only about 1000 on-line games. One of the most difficult hurdles confronting previous chess learning systems was getting an agent to avoid losing pieces without explicitly representing material. In this paper we have largely solved the material loss problem with this combination of modeling and learning techniques.

Section 2 gives some background on previous work. Section 3 describes the temporal difference learning algorithm in some detail, especially concerning its application to chess and game theory. Section 4 gives a general discussion of various linear and non-linear function approximation schemes. Section 5 ties this into the multi-layer network approach. Then Section 6 brings all of these elements together in describing our system's overall learning architecture. Section 7 presents the results of our experiments with off-line training on Grandmaster databases, on-line ICC, and bootstrap learning.

2   Previous Efforts

Since Samuel's checker playing program [21] there has been a desire to use TD methods for more sophisticated game-playing models. Tesauro's TD-Gammon represented a significant achievement in the fusion of neural networks and TD learning. Baxter and Tridgell created a program named KnightCap [2] for chess, which can improve the initial weights of a complex evaluation function that included other positional aspects besides simply material values. However, this approach was quite limited since it required good initial weights, human-supplied traditional chess features, and a knowledgeable opponent instead of relying on bootstrap learning. Morph, Korf and Christensen [9], and Beal and Smith [5] were also quite successful in applying temporal difference learning to learn piece values with no initial knowledge of chess and without the aid of a skilled adversary. Beal and Smith also report similar findings for Shogi [6] where piece values are less traditionally agreed upon. Another type of chess learning is rote learning in which exact positions that have been improperly evaluated are stored for future reference [23].

"At the time Deep Blue defeated the human World Champion Garry Kasparov, many laymen thought that computer-chess research had collapsed," until the publication of an extended report by "Ian Frank and Reijer Grimbergen [showed] that the world of games is still thrilling and sparkling, and that the game of chess is considered as the math reference point for many scientific and organizational problems, such as how to perform research, how to improve a search technique, how to handle knowledge representation, how to deal with grandmaster notions (cognitive science), etc" [13]. The Morph system uses a more complex model than previous approaches, which makes its goals "the central goals of artificial intelligence: efficient autonomous domain-independent machine learning for high performance" [22]. It uses graph representations of chess positions and pattern-oriented databases, in conjunction with minimax search to evaluate board states [5]. This approach can be limited in large domains since the number of patterns can become too large. Morph III and Morph IV [15,17] used nearest neighbor and decision trees to divide positions into equivalence classes and query them on-line in logarithmic time. However these approaches require a large amount of training data to achieve reasonable levels of play. "Morph is arguably the most advanced (not necessarily the strongest) temporal difference learning system in the chess domain...However, a major problem of Morph is that although it is able to delete useless patterns, it will still be swamped by too many patterns, a problem that is common to all pattern-learning systems" [10] (Fürnkranz, pg. 10). Compared to Morph I, we have shifted away from a bag of patterns towards a representation that stores fixed numbers of internal weights. This is more efficient in both space and time, and can lead to faster global optimization. Currently, to our knowledge, there is no completely adaptive autonomous program that is actually competitive on a tournament level.

3   Temporal Difference Learning

For adaptive computer chess, the problem is learning the values of board states or at least being able to accurately approximate them. In practice, there is seldom an exact actual value for a given model state but instead there may be an episodic task where an entire sequence of predictions receives only a final value. The actual values are typically approximated using some form of discounted rewards/penalties of future returns. This is the basic idea behind TD(λ), which has been successfully applied to many applications in game theory and others in machine learning. Temporal difference learning was originally applied to checkers by Samuel [21] but it has been less successful in more complex domains. KnightCap's variant of TD(λ) called TDleaf(λ) is used to evaluate positions at the leaves of a minimax search tree. The value of λ in TD(λ), provides a trade-off between bias and variance. Tesauro also successfully incorporated TD learning for backgammon, with a set of real valued features at the lowest level of representation. This differs from our model in the sense that there are no obvious numerical representations at the base level, so we learn a set of weights for each base feature. Our lowest level of representation is discussed further in Section 6.2.

The general structure of the TD learning paradigm, as described by Sutton [24], is based on learning the policy value function $V^\pi(s)$ given a sequence of predictions, s, and their associated reinforcement signals, r. Our system uses TD(0), which simplifies the update rule to

$$V(s_t) = V(s_t) + \alpha [r_t + \gamma V(s_{t+1}) - V(s_t)], \qquad (3.1)$$

where $\alpha\gamma$ is the learning rate parameter and γ is the discount rate.

4   Linear and Non-linear Optimization

Given a sequence of trials, the learner tries to accurately predict y_t, the value of the function f(x) given a vector x of attributes. The actual value y_t is returned to the learner, which attempts to minimize the loss $L(\hat{y}_t, y_t)$ between its prediction and the actual outcome. If the actual value is unknown it can be estimated using temporal difference learning methods described in the previous section, but obviously this makes convergence much more difficult.

The complexity of the learner, determined by its number of hidden nodes, should depend on the complexity of the target class. For example, we have considered the case where x is expanded into its power set with 2^n elements where n is the length of x. Therefore the learner stores 2^n internal weights—one for each of the non-linear combinations of x vector products—which are combined to predict with the equation

$$\hat{y}_t = \sum_{i=1}^{2^n} w_i C(x_i) + w_0, \qquad (4.1)$$

where n is the dimension of x and w_0 is a constant weight which corresponds to the empty set in the power set expansion of x. $C(x_i)$ represents the *ith* combination of x vector products.

The linear case is very similar except that the number of weights corresponds directly to the dimensionality of x and the prediction is of the form:

$$\hat{y}_t = \sum_{i=1}^{n} w_i x_i + w_0. \qquad (4.2)$$

Here w_0 is a constant factor used for scaling as in the non-linear case. This method performs much better for learning simple linear target classes of functions. Tesauro suggested the use of these linear combinations for learning rules at the bottom-level of a multi-layer network for TD-Gammon [26]. However chess may require the non-linear method described above or at least more complexity than the linear model. This is explored further in Section 6.

There are two other intermediate cases we consider, which allow us to work in higher dimensional spaces without relying upon a linear solution. These involve the use of only the paired or 3-tuple terms. This reduces the number of weights to $n(n+1)/2$ or $n(n^2 + 5)/6$ respectively, where n is the dimensionality of x. Therefore the prediction is

$$\hat{y}_t = \sum_{i=1}^{n}\sum_{j=1}^{i} w_{i,j} x_i x_j + w_0 \tag{4.3}$$

$$\hat{y}_t = \sum_{i=1}^{n}\sum_{j=1}^{i}\sum_{k=1}^{j} w_{i,j,k} x_i x_j x_k + w_0 \tag{4.4}$$

for the pairs, and similarly for the 3-tuples prediction. The practical difference between these two methods for chess is illustrated in Section 7.1.

4.1 Weight Update Policies

After each training example, all of the weights must be updated to move towards the objective function and minimize the loss of the learner on future predictions. In the words of Widrow: "The problem is to develop systematic procedures or algorithms capable of searching the performance surface and finding the optimal weight vector when only measured or estimated data are available" [29].

There are numerous methods for this regression problem, which make use of different loss functions. The two main algorithms considered here are gradient descent (GD), which is sometimes called the Widrow-Hoff algorithm [29], and exponentiated gradient (EG±) with positive and negative weight vectors. Kivinen and Warmuth showed the superiority of EG± for sparse target instances. In the non-linear case or even for the pairs, some of the additional expanded hidden terms will act like irrelevant attributes and therefore EG± has some advantages in that case. "For the EG± algorithm, the dependence on the number of irrelevant variables is only logarithmic, so doubling the number of irrelevant variables results in only a constant increase in the total loss" [14]. Complex games like chess can greatly benefit from ignoring many irrelevant attributes and focusing on only the relevant piece interactions, but situations may arise where GD is superior since "as one might expect, neither of the algorithms is uniformly better than the other" [12].

Both GD and EG± are presented briefly here since they are crucial to the proper convergence of any learning algorithm or predictor. Although the performance of the two methods is quite different, they both rely on a common framework that updates each of their weights from w_{old} to w_{new} in order to minimize

$$d(w_{new}, w_{old}) + \eta L(y_t, w \cdot x_t), \qquad (4.5)$$

where L is the square loss function $L(y,x) = (y - x)^2$, η is a positive learning rate, and $d(w_{new}, w_{old})$ is a particular distance measure between the two weight vectors. The only difference between them lies in the choice of this distance function, where the GD algorithm uses the squared Euclidean distance and the EG± algorithm uses the relative entropy, also called the Kullback-Leibler divergence [14].

4.2 Multiplicative Updates

"An alternative approach to account for the uncertainty in evaluating the strength of game positions is to translate these evaluations into estimates of the probability of winning the games from these positions, then propagate with estimates by the **product rule** as dictated by probability theory. In chess this translation can be based on statistical records. For example, a standard scale for scoring a chess position is the equivalent number-of-pawns that this position represents" [19]

- (Pearl, pg. 359)

In addition to the update rules considered above, we consider a second class of update rules that use a simple non-weighted product of the input vector components. This method offers some important advantages over other mappings from an arbitrary dimensional space to a real-valued prediction since each vector dimension has a greater effect on the overall prediction. It makes those inputs with the most extreme minimum values clearly stand out from other inputs, and therefore the learner is much more sensitive to small values and in the case of chess makes the learner more risk adverse. The prediction rule is of the form:

$$\hat{Y}_t = \prod_{i=1}^{N} X_{t,i}, \qquad (4.6)$$

where the prediction $\hat{Y}_t$ is based on the weighted product of input vector X_t in N dimensions. This product is proportional to the sum of the logs, which is oftentimes the preferred form since it is easier to calculate and its range is easier to bound, hence avoiding overflow or underflow. Therefore the prediction becomes

$$\hat{Y}_t = \sum_{i=1}^{N} \log_2(X_{t,i}). \qquad (4.7)$$

This method has the additional advantage/restriction of zero weights to be learned, which can add to the simplicity and stability of the learner. We also divide by the entropy to favor states with less variance or uncertainty giving the final prediction of

$$\hat{Y}_t = \frac{\displaystyle\sum_{i=1}^{N} \log_2(X_{t,i})}{\displaystyle\sum_{i=1}^{N} X_{t,i} \log_2\left(\frac{1}{X_{t,i}}\right)}. \tag{4.8}$$

During the course of our study we began by multiplying the prediction by the minimum of all the Chess Neighborhood vector products in order to induce a more conservative strategy. This naturally leads to the representation presented here, where the smallest values are valued the most.

4.3   Gaussian Normal Distribution

Since a winning position has an evaluation of 1 and a losing position has an evaluation of 0, TD will always return values in the range (0,1). This requires that the predictions of the on-line agent also be scaled into this range. Although this may seem trivial it has been a persistent problem in many of our complex learning models. Our solution lies in using the Gaussian probability density function or normal distribution to compress the predictions into the proper range before computing the loss for the regression network. This refers to approximating the integral,

$$\phi(x) = \int_{-\infty}^{\infty} \frac{e^{-\frac{1}{2}d^2(x,m,\sigma)}}{\sqrt{2\pi\sigma}} \quad , \tag{4.9}$$

where d^2 is the distance metric

$$d^2(x,m,\sigma) = \frac{(x-m)^2}{\sigma} \tag{4.10}$$

for input x, mean m, and standard deviation σ [3]. This can be approximated efficiently without any difficulty using a lookup table and interpolation. Beal and Smith [6] also suggest the use of this type of sigmoid squashing function to convert the prediction into a probability of winning. In particular they suggest that the function

$$S(v) = \frac{1}{1+e^{-v}} \tag{4.11}$$

can also be used effectively for game playing programs with a prediction v(x) that is a weighted combination of input x, but they only consider the linear case. Also, we have observed that by taking the variance into account, the system can adapt to tight ranges or wide oscillations that sometimes occur in the learning process.

5   Regression Networks

Neural networks and regression have been the focus of many studies in machine learning and statistics [7]. The traditional neural network has internal connections between each node, which creates a total of 2^n connections for n dimension input vectors. These paths are defined as hidden nodes since they directly correspond to the amount of weights that must be stored and heuristically they represent the relationships between all the different combinations of input spaces in a chess neighborhood. However, for large n this many connections can be impractical both for storage and for time-efficient calculation. Therefore we make use of the 2-tuples and 3-tuples of inputs as shown in (4.3) and (4.4). This improved the performance enormously from using linear terms at the lowest level, a method that already proved sufficient for backgammon. Tesauro motivated his linear approach at the base-level since "the neural network first extracts the linear component of the evaluation function, while nonlinear concepts emerge later in learning" [26]. Experiment 7.1 shows the relative performance change after including all the triples, which increases the number of internal nodes to 833 weights for our 17-length input. In general, the number of internal nodes for a k-tuple representation of an n-dimensional input vector is

$$\sum_{i=1}^{k} \binom{n}{i}, \qquad (5.1)$$

but since time is a crucial consideration in chess, we must restrict the learner to a maximum number of connections. The Chess Neighborhoods described below are of length 17 and each board contains 64 squares, so the entire power set of Chess Neighborhood terms would be 2^{17}, which is quite expensive. There is undoubtedly a trade-off between search depth and more internal network weights or connections. Our experiments show promising preliminary findings about the importance of nonlinear terms for chess evaluation.

5.1   Multi-layer Networks

Single layer regression networks are the easiest to train, but they are limited to a smaller class of problems. The multi-layer network has proven highly effective for small problems but it too can be very costly in higher dimensional spaces. A simple nested function learner example is given here where the learner tries to predict a function F(X), where F(X) is a non-linear weighted sum of g(x) terms, after receiving the vector x as input. The functions F(X) and g(x) are of dimension N and M respectively, which makes the number of internal weights on each level equal 2^N and 2^M where the internal nodes for g(x) must be calculated for each of N's components. For the case considered, N was set to 4 and M was set to 3. The functions F(X) and g(x) are set to

$$F(\overline{X}) = 0.2X_1X_2 + 0.6X_1X_2X_3X_4 + 0.1X_1X_3X_4 + 0.9$$

$$s.t. \quad \forall_{0 < i \leq N} \ X_i = g(\overline{x}_i) \ where$$

$$g(\overline{x}_i) = 0.1x_{i,1} + 0.2x_{i,1}x_{i,2} + 0.3x_{i,2} + 0.4\,x_{i,3} + 0.5x_{i,1}x_{i,2}x_{i,3}.$$

Using the on-line learning model described for the single-layer case presented in (4.1), the learner is able to effectively minimize its loss on the training data with weights that converge quite rapidly when applied to a 2-layer network with hidden nodes. Both of the update rules GD and EG± are compared using the same randomly drawn examples with the exact same target function. Figure 1 shows how both networks were able to accurately predict with a loss of 180 and 38 for GD and EG± respectively on 10,000 training examples.

Clearly the EG± algorithm outperforms the GD algorithm in this particular case. The value of η must be carefully tuned for both levels to ensure good performance. This problem is eliminated for our chess agent, with the use of a variable learning rate as shown in Section 6.3.

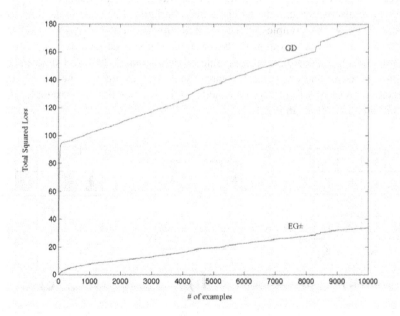

Fig. 1. Cumulative loss of EG± and GD for multi-layer function evaluation

6   Our Representation

"In more complex games such as chess and Go, one would guess that an ability to learn a linear function of the raw board variables would be less useful than in backgammon. In those games, the value of a particular piece at a particular board location is more dependent on its relation to other pieces on the board. A linear evaluation function based on the raw board variables might not give very good play at all – it could be substantially worse than beginner-level play. In the absence of a suitable recording of the raw board information, this might provide an important limitation in such cases to the success of a TD learning system similar to the one studied here."[26]

-Tesauro (TD-Gammon, pg. 10)

6.1   Chess Neighborhoods

The core of any learning system is the representation, since it provides the basis for anything it could possibly learn. Therefore for chess, we must choose a representation that accurately represents the geometry of the board. The naïve method of simply storing each of the 64 squares would suffer from lack of generalization and would therefore require too much training to be feasible in practice. In fact the geometry is counterintuitive for most humans due to the different movements of the pieces. Snyder and Levinson provided an abstraction in terms of "safe shortest path distances" [16] between the pieces but didn't indicate the best way to evaluate such paths.

We have discovered that the 64 chess neighborhoods that exist in each position can accurately show the complex piece relationships without being too specific to be worthwhile for on-line learning. A *chess neighborhood* is defined as a vector of length 17, where each dimension represents one of the: 2 ranks, 2 files, 4 diagonals, and 8 knight squares around one central square. Blank squares are also included in this representation so there will always be exactly 64 in any position. An example of such a neighborhood is shown in Fig 2.

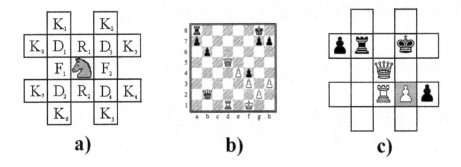

Fig. 2. a) This is a general example of a chess neighborhood with the 2 Files, 2 Ranks, 4 Diagonals, and 8 Knight distances away from one particular center square, which in this case happens to be a Knight. b) Shows a real game position our agent encountered. c) shows one of the 64 chess neighborhoods in the position in b). The darkened lower diagonal signifies its adjacency to the queen.

In a chess neighborhood the values for D_{1-4}, R_{1-2}, F_{1-2}, K_{1-8}, and the value of the center square from Figure 2a come from the lowest-level piece weights, which also take into account the chess neighborhood position, like upper left diagonal or left file. We represent the difference between adjacent and non-adjacent pieces in each neighborhood. This is illustrated for an actual game position in Figure 2b, in which our agent forked the opponent while simultaneously threatening mate.

6.2 Lowest Level Representation

On the lowest level, the weights are stored for each possible position where a piece could reside in a chess neighborhood, with respect to what piece is in the center of the position. For example, the value of having a black queen adjacent to the upper diagonal of a chess neighborhood is highly dependent on the piece in the center of that particular neighborhood. A white king in the center is in check, whereas an opponent's piece is supported by the black queen. Therefore our learner stores weights for each of these situations, which are updated proportionally to loss on each training example. In particular, the learned values of a piece being in the center square can be thought of as an approximation of the program's material piece values. The values obtained after approximately 2500 games of training, starting with randomly initialized values, are listed in Table 1 and they make a great deal of intuitive sense as material evaluation terms. This provides an adequate base onto which we can begin to develop a strong learning system, where there are no human-supplied bottom-level numerical features.

Table 1. The learned weight values after a mixture of on-line and off-line training with the 2-tuple agent using GD. These values are from white's perspective.

	King	Queen	Rook	Bishop	Knight	Pawn
Our Agent	0.51354	0.7158	0.63394	0.60757	0.61083	0.54027
Opponent	0.51354	0.29919	0.38528	0.41552	0.41471	0.47008

6.3 Bringing It All Together

The overall design of our system incorporates all of the features discussed in this paper. The weights on the lowest level are used as inputs to the 64 chess neighborhoods in each position. Each of the 17-vector chess neighborhoods uses a global set of weights for non-linear evaluation terms to make 64 predictions $\hat{y}_t$. These are then combined on the top-level multiplicatively to create an overall prediction for the board state. Figure 3 shows a graph of this learning hierarchy without including the table of base-level feature values presented in the previous section. The Chess Neighborhood weights are updated with the GD algorithm but we also consider EG± algorithm. For gradient descent, we use a learning rate η where $\eta = (Y_t - 0.5) / 200$ for actual value Y_t. Dividing by 200 decreases the magnitude that the weights move after each update, which therefore increases the stability of our entire model since the weights change less erratically over time. This puts a higher value on past experience.

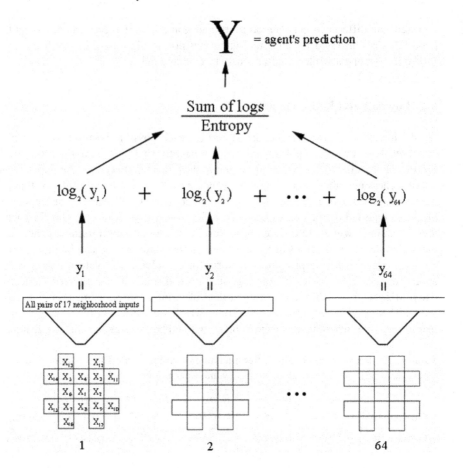

Fig. 3. This shows an overall model of our learner. The Chess Neighborhoods *(bottom)* of are expanded to create intermediate predictions $\hat{y}_t$. The sum of the logs of these predictions is then divided by the entropy to give the overall evaluation Y_t *(top)*.

6.4 Symmetry

One of the problems with building a learner based on, for example, a standard 64 square input representation is that the large number of symmetries on that board would not be exploited, leading to tremendous learning deficiencies. For example, a rook attacking a bishop should have similar significance regardless of where it occurs on the board. The Morph graph representation exploited some of these symmetries but was not able to exploit redundancies across graphs as we can with non-linear regression on the lower-layer [15]. After each training episode, the lowest level weights are averaged based on rank, file, diagonal, and knight movement symmetry. Pawns are a special case where only file symmetry applies.

7   Experimental Results

7.1   Varying Number of Internal Nodes

In the first experiment we studied the relative performance improvement with greater numbers of internal weights or nodes. Two versions of our system are trained against one another, where both versions have a different number of internal nodes. Games are conducted in an on-line fashion, where each agent begins with randomly initialized weights and doesn't learn after winning. Both agents use a 2-ply search. The results of Figure 4 show the improvement achieved by going from storing and updating all pairs of inputs to storing and updating all triples of the input vector. Tesauro showed for his TD net that performance increases monotonically with the number of hidden nodes [27]. We expect the performance to steadily increase in this fashion until the complexity of the internal representation is greater than or equal to the complexity of the problem. For example, a linear version of the regression model outperforms the non-linear version, when learning simple learner target classes, which is expected since the knowledge of the correct target starts the linear learner off with a much smaller universe of possible classifications. However, using a representation greater than the 2-tuples or 3-tuples will take a great deal of time, especially when combined with mini-max search. Ideally search can be eliminated entirely and replaced with positional knowledge in the form of internal weights and improved generalization among similar positions but minimal search may still be necessary to achieve competitive play and keep the program from throwing away material.

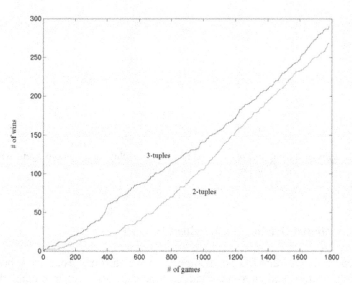

Fig. 4. GD learning agents with 2-tuples (pairs) and 3-tuples of input vector products

7.2   Replacing Minimax Search with Knowledge

A very important question is whether the representation and learning scheme we have illustrated in this paper can be used to effectively replace minimax tree search. Therefore we have tested the 3-tuple agent against a random agent with a greater search depth. Despite its random evaluation at the leaf nodes, the random agent is quite strong due to the importance of mobility in chess and the uniform distribution of the random numbers, causing those positions with the greatest number of leaf nodes to tend to have the most extreme evaluations [4]. Games are conducted on-line as in the previous experiment where the winner doesn't learn from the previous game. The actual learning agent is set to 2-ply of search, and the random agent is set to 5-ply. As the graph in Figure 5 expresses, the 3-tuple agent effectively learns to beat the 5-ply random agent at an increasing rate with 3-ply less search. This conclusion illustrates the importance of a good model with relevant features to accelerate learning and decrease the importance of search.

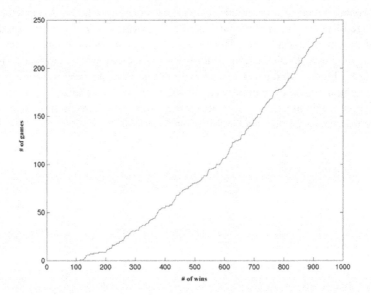

Fig. 5. The wins of a 3-tuple 2-ply GD agent over time playing a random agent with a 5-ply search. Initially the learner's weights are randomly chosen

7.3   Exponentiated Gradient vs. Gradient Descent

We previously showed that EG± can outperform GD in some instances, especially for sparse input vectors and non-linear target functions but neither method has been shown to be superior in every case. For this reason, it is important to compare the performance of both methods in this application. This might give us a greater understanding of the underlying target function, which represents the interactions between pieces in our model. As in the previous experiment both agents GD and EG± start with an empty database and play successive games against each other where only the loser learns from the previous trial. In this case our experiments up to now have proven inconclusive, since gradient descent is the clear winner initially but EG± is ahead at the conclusion of the experiment and appears to have a steeper learning curve. Gradient descent is better when "the weight vectors with low empirical loss have many nonzero components, but the instances contain many zero components," [12] which may be the case in our chess model since many of the 64 squares are blank squares, especially in the endgame positions which have the most extreme evaluations and the non-blank pairs of input vectors may stand-out. However, the nature of the multi-layer learning system with non-linear terms seems to favor the EG± algorithm, especially as the number of internal nodes increases. We intend to continue this experiment to compare their performance on a larger time scale and with different numbers of internal nodes.

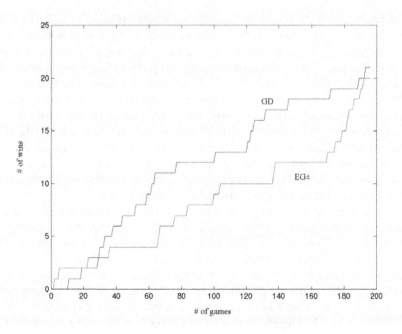

Fig. 6. The sum of wins for each of GD and EG± with a 2-ply search and initially random weights

7.4   Internet Chess Club and Grandmaster Datasets

Our new agent has competed for a few days on the Internet Chess Cub (ICC) and trained from several hundred Grandmaster games from online datasets. The former is by far the most useful training technique since the agent can actively explore the consequence of its own conclusions. As mentioned previously, the two initial random moves add enough variability to the games to make them interesting and productive for the learning process. Its rating continues to climb while playing on the internet chess server (ICC). So far the 2-tuple agent with a 4-ply search has achieved a rating of 1042 on ICC, which is a significant improvement over other learning programs such as Morph IV [17], which required months of training to reach the same level. Its learning and/or playing performance appears to be superior to that reported for other learning chess systems, such as NeuroChess [28], SAL [11], and Octavius [20]. We have also found bootstrap learning to be extremely effective for training, which contradicts the findings Baxter and Tridgell had for their agent KnightCap, which had difficulty acquiring knowledge autonomously [2]. Computers have certainly gotten faster in recent years, but our agent's improved learning ability is due to better representations rather than processing speed.

8   Conclusion

One of the enjoyable things about encountering a new opponent is to find out what "chess theory" they are consciously or unconsciously using. One difference between human and machine chess players is that the human's "theory" about chess evolves from game to game and sometimes move-to-move. However, for traditional chess computers, their "theory" is static, being built into their evaluation function and search heuristics - if any flexibility is retained it is style knobs that are alterable by the user. The program discussed in this paper represents a departure from the norm: the computer develops its own theory of chess evaluation and tests and evolves the theory through over-the-board experience. Another aspect of a theory is that its individual primitive components are either few (as in $E=MC^2$) or uniform. In this case we give the computer the framework of a theory based on uniform "chess neighborhoods" but leave it up to the system with its experience to fill in the details.

In order to participate properly in chess combat, the program must learn to evaluate various aspects of its position and then combine the values of these aspects into a single number that it attempts to maximize with the assistance of look-ahead search. To date, these two types of valuations: 1) Values of parts or features of a position 2) Whole positions based on the values of individual features - have been too combinatorially complex for machine learning programs to employ without making serious blunders such as needless tossing away of material. Unlike backgammon, where a probabilistic approach is highly appropriate, in chess, one bad move can be fatal. We believe that in this paper we have made significant steps in the resolution of the learned evaluation problem.

The contributions in this paper, by themselves, may not be original or novel, but when put together they represent a significantly new approach to chess evaluation. Contributions of this paper include:

1. The definition and use of chess neighborhoods to encapsulate local knowledge about a chess position.
2. The use of a regression network to learn non-linear combinations of the individual values of pieces that make up a piece neighborhood to arrive at a single value for

the entire neighborhood. The use of the regression network dramatically reduces the cost of learning the value of patterns over pattern systems such as Morph, that are unable to exploit the redundancy across the patterns.

3. The use of "exponential gradient descent" as opposed to traditional gradient descent in the learning of non-linear functions in which many sub-terms may be irrelevant.

4. The use of a symmetry updating phase to improve the speed of learning in a network by making nodes that "should be equal" be equal by taking an average of their values.

5. The use of a maximum product rule and minimum entropy rule to combine the 64 neighborhood evaluations in a position in a conservative risk-adverse way appropriate to good chess evaluation.

6. By starting the game by playing two random moves, increasing the exploration of the chess learned space.

In ongoing work, among other things, we are working on assessing the trade-offs between number of hidden (non-linear) nodes in the regression network, search depth and performance.

References

1. Allen, J., Hamilton, E., and Levinson, R. New Advances in Adaptive Pattern-Oriented Chess (1997). In H.J. van den Herik and J .W.H., Uiterwijk. Advances in Computer Chess 8, pp. 312-233., Universiteit Maastricht, The Netherlands.

2. Baxter, J., Tridgell, A., and Weaver, L. A chess program that learns by combining TD(λ) with game tree search. In *Proceedings of the 15th International Conference on Machine Learning (ICML-98)*, pages 28-36. Madision, WI. 1998. Morgan Kaufmann.

3. Ballard, D. H. An Introduction to Natural Computation. Cambridge: MIT Press.

4. Beal, D. F., & Smith, M.C. (1994). Random Evaluation in Chess. ICCA Journal, Vol. 17, No. 1, pp. 3-9 (A).

5. Beal, D. F., & Smith, M.C. Learning Piece Values Using Temporal Differences. *Journal of The International Computer Chess Association*, September 1997.

6. Beal, D. F., & Smith, M.C. First results from using temporal difference learning in Shogi. In H. J. van den Herik and H. Iida, editors, Proceedings of the First International Conference on Computers and Games (CG-98), volume 1558 of Lecture Notes in Computer Science, page 114, Tsukuba, Japan, 1998. Springer-Verlag.

7. Bishop, Christopher M. *Neural Networks for Pattern Recognition*, Oxford Univ. Press, 1998. ISBN 0-19-853864-2.

8. Bradtke, S. J., and Barto, A. G. (1996). Linear least-squares algorithms for temporal difference learning. *Machine Learning*, 22, 33-57.

9. Christensen, J. and Korf, R. (1986). A unified theory of heuristic evaluation functions and its applications to learning. Proceedings of AAAI-86 (pp. 148-152).

10. Fürnkranz, J., Machine learning in computer chess: The next generation. *International Computer Chess Association Journal*, 19(3):147-160, September (1996).

11. Gherrity, M. A Game-Learning Machine. Ph.D thesis. University of California, San Diego. San Diego, CA. 1993.

12. Helmbold, D. P., Kivinen, J., and Warmuth, M. K. (1996a), Worst-case loss bounds for sigmoided linear neurons, in "Advances in Neural Information Processing Systems 8," MIT Press, Cambridge, MA.

13. Herik, H.J. van den. A New Research Scope. International Computer Chess Association Journal 21(4), 1998.

14. Kivinen, J. and Warmuth, M. K. Additive versus exponentiated gradient updates for linear prediction. Information and Computation. Vol. 2, pp. 285-318, 1998.

15. Levinson, R. A., and Snyder, R. (1991). Adaptive pattern-oriented chess. In L. Birnbaum and G. Collins (Eds.), Proceedings of the 8th International Workshop on Machine Learning, pp. 85-89, Morgan Kaufmann.

16. Levinson, R. A., and Snyder, R., "Distance: Towards the Unification of Chess Knowledge", International Computer Chess Association Journal 16(3): 123-136, September 1993.

17. Levinson, R. A., and Weber, R. J. (2000). "Pattern-level Temporal Difference Learning, Data Fusion, and Chess". In SPIE's 14th Annual Conference on Aerospace/Defense Sensing and Controls: Sensor Fusion: Architectures, Algorithms, and Applications IV.

18. Littlestone, N., Long, P.M., and Warmuth, M. K. (1995), On-line learning of linear functions, *Journal of Computational Complexity* 5, 1-23.

19. Pearl, J. (1984). *Heuristics: Intelligent Search Strategies for Computer Problem Solving.* Addison-Wesley, Reading, Massachusetts.

20. Pellen, Luke. Neural net chess program Octavius: http://home.seol.net.au/luke/Octavius (1999).

21. Samuel, A. (1959. Some studies in machine learning using the game of checkers. *IBM J. of Research and Development*, 3, 210-229.

22. Scott, J. Machine Learning in Games: the Morph Project, Swarthmore College, Swarthmore, PA. http://forum.swarthmore.edu/~jay/learn-game/projects/morph.html.

23. Slate, D.J., A chess program that uses its transposition table to learn from experience. *International Computer Chess Association Journal* 10(2):59-71, 1987.

24. Sutton, R. S. (1988). Learning to predict by the methods of temporal differences. *Machine Learning*, 3, 9-44.

25. Sutton, R. S., & Barto, A.G. (1998). *Reinforcement Learning: An Introduction.* Cambridge: MIT Press.

26. Tesauro, G. Temporal Difference Learning and TD-Gammon. *Communications of the ACM*, Vol 38, No 3, March 1995.

27. Tesauro, G. Practical Issues in Temporal Difference Learning. *Machine Learning*, 8:257-278, 1992.

28. Thrun, S., 1995. Learning to Play the Game of Chess. In Advances in Neural Information Processing Systems (NIPS) 7, G. Tesauro, D. Touretzky, and T. Leen (eds.), MIT Press.

29. Widrow, B., and Stearns, S. (1985), "Adaptive Signal Processing," Prentice Hall, Englewood Cliffs, NJ.

Learning a Go Heuristic with TILDE

Jan Ramon, Tom Francis, and Hendrik Blockeel

Department of Computer Science, Katholieke Universiteit Leuven
Celestijnenlaan 200A, B-3001 Leuven, Belgium
{Jan.Ramon,Hendrik.Blockeel}@cs.kuleuven.ac.be

Abstract. In Go, an important factor that hinders search is the large branching factor, even in local problems. Human players are strong at recognizing frequently occurring shapes and vital points. This allows them to select the most promising moves and to prune the search tree. In this paper we argue that many of these shapes can be represented as relational concepts. We present an application of the relational learner TILDE in which we learn a heuristic that gives values to candidate-moves in tsume-go (life and death) problems. Such a heuristic can be used to limit the number of evaluated moves. Even if all moves are evaluated, alpha-beta search can be sped up considerably when the candidate-moves are approximately ordered from good to bad. We validate our approach with experiments and analysis.

Keywords: Machine Learning, Go, Decision trees, Inductive Logic Programming, Tsume-Go

1 Introduction

In Go, an important factor that hinders search is the large branching factor. Even in local problems, this can severely limit the ability of an algorithm to read far ahead and find a good solution.

Human players are strong at deep reading of Go positions. An explanation for this is that they easily learn and recognize frequently occurring shapes and vital points. This allows to select the most promising moves at each level and examine only these more thoroughly.

From a machine learning point of view, this motivates us to consider the task of learning a heuristic that evaluates candidate-moves. Such a heuristic would allow to select the most promising moves for further investigation. Alternatively, if one really wants to be sure that no good moves are missed, one could investigate all moves but in order of expected value. It is well-known that in an alpha-beta search the ordering of candidate moves can result in quite large performance gains.

In this paper we consider the problem of learning a theory that predicts the value of a move. Our goal is to learn a good heuristic in the tsume-go domain.

We believe that learning heuristics can be more efficient in the long run than hard coding them, as repeating a machine learning process on more data or with a better language bias can be done automatically while hand coded heuristics are difficult to improve as they become more and more complex. Therefore, in this paper we will try to get as much as possible without hard coding any non-basic knowledge.

T.A. Marsland and I. Frank (Eds.): CG 2000, LNCS 2063, pp. 151–169, 2001.

Most approaches learn patterns that describe a part of the board by assigning pieces to specific positions. This works well for many games, and also for describing elementary shapes such as 'a hane on the top of two stones' in Go. However, we argue that many concepts in Go are relational, and hence such propositional patterns are unsuitable for describing them. We will give an example of both attribute-value and relational representations of some concepts and discuss the advantages of both.

The remainder of this paper is organised as follows. In Section 2 we introduce the relational representation language we use. Next, in Section 3 we review the TILDE system. In Section 4 we describe our experiments and analyse the results. Finally, we present conclusions and ideas for further work in Section 5.

2   A Relational Representation Language

We will briefly illustrate rather than extensively introduce the first-order logic concepts used in the rest of the paper. For more theoretic foundations of Inductive Logic Programming we refer to [9]. We use the "learning from interpretations" setting of Inductive Logic Programming. Formal aspects of this induction paradigm are discussed in [4].

Most machine learning systems use a propositional or attribute-value setting. This means data can be represented in only one relation. What do we call a propositional pattern in the context of Go? Many machine learning approaches use board representations and patterns which assign *a black stone*, *a white stone*, *empty*, *edge* or *don't care* to an intersection. Here, *edge* means in fact 'off the board'. Such a pattern is given in Figure 1. When this pattern is applicable, the white stone on the second line can be captured by black by playing to the right of it on the intersection labeled '1'. This pattern can be represented in one relation as in Table 1 and therefore we call it 'propositional'. Notice that $(0, 0)$ is the coordinate of the move we want to make, in this case the intersection labeled '1'. Several approaches use fixed-size patterns (see e.g. [12], [8])) such as in Figure 2. Some do make the representation language more expressive by allowing the pattern to have other shapes (e.g. the flexible variant in [8]) or by using graph-based patterns (but then without board-matching part) (see [12], [7]), but not in a systematic way.

Fig. 1. A propositional pattern

However, in Go these approaches are often not natural. It is difficult to represent well-known shapes such as 'oi-otoshi', 'double atari', ... in a propositional representation language. Also, common tasks in playing Go such as counting liberties are difficult to implement (if possible at all) using only a propositional language. Therefore, both the board and the relations between groups (and between groups and empty spaces) need to be taken into account.

Table 1. Propositional representation of a pattern

black plays on (0,0) if		
-3	1	empty
-2	0	black
-2	1	empty
-2	2	edge
-1	-1	black
-1	0	white
-1	1	empty
0	1	empty
1	0	empty

Fig. 2. Fixed-size patterns square and diamond

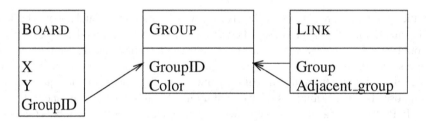

Fig. 3. A relational representation

We propose the relational representation in Figure 3. Every intersection on the board belongs to a group indicated by the BOARD relation. A group can be one empty intersection or can be a string of stones. This is defined by the GROUP relation which specifies the color of a group. Also, for each group the list of adjacent groups is stored in the relation LINK.

These relations give a view at both the board and the intergroup relations. On top of this additional implicit relations, called background knowledge, can be defined. Figure 4 gives some of these relations in PROLOG. Table 2 gives an overview of the predicates we used for learning on Go datasets. An important question is how much background knowledge to use. We decided to use sufficient knowledge to be able to express most tactical things easily. If more background knowledge was used, it could be expected that there was a chance of getting better results. However, this would increase the time needed to learn and the resulting heuristics would need more time to be executed when used in a program. Also, from the learning point of view perhaps not much more would be learned,

```
liberty(GroupID,Liberty) :-
      link(GroupID,Liberty), group(Liberty,empty).
liberty_cnt(GroupID,Nr_of_liberties) :-
      findall(Lib,liberty(GroupID,Lib),L),
      length(L,Nr_Of_liberties).
atari(GroupID) :- liberty_cnt(GroupID,1).
```

Fig. 4. Some background predicates

Table 2. Background theory for Go problems

Predicate	Meaning
liberty(G,L)	L is a liberty of the group G
liberty_cnt(G,N)	N is the number of liberties of G
stone_cnt(G,N)	N is the number of stones in the group G
group_on_pos(Pos,Displacement,G)	G is the group on the intersection Pos+Displacement
distance_to_edge(Pos,Dir,Dist)	Dist is the distance from Pos to the edge in the direction Dir
group_size(G,S)	S is the size of G
X < Y, X =< Y, X=Y	Comparison operators

as the system had already a higher level of knowledge to start with, and it would be harder to evaluate the results. Moreover, the same problems as when hard coding heuristics would arise. It is unclear what should be included and it is probable that simplicity would be lost. So we only provide background predicates which can be executed efficiently and follow directly from the rules of the game. From this, more complex queries can be composed by the learning system by combining several predicates into one test.

In PROLOG, variable names begin with a capital letter. Predicate symbols (names for relations) begin with a small letter. Bodies of the clauses should be read as existentially quantified conjunctions. E.g., the clause

```
p(X) :- q(X,Y),r(Y,Z).
```

should be read as $p(X) \Leftarrow (\exists Y, \exists Z : q(X,Y) \wedge r(Y,Z))$. In Figure 4, the liberty_cnt predicate contains a $findall$ predicate, which makes a list of all solutions of the call $liberty(GroupID, Lib)$. Since $GroupID$ is instantiated at the moment of this call, all liberties of the group $GroupID$ are collected in the list L. The length of the list is then the number of liberties.

Notice that the liberty predicate is also defined for a group consisting of one empty intersection. A liberty of an empty intersection is an adjacent empty intersection.

We deal with invariance due to the symmetry of Go with respect to mirror, rotation and color swap by doing all queries relative to some chosen axes. This causes all predicates which take as an argument a direction, some coordinate or a color to have an extra parameter. E.g., instead of using $board(X, Y, GroupID)$ we use $board(X, Y, Axes, GroupID)$. The $Axes$ variable can contain a triple $(Xmirror, Ymirror, XYmirror)$ where $Xmirror$, $Ymirror$ and $XYmirror$ are boolean values indicating whether hor-

```
double_atari((X,Y),A,B) :-
     liberty_cnt(A,2), liberty_cnt(B,2),
     liberty(A,L),liberty(B,L), board(X,Y,L).
geta_2((X,Y),Color,A) :-
     group_on_pos((X,Y),(1,1),A),group(A,Color),
     liberty_cnt(A,2),liberty(A,L1),liberty(A,L2),
     group_on_pos((X,Y),(1,0),L1),
     group_on_pos((X,Y),(0,1),L2),
     group_on_pos((X,Y),(1,-1),L3),group(L3,empty),
     group_on_pos((X,Y),(-1,1),L4),group(L4,empty).
```

Fig. 5. Definition of *double_atari* and *geta_2*

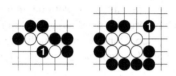

Fig. 6. Example of *double_atari* and *geta_2*

izontal, vertical and diagonal mirror should be done. Here, our prolog notation allows an elegant optimisation. The predicate translating from absolute coordinates to coordinates relative to the axes instantiates the *Axes* variable to one of its 8 possible values if it is uninstantiated, otherwise only performs the translation. In this way, no axes are chosen for queries such as *liberty_cnt(GroupID, Nr_of_libs)* which do not contain literals which require coordinates, and *Axes* remains uninstatiated such that no backtracking should be done.

Each example of the learning system has a key to identify the data belonging to that example. This also introduces an extra argument to many predicates. However, in this paper we make abstraction of this and drop the key-argument for simplicity of notation.

It could then be hoped that a machine learning algorithm could learn simple concepts such as *double_atari* and *geta_2* of which an approximate definition (omitting some constraints e.g. that surrounding groups should not be in atari) is given in Figure 5 and an example in Figure 6. In Figure 6, the intersection labeled '1' is the intersection on which black has to move.

Notice that all these definitions are short, natural and (for those used to reading PROLOG) simple while it would be very difficult to describe these concepts by propositional patterns. E.g. the definition for *geta_2* says: a group A of color *Color* could be captured in a geta by playing at $(0,0)$ if it has only two liberties, and those liberties $L1$ and $L2$ are at $(1,0)$ and $(0,1)$, and playing on them does not connect to another group, i.e. $L3$ and $L4$ are empty (at least, this is true if no group surrounding A will be captured etc.).

The representation of the data and the language bias presented here is not unique. E.g. the LINK relation is in fact redundant and can be computed from the BOARD relation. Also the GROUP relation could be eliminated by adding an attribute color to the board relation (though this would introduce redundant information). It is therefore possible to drop these relations and replace them by background knowledge, i.e. have them computed when necessary. This saves some memory in the case of dropping LINK, but involves a computational cost. Another factor that has to be taken into account is the time needed to update a position after a move has been done. In the approach presented here, we try to take the best of both sides.

Notice that even if both relations would be dropped, we would not call the resulting representation propositional because the background predicates still provide a substantial relational expressivity. This background knowledge could not easily be represented in a propositional way such that it could be used by a propositional learner. If this is tried anyway several complexity problems arise. These issues are described more thoroughly in [5].

Therefore, our approach is also different from these that use feature generation, which is frequently applied when using propositional learners or neural networks (see e.g. [7]). These approaches preprocess the data, generating some set of features which are then used in the learning step. Our system is not limited to such a predefined set of features. This implies that complex hypotheses can be generated if necessary.

3   The TILDE System

Tilde [1] stands for Top Down Induction of Logical Decision trees. This decision tree learner is an extension of the well-known TDIDT and C4.5 algorithms, in that it has the same basic algorithm. It enhances them by using a first order logic representation for both the examples and the tests in the nodes of the trees. In this section we briefly review the Tilde system.

Definition 1 (logical decision tree). *A logical decision tree T is either a leaf with class k, in which case we write $T = leaf(k)$, or it is an internal node with conjunction c, the decision tree l as left branch and the decision tree r as right branch, in which case we write $T = inode(c, l, r)$.*

Logical decision trees can be transformed into a set of rules in a straightforward way by following the paths from the top node to the leaves and making conjunctions of all encountered tests (or their negation, depending on which branch is chosen). A more thorough analysis of this is given in [1].

Figure 7 gives an example of a logical decision tree. It allows to classify black moves into normal moves, moves that capture an enemy (white) group and illegal moves (on non-empty intersections and suicides). Notice that nodes can share variables. At the top, $move(black, (X, Y))$ queries the move that is proposed. A move on an occupied intersection is illegal. If there exists a group E with only one liberty and the move fills that liberty, then E is captured. If the group containing the empty intersection on which is moved has a liberty, the move is legal. Otherwise there should be a friendly group adjacent to the move which has at least two liberties. Notice that the logical decision tree is built with both extensively defined predicates and background predicates in a transparent way.

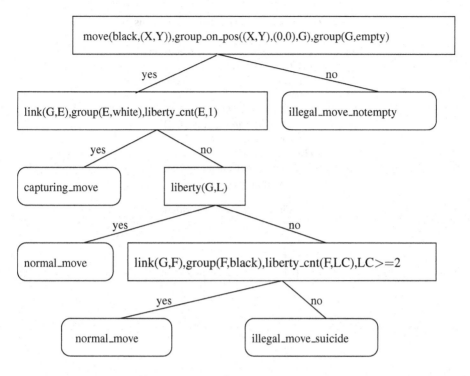

Fig. 7. Examples of a logical decision tree

The learning task performed by TILDE could be specified as:
Given:

- a set of classes C
- a set of classified examples E_{train}
- a background theory B

find:

- a logical decision tree (called the hypothesis H) that assigns (together with the background theory B) the correct class $c \in C$ to each example e of E_{train}.

In practice a more pragmatic view is adopted. E.g. to avoid overfitting a trade-off is made between accuracy and tree size. The induction of the tree is done using a recursive algorithm. An outline of the algorithm is given in Figure 8. When choosing the best test in a node, a heuristic such as gainratio is used.

TILDE has a separate mode, called regression mode, to predict real numbers instead of classes. In that case, a leaf is assigned the mean of the target value and the best test in a node is the test that minimizes the variance in its branches. This setting is described in detail in [2].

Learning decision trees have both advantages and disadvantages in the context of learning in games. An other frequently used approach is to learn sets of rules. One

procedure induce_tree(*E*:examples) **returns** tree
    Find best test T
    Let $E_+ = \{e \in E : T$ succeeds on $e\}$
    Let $E_- = E \setminus E_+$
    if partition $\{E_+, E_-\}$ good enough
    then
        Let $L = $ induce_tree(E_+)
        Let $R = $ induce_tree(E_-)
        return $inode(T, L, R)$
    else
        Let k be the majority class of E
        return $leaf(k)$

Fig. 8. The TILDE algorithm

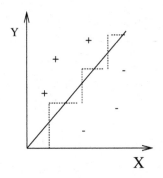

Fig. 9. A difficult to learn concept

rule then contains a pattern and a suggested move. The benefit of decision trees is that no conflicts between rules are possible, and deciding which move is best in case several patterns apply is not necessary. Distinction between any two patterns happens in a conceptual way (through a test) which is interesting from the point of view of understandability. Also, because of the structure of a decision tree, it can be executed faster than a set of rules followed by conflict resolution. For a decision tree, in worst case, only the longest pattern should be matched (i.e. all tests from the top node to the deepest leaf should be tested). For a set of patterns, usually several patterns must be matched.

However, decision trees are not so strong at certain kinds of numerical concepts. Consider e.g. Figure 9. Points above the (solid) line are positive, points below the line are classified negative. A decision tree learner that can only represent tests $X > c$ and $Y > c$ (c is a constant) would approach the line with a stepfunction-like concept. A neural network could easily adapt its weights to represent a linear function matching the concept. On the other hand, extending the language of the decision tree learner to include tests such as $aX + bY > c$ (with a,b,c constants) would also solve the problem.

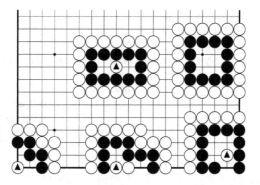

Fig. 10. Some nakade shapes

We feel in clearly defined logical problems such as tsume-go problems decision trees are not inferior to other learning settings. In large-scale strategic situations, sometimes things have to be traded off against each other. There, decision trees would probably need support of tools that are more oriented towards real-valued data.

4   Experiments

This section describes some experimental results of our approach. We first further illustrate the approach with a benchmark on a database of nakade shapes. Next, we compare different settings for learning on a database of tsume-go problems generated by Thomas Wolf's GoTools [13]. To increase repeatability of the experiments datasets and experimental setup can be obtained for academic purposes upon request.

4.1   Nakade Benchmark

A nakade shape is a big eyespace which can be reduced to one eye if the attacker plays on the vital point. Nakade shapes are important as elementary shapes in life- and death problems. Some nakade shapes are shown in Figure 10. In each case a triangled white stone is placed at the vital point. The square shape in the upper right is not a real nakade shape because it is dead whatever black does and therefore this shape has no vital point. We included it nevertheless to test the ability of the system to treat exceptions. We created a database of 166 nakade examples. As we are interested in tsume-go and only intend to use nakade shapes as benchmark, we did not use the large database of Dave Dyer [6] for this purpose. The task of trying to compress this database into a small theory is a possibility for further work. An example is a tuple (nakade shape, move). The class of an example is positive if the move is good (on the vital point), otherwise the class is negative.

We did a tenfold crossvalidation on this database. This means we did 10 folds where each time 90% of the examples were used as training set and 10% as test set. The total predictive accuracy was 100%. This is not really impressive since these shapes are relatively easy to learn. On the other hand, it is interesting to look at the induced trees.

```
move((X,Y)),board((X,Y),G),
   libertycnt(G,Libs), Libs<2 ?
   --yes : group_on_pos((X,Y),(1,0),D), group(D,defender) ?
   |           +--yes : group_on_pos((X,Y),(0,1),E), group(E,defender) ?
   |           |           +--yes : negative
   |           |           +--no : liberty(B,H,T) ?
   |           |                         +--yes : group(E,attacker) ?
   |           |                         |           +--yes : positive
   |           |                         |           +--no : negative
   |           |                         +--no : negative
   |           +--no : negative
   +--no : group_on_pos(G,(1,1),F), group(F,empty) ?
               +--yes : libertycnt(G,Libs), Libs=3 ?
               |           +--yes : positive
               |           +--no : negative
               +--no : positive
```

Fig. 11. An induced tree for determining vitality of points in nakade shapes

The ten experiments resulted in very similar trees. An example is given in Figure 11. It contains only 7 nodes, providing a compact and understandable description of the dataset. The test in the top node is $liberty\_cnt(Move, N), N < 2$. This is indeed an important test: in all nakade shapes shown in Figure 10, the vital point is also the point with the most liberties.

4.2   Tsume-go Problems

The tsume-go database we use for these experiments contains 3600 tsume-go problems generated by the GoTools program of Thomas Wolf. Of these, we use 2600 for training and 1000 for testing the performance of the learned theory on unseen problems. Each problem contains a board situation and one or more solutions ordered from high to low quality. Some problems have both solutions for black to move and white to move and can therefore be split into two separate problems. A solution contains the result (dead/alive), a representative sequence of moves (showing the failure of some resistance of the opponent) and some GoTools-specific information. The result info could allow us to determine if the task is to make a group alive or to kill an enemy group. However, in this paper we do not use this information. The learned theory will have to determine the goal itself, just as would be necessary in a real game where it is not known if it is possible to kill the enemy or not. This dataset covers a broad range of tsume-go problems, and though some types of problems are not included, we believe it is sufficient to evaluate the performance of our approach.

A problem can have several correct answers, which all solve the problem (live or kill) but perhaps differ a little (e.g. in endgame possibilities). In that case the answers are awarded 1, 0.9, 0.8, ... points in descending order of optimality. Wrong moves are awarded 0 points.

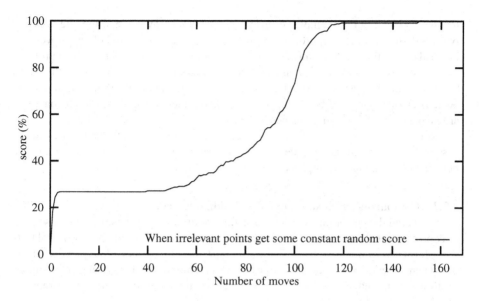

Fig. 12. Learning only the database moves

Running TILDE in regression mode on the 2600 training examples produces a logical decision tree that can be used to assign expected rewards to new problem - candidate move pairs. We evaluate our results as follows. The induced tree is used to predict the expected reward of all possible moves in the 1000 test problems. These moves are ordered from high to low expected reward. Then, a curve is drawn giving for each number n the total effective reward (given in the answers of the test problems) of the n first moves from the heuristic ordering in each problem.

4.2.1 Learning Is not Robust to Changes in the Environment

The database gives solutions to problems, in the same way it is done in problem books for human players. The best solution is given, together with some wrong solutions which could look as good moves. Normally, the problem is located in a corner or at the edge of the board and the rest of the board is empty. No human player would think of playing somewhere on that empty part of the board, 'far away' from the problem. However, when an algorithm does not get examples of playing on such uninteresting points, it will not necessarily classify such moves as bad. In fact, it will give them all the same score (their neighbourhood is empty and hence identical) but it is unpredictable what that score will be. If this score is higher than that of possible good moves, we get a curve like the one given in Figure 12. As explained earlier, the score of the first n moves in the heuristic ordering are set out against n. There is a plateau at the point where the expected value of the moves outside the problem is reached.

Several solutions are possible. First, one could predict the value of moves inside the problem only. However, it is not always easy to delimit a problem. Also, this is not a complete solution because there are still some classes of moves that will never be mentioned in a training example because they are stupid moves, e.g. filling the eyes of a two-eyed group. A better approach is to provide examples of all kind of bad moves, not just bad moves that could be mistakenly seen as good ones. Therefore, for each problem in the database, we added also an example of a random (allowed) move, with reward 0. As a result, the system learns that all these moves are also bad (though it gets some noise because some good but suboptimal moves not mentioned as such in the database are labeled as bad) and performs better.

Notice that it would not be a good idea to provide all possible bad moves to the system as that would let explode the size of the training set and disturb the balance between good and bad examples too much.

4.2.2   Comparing Propositional and Relational Learning

We compared the performance of the trees for two different settings. In the first setting, only propositional patterns are considered. This means we allow only propositional tests of the form $group\_on\_pos(Move, displacement, Group)$, $group(Group, color)$ in the tree where $displacement$ and $color$ are constants. This approach generates rules similar to those considered in [8]. This becomes clear when we introduce the exists predicate which is used in this work:

```
exists(RelativeCoordinate,Color) :-
     move(Side,Move),
     group_on_pos(Move,RelativeCoordinate,Group),
     group(Group,Color).
```

Indeed, using this predicate, rules of the form "IF $exists(xy_1, color_1) \land .... \land exists(xy_n, color_n)$ THEN play((0,0),value)" will be generated. Since we do not further restrict the rules, the hypothesis space is the same as used by the Flexible variant in [8].

In the second setting, we use a relational language. All relations $board$, $group$, $link$ and background predicates given in table 2 are allowed.

Figure 13 shows the curves obtained when learning on only 10% of the training set. It can be observed that the relational tree already gives quite good results. The results of the propositional tree are clearly inferior. Figure 14 shows the results obtained by learning on the full training set. Both settings benefit from having more training data. However, the propositional setting is still clearly below the relational setting.

The propositional curves still have one or more plateaus. This could be caused by the fact that at some points no suitable tests can be found distinguishing good from bad moves, because the system is unable to represent tests that could do so in a propositional way.

The top of the relational tree learned on the full training set is shown in figure 15. It can be seen that at the top two things seem to be important to have a fast idea of the value of the move. First, there are tests that investigate locally the number of liberties of surrounding groups. Second there are tests that look if the move is in a corner, along the edge or in the center. Both are indeed important parameters to evaluate a move.

If we record the position in the heuristic ordering of all moves that are correct answers and average over these numbers, we get 2.44. This also applies when there are several good (winning) moves. This seems to indicate that the quality of our learned heuristic

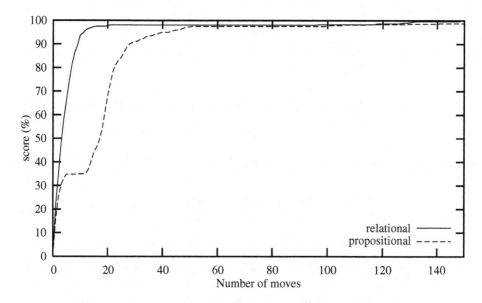

Fig. 13. Curves learned on 10% of the training set

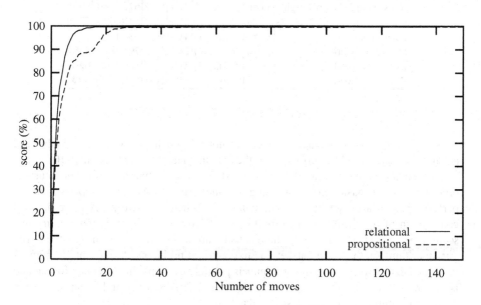

Fig. 14. Curves learned on the full training set

is comparable to the handcrafted heuristic in GoTools (of which [13] says: "If there is one or more winning moves in a given situation, then such a move comes currently on average on better than the second place in the heuristic ordering").

```
example((X,Y),Color), group_on_pos((X,Y),0,GXY),
liberty(G1,(X,Y)),liberty_cnt(G1,Libs1), Libs1>4 ?
   yes: liberty(GXY,LibXY) ?
      yes: groupsize(G1,Size1), Size1=2 ?
         yes: 0.2
         no: .....
      no: liberty(G2,(X,Y)), groupsize(G2,Size2), Size2=1 ?
         yes: ...
         no: liberty(G3,(X,Y)), group(G3,friend) ?
            ...
   no: distance_to_edge((X,Y),Dir,2) ?
      ...
```

Fig. 15. Top of the relational tree

Table 3. Score of some systems for the first five moves

System	Testset(size)	1	2	3	4	5
Propositional decision tree	Gotools(1000)	29%	49%	61%	69%	73%
Relational decision tree	Gotools(1000)	35%	58%	73%	79%	87%
Flexible rules with Weights	Basic(100)	36%	51%	63%	73%	79%
Flexible rules with Weights	3 dan(100)	31%	57%	67%	74%	80%
Flexible rules with Weights	5 dan(100)	26%	46%	55%	69%	77%
Neural network	Subset Basic(61)	41%	60%	63%	68%	75%
Neural network	Subset 3 dan(43)	27%	56%	79%	81%	83%
Neural network	Subset 5 dan(47)	21%	43%	55%	60%	62%
Neural network	A-E (1000)	35%	50%	59%	65%	69%

On the other hand, average numbers do not tell much. Even the commonly used accuracy measure has been criticized in the machine learning field recently [10].

[8] reports having found 79% of all good moves after five moves in the heuristic ordering, also for 'basic' problems, using weighted patterns. Also, [11] reports a similar performance on a subset of the same dataset with neural networks. A more detailed comparison of the results is given in Table 3. So our results (87% after five moves) seem to be at least comparable to the results mentioned there. It was impossible to run our algorithms on the same dataset due to copyright problems. At present, no human-generated database of tsume-go problems is publicly available in electronic format and is big enough to train a program and evaluate it. Therefore, it would be an interesting project for future work to create such a database.

It would be interesting to compare our results with other learning approaches and handcrafted heuristics more thoroughly, but we are not aware of many published and detailed studies. In particular, not much work mentions the number of candidate moves that can be eliminated safely as bad moves, while this is an important characteristic as shown below. Also, the lack of good and sufficiently large databases is a problem. The use of test sets smaller than a few hundred examples makes it difficult to get statistical significant results.

Table 4. Size and induction time of the tree

	on 10% of training set		on full training set	
	tree size	induction time	tree size	induction time
relational	53	12h	152	44h
propositional	53	5h	205	17h

Table 4 gives the complexity of the tree for each setting. When training on only 10% of the training set, the tree sizes are similar. For the trees induced on the full training set, it can be seen that the relational one is smaller.

Table 4 also gives the times used by Tilde to induce these trees. Recently significant improvements are made in the technology of first order learning algorithms (see e.g. [3]) sometimes increasing their speed by a factor of 40. Hence, the absolute numbers do not say much and will soon be reduced further. Their relative size is interesting from the machine learning point of view as inducing relational trees takes more time.

More important is the time needed to execute the resulting heuristic when used in a program. Evaluating a move with the relational tree in PROLOG on a pentium166 computer takes about 6 ms. Translating the heuristic to C would probably give a speedup of at least 10, which means that per second a few thousand moves can be evaluated. The propositional tree is faster, and requires about 2.5ms in PROLOG to evaluate a move. However, as will become clear in the next section, this is outweighted by the time saved because for relational trees the reduction of the branching factor that can be achieved is much higher.

The results presented here are obtained for the first move of the problems. We noticed that also for the next few moves the learned heuristics perform well. However, this is not sufficient to finish the task. As soon as it comes down to the easy but unseen part of e.g. filling liberties and effectively capturing, the heuristic fails, in the same way as in Section 4.2.1 where it was never told that moves far away from the problem are bad. However, it is well-known that the first move in a tsume-go problem is usually the most difficult to find. Therefore, we expect that the same learning strategy can be used to learn a heuristic that can be used deeper in the search tree.

4.2.3   Implication on the Search Complexity

The complexity of searching in a minimax-tree is an exponential function b^k in the depth k of the tree. b could be called the average effective branching factor. While the exponentiality usually can not be avoided, optimizations such as alpha-beta pruning and a heuristic can reduce the value of b.

How good are our results in terms of reducing this average branching factor? Figure 16 shows a detail of the relational curve learned on the full training set. 'Cumulative' is the same curve from Figure 14. 'Score' gives for each n the average reward of the n-th best classified move.

Such a heuristic can reduce the average branching factor in two ways:

– Moves that score badly for the heuristic, have a low probability of being the best move. When finding the best move with a high probability is sufficient (e.g. when time is limited and one wants to take the risk of missing a good move), these candidate

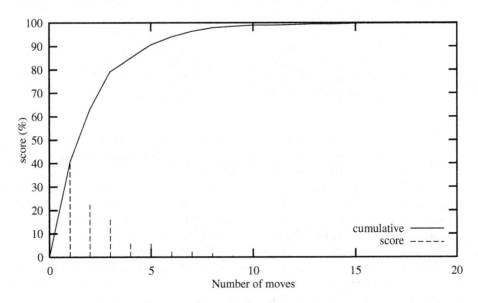

Fig. 16. Detail of the results of the relational tree

moves can simply be omitted. When a fraction k of all moves is eliminated in this way, the branching factor becomes $b' = (1 - k)b$. The risk of missing a good move should be compared with other factors causing the search to be suboptimal, e.g. errors in the evaluation function of the leaves. Moreover, if the search is done to achieve a particular goal (kill or live), failure to do so can still be a reason for the evaluation of the originally eliminated moves.

From Figure 16, it can be seen that if only the 5 best classified moves are examined, 10% of the good moves are missed, while if the best 10 classified moves are examined, only 1% of the good moves are missed. When compared with about 150 available moves on the 13x13 problem boards, or the 30 moves available on average inside the tsume-go problem, it becomes obvious that quite good branching factor reductions are possible.

– The moves that are left as candidates in the search tree can be investigated more efficiently because they are ordered from high to low expected value. Here alpha-beta pruning comes into play. In the worst case (reverse ordering) this still gives the average branching factor of b' of the minimax algorithm. However, in the case of perfect ordering, the average branching factor is reduced to $\sqrt{b'}$. While perfect ordering can never be achieved in practice (because it implies already knowledge of the best move), it is possible to approximate perfect ordering very closely.

While it is difficult to give an exact estimate for non-optimal cases, we can give a good upper bound. Let $T(n)$ be the complexity of searching a tree of depth n. Let p_i be the probability that the best move is the i-th move in our ordering. Suppose we have a tree of depth n (see Figure 17). If the i-th move is the best move, then in worst case, the moves $1, \ldots, i$ are ordered from bad to good and no pruning is possible. So the complexity to examine them could be $i.T(n - 1)$. For the other

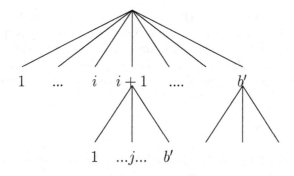

Fig. 17. A search tree

Table 5. Branching factor reductions

Heuristic	1% b'	1% upper bound	5% b'	5% upper bound
Relational on full training set	10	5.7	5	3.6
Relational on 10% of training set	127	28.4	10	7.3
Propositional on full training set	25	12.4	16	8.4
Propositional on 10% of training set	151	55	28	21.8

$b' - i$ moves pruning is possible. This can certainly be done as soon as the best answer of the opponent is encountered. Since with probability p_j this is the j-th move examined, The complexity is at most $\sum_j p_j.j.T(n-2)$. We can conclude that

$$T(n) = \sum_{i=1}^{b'} p_i \ i.T(n-1) + (b'-i). \sum_j p_j.j.T(n-2).$$ Solving this recurrent equation yields a function of exponential order, of which the base is an upper bound on the average branching factor.

Table 5 illustrates this by giving b' (with 1% and 5% of good moves missed) and the corresponding upper bound on the average branching factor we obtained by solving the equation mentioned above using the ordering obtained with the heuristics presented in Section 4.2.2.

We can conclude that the branching factor is effectively reduced. Here also, the logical decision trees are superior to the propositional decision trees.

5   Conclusions and Further Work

In this paper we presented the application of the logical decision tree learner TILDE on Go and discussed the relation to other learning algorithms. We have argued that it is useful to use a relational representation language to describe the learned concepts. We have presented experiments of our application and analysed the results, both in terms of accuracy and in terms of average branching factor reduction. The results support our claim that more expressive representations are superior to propositional ones. More-over, the performance of our learning technique is comparable to that of other learning approaches and to handcrafted heuristics. Also, we argued in favor of more thorough evaluations of machine learning approaches to go.

For further work we see several lines of possible research. First, the heuristic could be extended to be able to give predictions for all moves in a tsume-go sequence, including the final stage of effectively capturing or timely defending when dames are filled. This would enhance the usefulness of it in a real playing program. Next, the same techniques could also be tried in other parts of the game, such as endgame or opening patterns.

Acknowledgements

Jan Ramon is supported by the Flemish Institute for the Promotion of Science and Technological Research in Industry (IWT). Hendrik Blockeel is a post-doctoral fellow of the Fund for Scientific Research (FWO) of Flanders. We thank Thomas Wolf for making available his dataset of tsume-go problems. We also thank the reviewers for their comments and Maurice Bruynooghe and Johannes Fürnkranz for proofreading the paper.

References

1. H. Blockeel and L. De Raedt. Top-down induction of first order logical decision trees. *Artificial Intelligence*, 101(1-2):285–297, June 1998.
2. H. Blockeel, L. De Raedt, and J. Ramon. Top-down induction of clustering trees. In *Proceedings of the 15th International Conference on Machine Learning*, pages 55–63, 1998. http://www.cs.kuleuven.ac.be/~ml/PS/ML98-56.ps.
3. H. Blockeel, B. Demoen, L. Dehaspe, G. Janssens, J. Ramon, and H. Vandecasteele. Executing query packs in ILP. In J. Cussens and A. Frisch, editors, *Proceedings of the 10th International Conference in Inductive Logic Programming*, Lecture Notes in Artificial Intelligence, London, UK, July 2000. Springer.
4. L. De Raedt. Logical settings for concept learning. *Artificial Intelligence*, 95:187–201, 1997.
5. L. De Raedt. Attribute-value learning versus inductive logic programming: the missing links (extended abstract). In D. Page, editor, *Proceedings of the Eighth International Conference on Inductive Logic Programming*, volume 1446 of *Lecture Notes in Artificial Intelligence*, pages 1–8. Springer-Verlag, 1998.
6. D. Dyer. An eye shape library for computer go. http://www.andromeda.com/people/ddyer/go/shape-library.html.
7. M. Enzenberger. The integration of a priori knowledge into a go playing neural network, 1996. http://home.t-online.de/home/markus.enzenberger/neurogo.html.
8. T. Kojima and A. Yoshikawa. A two-step model of pattern acquisition: Application to tsume-go. In H. van den Herik and H. Iida, editors, *Proceedings of the 1st International Conference on Computers and Games*, volume 1558 of *Lecture Notes in Computer Science*, pages 146–166. Springer-Verlag, 1998.
9. S.-H. Nienhuys-Cheng and R. Wolf. *Foundations of inductive logic programming*, volume 1228 of *Lecture Notes in Computer Science and Lecture Notes in Artificial Intelligence*. Springer-Verlag, New York, NY, USA, 1997.
10. F. Provost, T. Fawcett, and R. Kohavi. The case against accuracy estimation for comparing induction algorithms. In *Proceedings of the 15th International Conference on Machine Learning*, pages 445–453. Morgan Kaufmann, 1998.

11. N. Sasaki, Y. Sawada, and J. Yoshimura. A neural network program of tsume-go. In H. van den Herik and H. Iida, editors, *Proceedings of the 1st International Conference on Computers and Games*, volume 1558 of *Lecture Notes in Computer Science*, pages 167–182. Springer-Verlag, 1998.

12. D. Stoutamire. Machine learning applied to go. Master's thesis, Case Western Reserve University, 1991.

13. T. Wolf. The program GoTools and its computer-generated tsume go database. Report, School of Mathematical Sciences, Mile End Road London E1 4NS, November 1996. http://www.qmw.ac.uk/~ugah006/gotools/.

Learning Time Allocation Using Neural Networks

Levente Kocsis, Jos Uiterwijk, and Jaap van den Herik

Department of Computer Science, Institute for Knowledge and Agent Technology,
Universiteit Maastricht, P.O. Box 616, 6200 MD Maastricht, The Netherlands
{l.kocsis,uiterwijk,herik}@cs.unimaas.nl

Abstract. The strength of a game-playing program is mainly based on the adequacy of the evaluation function and the efficacy of the search algorithm. This paper investigates how temporal difference learning and genetic algorithms can be used to improve various decisions made during game-tree search. The existent TD algorithms are not directly suitable for learning search decisions. Therefore we propose a modified update rule that uses the TD error of the evaluation function to shorten the lag between two rewards. The genetic algorithms can be applied directly to learn search decisions. For our experiments we selected the problem of time allocation from the set of search decisions. On each move the player can decide on a certain search depth, being constrained by the amount of time left. As testing ground, we used the game of Lines of Action, which has roughly the same complexity as Othello. From the results we conclude that both the TD and the genetic approach lead to good results when compared to the existent time-allocation techniques. Finally, a brief discussion of the issues that can emerge when the algorithms are applied to more complex search decisions is given.

Keywords: temporal difference learning, genetic algorithms, search decisions, time allocation, Lines of Action

1 Introduction

Most computer game programs rely on full-width game-tree search. The heuristically estimated values of the leaf positions are propagated up to the root using the basic minimax principle. This simple skeleton, together with its enhancements, led to great performances, such as DEEP BLUE's success against Kasparov [29,30]. Nevertheless, roughly speaking, the two players were equally strong, which proves the effectiveness of both game-playing approaches, viz. the brute-force approach and the human game-playing approach. Two key features of the latter approach are the amount of knowledge acquired by learning and an intelligent search procedure (compared to minimax-like algorithms). It is a challenging idea, whether the human techniques can improve the brute-force approach for chess-like games.

In game programs, learning is mostly used to improve the parameters of the evaluation function. In the last two decades, a wide range of learning techniques has been employed for this purpose, such as temporal difference learning [3,4,5,31,33], genetic algorithms [10,25], explanation-based learning [23] and Bayesian methods [2]. These techniques lead to significant improvements in many games including backgammon [33], chess [1] and Othello [7].

T.A. Marsland and I. Frank (Eds.): CG 2000, LNCS 2063, pp. 170–185, 2001.

Beside the evaluation function, the other main component of a game program is the search algorithm. Although most of the game programs rely on alpha-beta search, a number of additional decisions has to be made. These 'search decisions' include time-allocation details, the maximum depth of the search tree, move ordering, narrowing the search tree in certain positions, stopping the search in nodes considered 'safe', and various other enhancements. In almost all game programs the conditions for these decisions are set by the programmer. It is an interesting issue, whether these decisions can be improved upon by some learning methods. So far, this issue received only little attention. One of the few exceptions in this context is the influence of information about the opponent on the search, i.e., opponent modelling [8,14].

In this paper, we focus on the research question how search decisions can be learned using the final outcome of games.

For the learning of the evaluation functions and search decisions we use neural networks, but the learning techniques involved can be applied to any other parameterized function approximator as well.

The topic of our choice is the time-allocation problem. Although this is a major issue in tournament programs, not much has been published on this subject, mainly because simple ad-hoc heuristics seem to work reasonably well in practice [2,7,13] (see also subsection 1.1).

In order to study search-decision learning techniques, we need some evaluation functions. Two methods of learning evaluation functions using neural networks are described in section 2. Subsequently, a new actor-critic architecture for learning search decisions is presented in section 3. Section 4 provides experimental results for learning time allocation in the domain of Lines of Actions, a game comparable in complexity with Othello. In section 5 we provide some conclusions and section 6 contains a brief discussion of the issues that can occur when the proposed algorithms are applied to learn more complex search decisions.

1.1 Related Work

This subsection provides an overview of related work. We emphasize on recently published novel results of tuning evaluation functions using TD learning and genetic algorithms, and of algorithms dealing with time allocation.

The use of TD learning for tuning evaluation functions dates back to Samuel's checkers program [28]. Significant results in this domain of learning were obtained by Tesauro's master-level backgammon program, TDGAMMON [33]. Although the success of TDGAMMON was not replicated for other games[1], some interesting results have been obtained. In NEUROCHESS [34] a combination of explanation-based and TD learning raised the playing strength from zero knowledge to a score of 25% against GNUCHESS. Better results were obtained in KNIGHTCAP [3]. Using the TDleaf algorithm Baxter *et al.* tuned the weights of the evaluation function to reach a rating of 2150 ELO points in just three days. Moreover, Beal and Smith [4,5] applied TD learning in chess and shogi to find automatically the values of the pieces and piece-squares. The implementation of

---

[1] Apparently the strong chess program CILKCHESS [20,21] also uses some form of TD learning.

these values showed to improve competitive play. As an experimental measure, they are interesting for human players too.

Although evolving neural networks using genetic algorithms did not repeat the success of TDGAMMON, promising results have been obtained in checkers [9] and Go [26].

The first major approach to time allocation is Hyatt's publication [13] considering some manually programmed enhancements to a time-allocation scheme on the basis of the backed-up minimax value. Markovitch and Sella [22] had a more adaptive approach to the problem of time allocation. For the game of checkers the authors trained their system to recognize the class of positions which gain from extra search.

Some specific search algorithms have built-in time-allocation mechanisms. Notable examples for this category are BPIP [2] and Multi-ProbCut [7]. BPIP uses a probability distribution over the evaluation function to predict the gain of expanding a given leaf. The algorithm allocates extra time only if the expected gain from the extra search is higher than the "cost of time". In Multi-ProbCut a sequence of shallow searches is performed to detect whether a certain subtree will affect the minimax value of the root. If it is unlikely to affect the root value, the subtree is pruned, saving the time for probably more relevant lines. The use of shallow searches for forward pruning appears also in adaptive null-move pruning [12]. In this case one side can make two moves in a row, and if in this way (s)he still cannot achieve anything, this line will be pruned. The depth of the null-move search can vary with the depth of the position in the search tree or with some specific features of the position.

The idea of learning search decisions appears also in [6]. The authors propose a framework for different search extension techniques using a parameterized approximator for the search depth of the nodes. To improve the parameter vector an adaptive cost model is used to predict how many nodes it takes to search a certain position for a given parameter vector.

2   Learning Evaluation Functions

Using the final outcome of games, we deal with two approaches to find the weights of a neural network that represents an evaluation function, viz. TD learning and genetic algorithms. In both cases, the neural network receives as input the position in some encoded form, and provides as output the estimated value of the position. Further structure and functionality of the network is a choice of the designer.

2.1   Temporal Difference Learning

An attractive approach to learn an evaluation function is TD learning (see, e.g., [31]). Using this approach, each state s has associated a value V, representing the estimation of the expected outcome of the game. The state-value can be used as an evaluation function in the search tree.

In the learning phase, the state-values are updated so that they approach a target value. Let us consider a sequence of game positions $s(t_0), s(t_1), \ldots, s(t_{final})$. The target value for the final position, $s(t_{final})$, is given by

$$V^{target}(s(t_{final})) = \begin{array}{l} 1, \text{ if } s(t_{final}) \text{ is a win for Black,} \\ -1, \text{ if } s(t_{final}) \text{ is a win for White} \end{array}$$

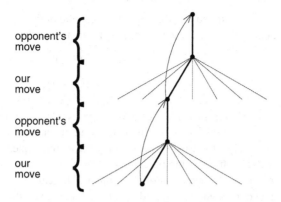

Fig. 1. TD back-ups in game trees. The solid lines represent the moves made during the game; the dashed lines represent moves that the player considered but did not play. The curved arrows denote the back-up caused by each move.

The target value for $s(t_{final-1})$ is equal to the target value of $s(t_{final})$ since the opponent is not able to reply after the move made in $s(t_{final-1})$. The target values for the non-terminal positions $s(t_0), s(t_1), \ldots, s(t_{final-2})$ are given by

$$V^{target}(s(t_k)) = \gamma \cdot V(s(t_{k+2})) \qquad \text{(with } 0 < \gamma \le 1) \qquad (1)$$

The discount factor γ is used to favour early over late success. According to equation 1, after each move we 'back up' the state-value to the position two plies before. The TD back-up mechanism, together with a sequence of moves made and considered during a game is illustrated in Fig. 1. Using only the 'even' positions for back-up complies with the general idea in TD learning that the state-values are propagated through states where the agent (the player to move) has to take an action (followed by a change in environment – in our case the opponent's move).

To speed up TD learning, we can average toward future target values using a trace-decay parameter[2], λ (with $0 \le \lambda \le 1$):

$$V^{target}(s(t_k)) = (1 - \lambda)(V(s(t_{k+2})) + \lambda V(s(t_{k+4})) + \lambda^2 V(s(t_{k+6})) + \ldots) \quad (2)$$

In game programs using the TD approach, V is typically represented by an artificial neural network, which is trained to fit the target values V^{target} obtained by the equations above.

With the TD error $\delta(t_k) = \gamma \cdot V(s(t_{k+2})) - V(s(t_k))$, the gradient updating rule for the weights w of the neural network will be:

$$\Delta w_i(t_k) = \alpha \cdot \delta(t_k) \cdot \frac{\partial V(s(t_k))}{\partial w_i} \qquad (3)$$

---

[2] In equation 2, for simplicity we omitted the discount factor, γ. The precise combination of γ and λ is left to the reader.

In equation 3 we considered for back-up only the value of the next state (i.e., $\lambda = 0$). For $\lambda > 0$, we can use eligibility traces (ew) [31] to memorize the previous derivatives:

$$\Delta w_i(t_k) = \alpha \cdot \delta(t_k) \cdot ew_i(t_k)$$

$$ew_i(t_k) = \lambda \cdot ew_i(t_{k-2}) + \frac{\partial V(s(t_k))}{\partial w_i}$$

The formulae presented should be perceived only as a skeleton for a game-playing program using TD learning. Further details include the mechanism of generating game sequences and the way how the positions are selected from the sequence, the architecture of the neural network or the learning parameters.

A possible choice on the selection of positions is TDLeaf [3]. In TDLeaf the state values are propagated through the leaf nodes of the principal variations and not through the root nodes of the search trees.

Increasing the size of the hidden layer increase the representational power of the neural network. The typical drawback is the increased difficulty of the training (since there are more weights to be trained), and the time used by propagating the activation from the input to the output. A very fast solution is not to use hidden layers at all. In this case, however, it is not possible to detect nonlinear dependencies. Some experimental results on the size of the hidden layer are presented in [33].

The gradient updating rule given above suggests a 'plain' back-propagation-like adaptation, but some of the improvements developed for supervised learning are likely to work for TD learning also. One of the few experimental results on this issue is presented in [5].

A discussion on our experimental details is given in subsection 4.2.

2.2   Evolving Evaluation Functions with Genetic Algorithms

A genetic algorithm (e.g., [10,25]) performs a global search in the space of possible neural networks. This search is guided by the individual performances in the game of the neural networks. There is no punishing or rewarding of individual moves through a gradient procedure.

The genetic algorithm maintains a population of neural networks that are encoded in so-called chromosomes. In each iteration (generation) all the neural networks are evaluated assigning to each individual a certain fitness depending on its performance (see below). Then, pairs of individuals are selected depending on their fitness. Subsequently, they are combined using crossover and mutation operators, and inserted in a new generation.

The networks are compared either in a tournament environment including all the individuals from a generation, or by playing a number of games against an external opponent. The assigned fitness of each player is a function of the number of points obtained.

3   Learning Search Decisions

In the time available for a move, searching the whole search tree can usually not be completed. With an incomplete search tree and an imperfect evaluation function it is important to emphasize on answering the question where and when are the search efforts most profitable. The underlying idea is to enhance the *quality* of the search decisions. The decisions include next to the one mentioned in section 1, for instance, selecting the next node for expansion, pruning a move (and thus a subtree), and stopping the search in nodes where the evaluation function is reliable.

Below we distinguish between the actual search and the search control. The original problem of selecting a certain move is usually termed an 'object-level' problem, and the problem of the search control a 'meta-level' problem. The object-level search which is carried out as a consequence of a meta-level decision (i.e., a search decision) is termed *computational action*.

The main property of a meta-level problem is its increased complexity compared to that of the object-level problem. The high complexity makes it intractable to compute the utilities of the search decisions. Even the currently available approximation methods are insufficient (e.g., MGSS [27] fails for search trees with conspiracy numbers higher than 1). An attractive alternative to these algorithms is improving the selection between the computational actions through learning. This is the motivation of our research.

Another important property of meta-level problems is that the computational actions between two object-level actions are approximately commutative [27] (e.g., it is not important which subtree[3] is generated first if each of them will be investigated).

The high complexity of the meta-level space influences also the encoding of the search state, enforcing us to drop some of its features. This approach transforms the original problem (for the learning program) to a Partially Observable Markov Decision Process (POMDP; e.g., [16]). The possibly high number of available search decisions can form an obstacle in learning. If it is the case, some of the decisions should be omitted or the learning environment should be restructured.

At the meta-level, test-expansion of a search decision is usually too expensive. This requires the explicit computation of the state-action values [11].

When using a gradient method with neural networks to learn the values of the search decisions, a basic idea is to enforce the search decisions leading to a won game. Then the usual problem is how to distribute the credit over the decisions taken in the game. Another problem is still the long lag before the reward.

A solution to these problems may come from the use of the TD error of the evaluation function as a reward. In this case, we will enforce those decisions that improved the position. Since the computational actions between two moves are approximately commutative the credit assignment can be made uniformly.

The use of TD error as a reward leads us to an actor-critic architecture, where the evaluation function is the critic and the part which learns the search decisions is the actor. There are two main differences between the classic actor-critic (e.g., [19,31,32]) (Fig. 2a) and the resulting search-decision architecture (Fig. 2b). First, the state space where the actor operates can differ from the state space of the critic (usually the first includes the latter). Second, there can be more than one decision of the actor between two 'signals' from the critic. The first difference does not create a problem since the evaluation function still predicts the overall performance of the complete program. The second one is solved by the commutative nature of the computational actions.

---

[3] The subtree is identified in this point by all constituent nodes and not only by its root.

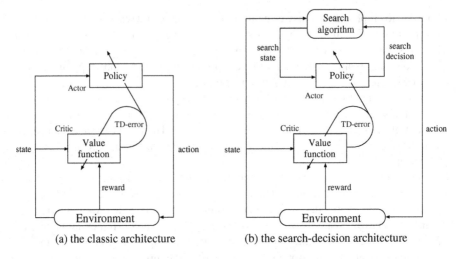

(a) the classic architecture   (b) the search-decision architecture

Fig. 2. Actor-critic architectures. In both architectures the critic learns the state-values of the object-level problem using the reward provided by its environment.

To avoid the influence of the inaccuracy in the evaluation function on the final search policy we can use eligibility traces with the same eligibility factor as the discount rate of the critic [17]. The use of the eligibility also helps if the original problem is transformed to a POMDP [15].

Consequently, a slight adaptation of the update rule to learn a search policy is proposed. It reads

$$\Delta v_i(t_k) = \alpha \cdot \delta(t_k) \cdot ev_i(t_k) \tag{4}$$

$$ev_i(t_k) = \beta \cdot ev_i(t_{k-2}) + \sum_{\tau=t_{k-2}}^{t_k} \frac{\partial}{\partial v_i}(\mathbf{y}(\tau) \cdot \mathbf{d}(\tau)) \tag{5}$$

where v_i denotes a weight of the neural net implementing the search decision, ev_i is the associated eligibility, t_k is the time sequence of the external actions, τ denotes the moments of the search decisions, α is the learning rate of the actor, δ is the TD error[4] of the critic, β is the discount factor for the eligibility trace, $\mathbf{y}$ is the output vector, and $\mathbf{d}$ is the direction of the performed search decision associated to $\mathbf{y}$ with:

$$d_i = \begin{cases} 1, & \text{if a high value of } y_i \text{ is required} \\ -1, & \text{if a low value of } y_i \text{ is required} \\ 0, & \text{if the ith output unit does not influence the decision} \end{cases}$$

To illustrate how the values for $\mathbf{d}$ are set, let us consider a meta-level with five search decisions ($sd_1..sd_5$) and a search state where only the first three (sd_1, sd_2, sd_3) are possible. If we encode the search decisions using five output units ($y_1..y_5$), and assume

---

[4] If the critic estimates the outcome of the game and not the 'reward', then we must use the negamax value of the evaluation function.

that the choice among them is made with the use of the maximum operator, then the direction vector associated to sd_1 is $(1, -1, -1, 0, 0)$, to sd_2 is $(-1, 1, -1, 0, 0)$ and to sd_3 is $(-1, -1, 1, 0, 0)$.

Although actor-critic algorithms are not very new, theoretical results on optimality and convergence properties have been obtained mostly for the tabular and linear case (e.g., [18]). Since we are more interested in the nonlinear case (as produced by multilayer neural networks), and the original problem is moved to the meta-level space (which makes a formal analysis even more difficult), we chose to validate the proposed method by experiments. The experimental setting and results are described in subsection 4.3.

If we use genetic algorithms to evolve the neural network that implements the search policy, then the fitness assignment to a certain player is not different from the case of learning evaluation functions.

4 Experimental Results

In this section we describe the experimental results. For the experiments, we have chosen the domain of Lines of Action (LOA), which is a two-person zero-sum perfect-information game, of the same general type as chess. The preference for LOA was based on the observation that LOA positions and rules are well suited for pattern-recognition techniques (such as neural networks). In LOA there is no need for recoding schemes as is the case in chess. Such a recoding may concern the set of features of a classic evaluation function or the input board when enhanced with relations derived from the elementary concepts of the game (e.g., 'attack' in chess).

4.1 The Game of LOA

LOA is played on a traditional 8x8 board by two players: Black and White (Black to play first). Twelve black stones are placed in two rows along the top and bottom of the board (not in the corners), while twelve white stones are placed in two files at the left side and right side of the board (also not in the corners) (Fig. 3, left).

A player can move one of his/her pieces, in a straight line (horizontally, vertically or diagonally), exactly as many squares as there are pieces of either colour anywhere along the line of the move. It is possible to jump over own pieces, but not over opponent's pieces. If the move lands on a square occupied by an opponent stone, that stone is removed. If a player is unable to move (s)he loses[5]. The goal of the game is to connect all own pieces. In Fig. 3 (right) Black has connected all its pieces.

The average game length is some 38 plies. Only a very small number of games terminates in a draw (repetition of moves). The average branching factor is around 30 [35].

4.2 The Evaluation Function Population

In order to test the performance of the learning algorithms for search decisions we need an evaluation function. To ensure, that our player does not become specified against a single

---

[5] More details about LOA can be found at http://www.andromeda.com/people/ddyer/loa/loa.html

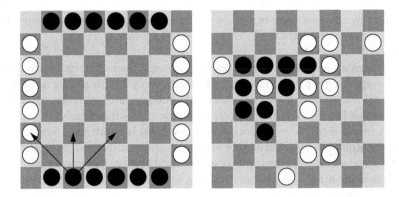

Fig. 3. Start position (on the left) and a sample end position (on the right) in LOA. The latter position is won by Black since all black pieces are connected.

opponent we need a large number of different evaluation functions. In the experiments we used three variations of the TD-learning algorithm (*STD, MTD, SANE-TD*) and a specific genetic algorithm (*SANE*).

All the evaluation functions used a feed-forward neural network with a hidden layer consisting of 40 neurons. Both the output and the hidden neurons used a sigmoidal activation function. A position was encoded by 65 input units. Each square was associated to a neuron (+1 for a black piece, -1 for a white one, 0 for an empty square) with an additional neuron indicating the side to move. The output neuron represents the minimax estimation of the position.

To evolve the weights of the evaluation functions with genetic algorithms, we used a method called SANE (Symbiotic, Adaptive Neuro-Evolution) [24]. The main feature of SANE is that it evolves two separate populations: a population of neurons (defined by their input and output connection), and a population of neural networks using, in the hidden layer, neurons from the first population. With the two populations the algorithm manages to keep a high level of diversity throughout the evolution.

To generate evaluation functions through TD learning, we used the equations from subsection 2.1, with $\lambda = 0.8$ and $\gamma = 0.9$. Although the implementation of the described method seems straightforward at first glance, there are some details which are to be discussed. These include (1) the way of generating the game sequence, (2) which positions are used from the sequence, (3) variations on the size of the hidden layer, and (4) the values of the learning parameters, such as λ and γ.

In general, game sequences can be generated using game databases or games played by the learning program itself. In the latter case, a further choice can be made on the opponent. Possible options include using an already existing game program, playing against players with similar strength on the Internet, playing against itself, or using more learning players which improve their skill by playing together. In LOA, game databases and an Internet server are not available. The remaining option is to use games played by the program which use the evaluation function to be trained. Three training strategies were tested: *STD* (Single TD player), *MTD* (Multiple TD player) and *SANE-*

Table 1. Gathering the evaluation function population: 1st tournament. The players for *std1* are generated using *STD*, for *mtd1* using *MTD*, for *sane-td* using *SANE-TD*, and for *sane* using *SANE*.

method	# of players	result	std. dev.
std1	50	0.276	0.024
mtd1	50	0.288	0.025
sane-td	50	0.237	0.011
sane	50	0.199	0.027

TD (this name is connected to the opponent). In the first strategy (*STD*) the evaluation function was trained using games played against itself. The second strategy (*MTD*) used 10 TD players each being trained by playing against the rest. The third one (*SANE-TD*) was playing against genetic opponents.

Given a game sequence, with the search trees associated to the positions from the sequence, a further choice can be made on which positions are selected for the learning process and thus for back-up. Equation 2 suggests that the back-up should be made using the game positions where the learning player is to move in the actual game. There are, however, variations on this scheme. E.g., TDLeaf [3] propagates the values between the leaf nodes of the principal variations. Another possible idea is to use as target value for the root node, the value obtained by search. Although in our experiments we found no significant difference in performance between the implementation according to equation 2 and TDLeaf, an extensive comparative study of these methods would be welcome. The alternative with the search-computed value as target for the the root node can be implemented only with $\lambda = 0$, and consequently it has a much slower convergence.

Varying the size of the hidden layer in our experiments did not influence significantly the performance, except in the range of 0 to 10 units, when the drop in performance was drastic. The choice for the value 40 was apparently a good compromise between performance and speed of evaluation. The failure to increase the performance for larger networks may be due to the increased difficulty of finding a good weight configuration. Although the experiments made with different learning parameter sets were not very extensive, the values given above are likely to be a good choice.

We compared the four methods (*STD, MTD, SANE-TD, SANE*) in a tournament with 50 players for each method (a total of 200 players). The tournament used the swiss pairing system, where the players are paired with opponents with approximately the same amount of points, and lasted 50 rounds. The results are summarized in Table 1. The result for each method is given as the ratio between the total amount of points obtained by the players representing the method, and the total number of points in the tournament. The standard deviation reflects the deviation in performance inside a method. The tournament resulted in a clear superiority of the TD players over the genetic ones. Out of the three TD training strategies, *SANE-TD* had the worst result, due to the weaker opponents (provided by the genetic algorithm). The values of the standard deviation indicate that the advantage of *STD* and *MTD* compared to the other two methods is significant, whereas the difference between *STD* and *MTD* was not significant. To obtain a better set of evaluation functions, we generated 60 more players using the two 'pure' TD methods (*STD* and *MTD*) and

Table 2. Gathering the evaluation function population: 2st tournament. The players for *std1* and *mtd1* are from the 1st tournament, and the players for *std2* and *mtd2* originates from a new set generated using *STD* and *MTD*.

method	# of players	result	std. dev.	final set
std1	30	0.228	0.049	20
mtd1	30	0.261	0.037	28
std2	30	0.238	0.038	24
mtd2	30	0.273	0.051	28

included them in a second tournament (Table 2), together with 60 top players (30 *STD* and 30 *MTD*) from the first tournament.

The final set of evaluation functions were selected as the top 100 players from the second tournament based on their performances. The resulting evaluation function population was subsequently used in learning time allocation.

4.3   Learning Time Allocation

Experimental results for learning search decisions have been obtained for the problem of time allocation. The problem consists of deciding on the depth of the search tree given a certain amount of time for a game.

The time-allocation strategies can be divided into three groups [22]:

- static strategies (strategies that decide the depth before starting the game);
- semi-dynamic strategies (strategies that decide before each move);
- dynamic strategies (strategies that can change the depth during the search process).

In our experiments we compared the performances of four algorithms (one static and three semi-dynamic). The semi-dynamic algorithms included the algorithms proposed in section 3, and a theoretical bound on a class of existing algorithms. Major examples for dynamic time allocation are represented by the stopping criteria in BPIP [2] and the probe search in the Multi-ProbCut [7] search algorithm. The time-allocation strategies in these search techniques are inherently interwoven with the search strategies, and therefore the results for time allocation cannot be separated easily from the search. This is the reason why we did not include them in our experiments.

The algorithms were compared using a test player (taken from the final set of evaluation functions described in subsection 4.2) playing with both colours against the whole population (a total of 200 games). Each player used a plain alpha-beta algorithm, the opponents searching to a depth of 2. At each move three search depths were available (for decision) for the test-player: 1, 2 or 3. Time constraints on the experiments did not allow us to have a greater variety in search depths.

The time constraint was enforced not in seconds but in the number of nodes evaluated. The amount of nodes available for searching is incremented with a fix number ($incr$) after each move and diminished by the amount spent to decide on the move[6]. The starting

---

[6] The incremental timing is widely used on Internet game servers and, with the help of electronic clocks, becomes more popular in human rapid tournaments too. Its main benefit is that it avoids losing on time if a game lasts too long.

resource is $5 \cdot incr$. The constraint influenced only the test player. If the available resource was consumed, the test player was forced to move using the current search tree, without any further evaluations. The advantage of incremental timing for our experiments is that it already contains a uniform division of time per move. Without such a time division, a hand-tuned time division method would have been necessary for the static strategy. The task for a 'wiser' time-allocation method (such as the semi-dynamic ones), is to shorten the search at certain moves to save the time for later when, possibly, it is more helpful.

Below we briefly describe the four algorithms used in our experiments.

The first algorithm is called *static*, and it contains a static strategy. For each of the three possible search depths a performance was obtained. The resulting performance of the algorithm is the best performance for the three individual depths.

The second algorithm, *SD-ES* (Semi-Dynamic Extra-Search), implements a theoretical bound on the performance of semi-dynamic strategies which learn the "boards worth extra search" [22]. More precisely, the algorithm searches to depth $k + n$ (instead of k) only if the extra search alters the move choice. This information is not available before the $k + n$-ply search, so it is necessary to learn which boards will gain from extra search. For our experiments we assumed that we have such an oracle[7]. For k we used depth 1 or 2, and n was set to 1. The choice between the two values for k is made in a similar way as for the static strategy.

The third algorithm, *SD-TD* (Semi-Dynamic TD), involves a neural net to decide the search depth (1, 2 or 3) before each move. The neural net was trained using equations 4 and 5, with an additional momentum term in the update rule. The input of the net consisted of the position and the available resources. Additional input could include the average branching factor or node consumption for the type of position, some measure of the evaluation volatility, and so on. Most of these informations, however, cannot be computed easily, and if it is to be learned, then why not let the neural network itself discover them in an implicit way (those which are relevant). The network had three outputs corresponding to the three search depths. The depth was decided in a winner-takes-all manner. In the training phase β was 0.9 (same as γ for the evaluation functions).

The fourth algorithm, *SD-SANE* (Semi-Dynamic SANE), is a genetic one, using the final outcome of the game for the fitness value. The structure of the evolved neural network is the same as in the previous algorithm.

The performances of the four algorithms are plotted in Fig. 4. A comparison between the time consumed by the two learning algorithms is given in Fig. 5.

Comparing the tournament performances of the four time-allocation algorithms, the static one had the worst results. The TD learning algorithm had roughly the same overall performance as the *SD-ES* algorithm. Finally, the genetic algorithm outperforms each algorithm for almost all the values of the increment.

The time used by the genetic method, however, was significantly higher compared to the method using TD learning. To have a more precise idea about time, the experiment for one tournament (200 games) can be performed in approximately 10 minutes,

---

[7] The oracle compares the move choices for both search depths. If the two choices are identical, it will suggest depth k, otherwise $k + n$. The search resource consumed by the oracle is not counted for the player.

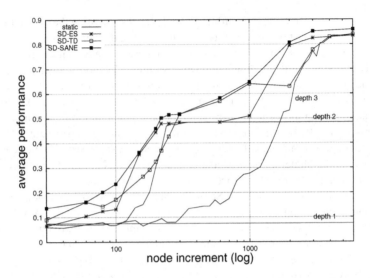

Fig. 4. The performance of the four time-allocation algorithms, for varying available nodes/move. The performance of the static method should be interpreted as the maximum of the three curves (corresponding to the three possible depths). Each data point is an average of 200 games.

consequently the last data points for the genetic algorithm took in the order of some weeks.

5   Conclusions

The experiments with learning evaluation functions suggest a superiority of TD learning over the genetic algorithm in the game of LOA. For TD learning the strength of the opponents seems to be an important factor. Another factor which had a slight influence on the final performances of the evaluation functions was the diversity in the learning programs, and consequently the diversity of the opponents.

The experimental results on time allocation provide an indication that the algorithms proposed to learn search decisions are promising. Both the TD and the genetic approach lead to competitive results with the existent time-allocation strategies. Although the genetic algorithm proved to be better than the TD approach, the time used for learning appears to be a critical constraint on its applicability.

We note that the TD approach performed better in building evaluation functions, and the genetic algorithm proved to be more useful in learning decisions. In the latter case TD learning still can be very useful when the learning time is critical. Naturally, if online learning is needed, then the TD approach is a more appropriate candidate.

6   Summary and Future Research

This paper presents some attempts to improve search decisions in games by using TD learning and genetic algorithms. The suggested algorithms were tested for the problem

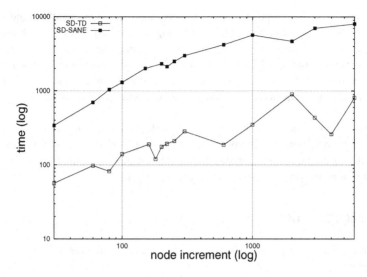

Fig. 5. The amount of time used by the two learning algorithms to reach the performance plotted in Fig. 4. The 'time' is measured as the number of tournaments (i.e., 200 games) scaled with $incr/300$.

of time allocation in the game of LOA. Both approaches obtained very good results compared to the existent algorithms.

Although the algorithms discussed proved to be useful for learning time-allocation strategies, a number of issues have to be investigated before deciding whether they can be applied for a more general class of search decisions.

A major issue, which already arose in the presented experiments, is the time required for learning. The time needed to find the suitable weights is roughly linear with the time required for a game or with the number of nodes to be searched. The factor is reasonable for TD learning, but could be prohibitive for the genetic approach. It is an important issue how the number of iterations or generations scales up with more complex search decisions to be learned.

For time allocation we were able to encode in the input of the neural net all the relevant information. For more local search decisions however this is not feasible (e.g., in the case of forward pruning a natural encoding would include the current position, the search depth, the remaining time and certainly not the whole search tree). It will be interesting to see how the performance of the learning algorithms (especially, the TD learning) will be influenced by the transformation of an MDP to a POMDP.

A potential sensible issue for the TD learning is the increasing lag when the number of search decisions between two moves is high. If the learning algorithm will not be able to overcome this problem, special (possibly problem-dependent) credit-assignment techniques need to be developed.

Finally, an important issue for the meta-level techniques is the time used by the extra computation needed for the more elaborate search decisions. If we assume that the time required for a search decision is of the same order as the time required to evaluate a board position, and the number of search decisions is not significantly higher than the number of nodes in the tree (it is difficult to imagine search decisions which grow more than linear with the number of nodes), then we can expect that it will not be necessary to decrease the size of the searched tree significantly. The time parameters possibly changed need to be considered when performances are compared. In the case of semi-dynamic time allocation this was not necessary, since there was only one search decision for a search tree.

Further experiments with other types of search decisions will reveal whether one of the four issues represent a major obstacle for the proposed learning algorithms. Experiments with time allocation for higher depths can also test the issue of how the learning time and efficiency scales up for more difficult problems.

Acknowledgements

The authors thank the referees for their constructive comments and suggestions for improvements.

References

1. T.S. Anantharaman. Evaluation tuning for computer chess: Linear discriminant methods. *ICCA Journal*, 20(4):224–242, 1997.
2. E.B. Baum and W.D. Smith. A bayesian approach to relevance in game playing. *Artificial Intelligence*, 97(1-2):195–242, 1997.
3. J. Baxter, A. Tridgell, and L. Weaver. Experiments in parameter learning using temporal differences. *ICCA Journal*, 21(2):84–99, 1998.
4. D.F. Beal and M.C. Smith. Learning piece values using temporal difference learning. *ICCA Journal*, 20(3):147–151, 1997.
5. D.F. Beal and M.C. Smith. Temporal difference learning for heuristic search and game playing. *Information Sciences*, 122(1):3–21, 2000.
6. Y. Björnsson and T. Marsland. Learning search control in adversary games. In H.J. van den Herik and B. Monien, editors, *Proceedings of the Advances in Computer Games 9 Conference*, 2000.
7. M. Buro. Experiments with Multi-ProbCut and a new high-quality evaluation function for Othello. In H. J. van den Herik and H. Iida, editors, *Games in AI Research*. 1999.
8. D. Carmel and S. Markovitch. Incorporating opponent models into adversary search. In *Proceedings of the Thirteenth National Conference on Artificial Intelligence (AAAI-96)*, pages 120–125, 1996.
9. K. Chellapilla and D.B. Fogel. Co-evolving checkers playing programs using only win, lose, or draw. In *Proceedings of SPIE's AeroSense'99: Applications and Science of Computational Intelligence II*, 1999.
10. D.E. Goldberg. *Genetic Algorithms in Search, Optimization, and Machine Learning*. Addison-Wesley, Reading, MA, 1989.
11. D. Harada and S. Russell. Extended abstract: Learning search strategies. In *AAAI Spring Symposium on Search Techniques for Problem Solving under Uncertainty and Incomplete Information*, 1999.
12. E.A. Heinz. Adaptive null-move pruning. *ICCA Journal*, 22(3):123–132, 1999.
13. R.M. Hyatt. Using time wisely. *ICCA Journal*, 7(1):4–9, 1984.

14. H. Iida, J.W.H.M. Uiterwijk, H.J. van den Herik, and I.S. Herschberg. Potential applications of opponent-model search. Part 1: The domain of applicability. *ICCA Journal*, 16(4):201–208, 1993.

15. T. Jaakkola, S. Singh, and M. Jordan. Reinforcement learning algorithm for partially observable markov problems. In *Advances in Neural Information Processing Systems 7*, pages 345–352, 1994.

16. L.P. Kaelbling, M.L. Littman, and A.R. Cassandra. Planning and acting in partially observable stochastic domains. *Artificial Intelligence*, 101(1-2):99–134, 1998.

17. H. Kimura and S. Kobayashi. An analysis of actor/critic algorithms using eligibility traces: Reinforcement learning with imperfect value function. In *Proceedings of the 15th International Conference on Machine Learning*, pages 278–286, 1998.

18. V.R. Konda and V.S. Borkar. Actor-critic type learning algorithms for markov decision processes. *SIAM Journal of Control and Optimisation*, 38(1):94–133, 1999.

19. V.R. Konda and J.N. Tsitsiklis. Actor-critic algorithms. In *Advances in Neural Information Processing Systems 12*, 2000.

20. B.C. Kuszmaul. The StarTech massively parallel chess program. *ICCA Journal*, 18(1):3–19, 1995.

21. C. Leiserson. Using the Cilk multithreaded programming language to implement a multiprocessor chess program. In H.J. van den Herik and B. Monien, editors, *Proceedings of the Advances in Computer Games 9 Conference*, 2000.

22. S. Markovitch and Y. Sella. Learning of resource allocation strategies for game playing. *Computational Intelligence*, 12(1):88–105, 1996.

23. T.M. Mitchell, R.M. Keller, and S. Kedar-Cabelli. Explanation-based generalization: A unifying view. *Machine Learning*, 1(1):47–80, 1986.

24. D.E. Moriarty and R. Miikkulainen. Hierarchical evolution of neural networks. In *Proceedings of the 1998 IEEE Conference on Evolutionary Computation*, pages 428–433, 1998.

25. D.E. Moriarty, A.C. Schultz, and J.J. Grefenstette. Evolutionary algorithms for reinforcement learning. *Journal of Artificial Intelligence Research*, 11:241–276, 1999.

26. N. Richards, D. Moriarty, and R. Miikkulainen. Evolving neural networks to play Go. *Applied Intelligence*, 8:85–96, 1998.

27. S. Russell and E.H. Wefald. *Do the Right Thing: Studies in Limited Rationality*. MIT Press, 1991.

28. A.L. Samuel. Some studies in machine learning using the game of checkers. *IBM Journal of Research and Development*, 3(3):211–229, 1959.

29. J. Schaeffer and A. Plaat. Kasparov versus DEEP BLUE: The rematch. *ICCA Journal*, 20(2):95–101, 1997.

30. Y. Seirawan. The Kasparov - DEEP BLUE games. *ICCA Journal*, 20(2):102–125, 1997.

31. R.S. Sutton and A.G. Barto. *Reinforcement Learning: An Introduction*. MIT Press, 1998.

32. R.S. Sutton, D. McAllester, S. Singh, and Y. Mansour. Policy gradient methods for reinforcement learning with function approximation. In *Advances in Neural Information Processing Systems 12*, pages 1057–1063, 2000.

33. G.J. Tesauro. Practical issues in temporal difference learning. *Machine Learning*, 8:257–277, 1992.

34. S. Thrun. Learning to play the game of chess. In *Advances in Neural Information Processing Systems 7*, pages 1069–1076, 1995.

35. M.H.M. Winands. Analysis and implementation of Lines of Action. Master's thesis, Department of Computer Science, Universiteit Maastricht, 2000.

The Complexity of Graph Ramsey Games

Wolfgang Slany

Institut für Informationssysteme
Technische Universität Wien
Favoritenstr. 9–11, A-1040 Wien, Austria
wsi@dbai.tuwien.ac.at

Abstract. We consider combinatorial avoidance and achievement games based on graph Ramsey theory: The players take turns in coloring edges of a graph G, each player being assigned a distinct color and choosing one so far uncolored edge per move. In avoidance games, completing a monochromatic subgraph isomorphic to another graph A leads to immediate defeat or is forbidden and the first player that cannot move loses. In the avoidance$^+$ variant, both players are free to choose more than one edge per move. In achievement games, the first player that completes a monochromatic subgraph isomorphic to A wins. We prove that general graph Ramsey avoidance, avoidance$^+$, and achievement endgames and several variants thereof are **PSPACE**-complete.

Keywords: combinatorial games, graph Ramsey theory, Ramsey game, PSPACE-completeness, complexity, edge coloring, winning strategy, achievement game, avoidance game, the game of Sim, Java applet, endgames

1   Introduction and Overview

To illustrate the nature of combinatorics, [5] uses the following simple game: Two players, Red and Green, compete on a game board composed of six vertices and all $\binom{6}{2} = 15$ possible edges between these vertices. The players alternate in coloring at each move one so far uncolored edge using their color, with the restriction that building a complete subgraph with three vertices whose edges all have the same color (a monochromatic triangle) is forbidden. The loser is the first player that cannot legally move or, in case of a fallible human player, that builds a triangle by mistake.

This game was first described under the name Sim by [49] in 1969. Since then, it has attracted much interest [2], [3], [4], [1], [5], [7], [8], [9], [10], [12], [13], [15], [19], [20], [21], [22], [25], [26], [27], [29], [30], [32], [33], [34], [36], [39], [41], [43], [45], [46], [47], [48], [50], [51], [53]. Figure 1 shows a typical play sequence[1].

Besides their value as motivational examples for the field of combinatorics and the psychological reasons why humans may be attracted by combinatorial games such as

---

[1] Considering that a hands-on session with an interactive system often is worth more than a thousand images, you might want to challenge a Java applet at http://www.dbai.tuwien.ac.at/ proj/ramsey/ that plays Sim, playing perfectly when possible and improving its strategy by probabilistic reinforcement learning from playing over the Internet when perfect play is impossible. In case you win, you will be allowed to leave your name in our hall-of-fame!

T.A. Marsland and I. Frank (Eds.): CG 2000, LNCS 2063, pp. 186–203, 2001.
© Springer-Verlag Berlin Heidelberg 2001

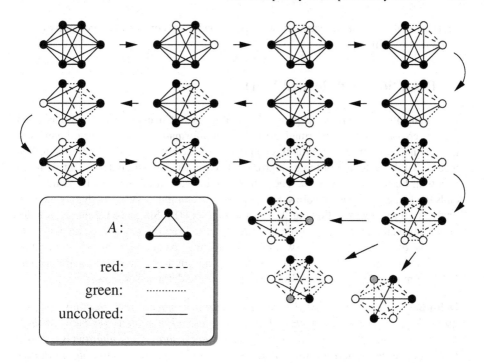

Fig. 1. Sample play sequence of Sim. The initial, uncolored game board is shown at the top left corner. Player Red (= dashed lines) starts by coloring some edge, then player Green (= dotted lines) colors another one, etc. The "highlighting" of vertices just serves to indicate the current move. Finally, Red is forced to give up since any further coloring would complete a red triangle (= a monochromatic subgraph isomorphic to A).

Sim, these games are also of practical interest because they can serve as models that simplify the analysis of competitive situations with opposing parties that pursue different interests, or for situations where one is faced with an unforeseeable environment such as Nature. It is easy to see that playing against a perfectly intelligent opponent with unlimited computational resources is the worst case that can happen. If the problem of winning against such an opponent can be solved, one will also be able to handle all other eventualities that could arise. Finding a winning strategy for a combinatorial game can thus be translated into finding a strategy to cope with many kinds of real world problems such as found in telecommunications, circuit design, scheduling as well as a large number of other problems of industrial relevance [18,24,37]. Proving or at least providing strong evidence that finding such a winning strategy is of high complexity helps to explain the great difficulties one often faces in corresponding real world problems [17,23,38].

   In Sect. 2, we define the notions from combinatorial games, computational complexity and Ramsey theory that we use later, formally define all mentioned games and discuss previous work. Section 3 contains our main complexity results. These results imply that the unrestricted graph Ramsey games are at least as hard as a large number of well-known games (e.g., Othello [28]) and problems of industrial relevance (e.g., decision-making under uncertainty such as stochastic scheduling [37]) generally recognized as very difficult.

2   Definitions and Related Work

Like many other combinatorial games, including Chess, Checkers, and Go, Sim is a two-player zero-sum perfect-information (no hidden information as in some card games, so there is no bluffing) game without chance moves (no rolling of dice). Zero-sum here means that the outcome of the game for the two players is restricted to either win-loss, loss-win, tie-tie, or draw-draw. The distinction between a draw and a tie is that a tie ends the game, whereas in a draw, the game would continue forever, both players being unable to force a win, following the terminology in [18]. Sim is based on the simplest nontrivial example of Ramsey theory [24,35], the example being also known under the name of "party-puzzle": How many persons must be at a party so that either three mutual acquaintances or three persons that are not mutual acquaintances are present? More formally, classical binary Ramsey numbers are defined as follows:

Definition 1. $\mathrm{Ramsey}(n, m)$ *denotes the smallest number r such that any complete graph K_r (an undirected graph with r vertices and all possible edges between them) whose edges are all colored in red or in green either contains a red subgraph isomorphic to K_n or a green subgraph isomorphic to K_m. In classical* symmetric *binary Ramsey numbers, n equals m.*

The classical result of F. P. Ramsey [40], a structural generalization of the pigeon-hole principle, tells us that these numbers always exist:

Theorem 1 (Ramsey [40]). $\forall (n, m) \in \mathbb{N}^2$ $\mathrm{Ramsey}(n, m) < \infty.$

A simple combinatorial argument that $\mathrm{Ramsey}(3, 3) = 6$ is shown in Fig. 2, and so the minimal number of persons satisfying the "party-puzzle" from above is six. Theoretically, Sim ends after a maximum of 15 moves since this is the number of edges in a complete graph with six vertices. If we define Sim such that monochromatic triangles are not allowed, and since Ramsey theory says that any edge-2-colored K_6 will contain at least one monochromatic triangle, we know that the second player will not be forced to give up simply because all edges are colored after 15 moves, as all games will end before the 15th move. The game of Sim as it is usually described and played ends when one of the players completes a triangle in his color, whether forced or by mistake (i.e., the last player to *move* loses), with no winner, that is, a tie, defined for the case when all edges are colored without a monochromatic triangle having been completed. For this variant of Sim, the $\mathrm{Ramsey}(3, 3) = 6$ result implies that no game will ever end in a tie.

It is easy to see that in finite, two-player zero-sum perfect-information games with no ties and no chance moves, either the player who starts the game or his opponent must have the possibility to play according to a *winning strategy*: A player who follows such a strategy will always win no matter how well the opponent plays (for the existence of such a strategy, see for instance the fundamental theorem of combinatorial game theory in [17]). Clearly, this means that one of the players will have an

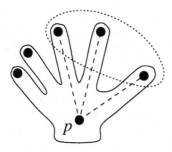 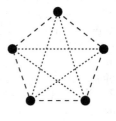

Fig. 2. Visual proof that Ramsey$(3,3) = 6$. The drawing on the left shows that six vertices are enough, as follows: Take any vertex p (as in "palm") of an edge-2-colored K_6. At least three edges connected to p will have the same color. Without loss of generality, assume that this color is the dashed one. Consider the three vertices connected to p through these three edges: Either one of the edges that connects two of these vertices is of the dashed type (and then there is a dashed triangle with the edges connected to p), or not (and then the three top edges form a triangle in the other color). The edge-2-colored K_5 on the right serves as a counter-example, showing that five vertices are insufficient to force a monochromatic triangle. Thus, six is the smallest number with the required property.

a-priori upper-hand in Sim, so the answer to the following question is of central interest: Which of the two players has a winning strategy, the first or the second to move? [32] have shown that the second player can always win. Nevertheless, a winning strategy that is easy to memorize for human players has so far eluded us, despite much effort [7,10,19,20,21,22,32,33,34,36,43,45,46,47,48,50,53]. While several implementations of perfect winning strategies for Sim exist, it is straightforward to define games based on slightly larger Ramsey numbers that will defy any brute force attack. In the rest of this paper, we will show that natural generalizations of Sim and some variants thereof are of high computational complexity, thus trying to abductively shed some light on the difficulties to formulate "simple" winning strategies for these games.

A game being finite means that it should theoretically be possible to solve it. However, the trouble is that it might take an astronomical amount of time and memory to actually compute the winning strategy. Thus, we turn to the next best thing, which is to classify the games in terms of computational complexity classes, that is, to find out how the function bounding the computational resources that are needed in the worst case to determine a winning strategy for the first player grows in relation to the size of the game description.

To be able to speak about the computational complexity of game problems, the games must be scalable instead of having a fixed size. Generalizations to boards of size $n \times n$ of well-known games such as Othello, Checkers, and Go have been classified as **PSPACE**-complete and **EXPTIME**-complete [28,16,31]. **PSPACE** in particular is important for the analysis of these and large classes of more formal combinatorial games [17,18,23,38,44]. **PSPACE** is the class of problems that can be solved using memory space bounded by a polynomial in the size of the problem description. **PSPACE**-complete problems are the hardest problems in the class **PSPACE**: Solving one of these problems efficiently would mean that we could solve *any* other problem in **PSPACE** efficiently as well. While nobody so far was able to show that **PSPACE** problems are inherently difficult, despite much effort to show that the complexity class **P** containing the tractable problems solvable in polynomial time is different from **PSPACE**, it would be very surprising if they were not. Indeed, the well-known complexity class **NP** is included in **PSPACE**, so problems in **PSPACE** are at least as difficult as many problems believed to be hard such as the satisfiability of boolean formulas or the traveling salesman problem. This means that it is rather unlikely that efficient general algorithms to solve **PSPACE**-complete combinatorial games do exist. Details on computational complexity theory can be found in [23] and [38]. Obviously, the high complexity of many combinatorial games contributes to their attractiveness.

So the question is, what could be a generalization of Sim to game boards of arbitrary size? We first introduce graph Ramsey theory which generalizes classical Ramsey theory in a natural way to arbitrary graphs. Graph Ramsey theory was pioneered around 1973 independently by [6], [11], by [14], and by [42]. One possible formulation of the central concept of "arrowing" is the following:

Definition 2. $(G, E^r, E^g) \to A$: *A partly edge-colored graph* (G, E^r, E^g), *where some edges* E^r *of G are precolored in red and some other edges* E^g *of G are precolored in green, "arrows" a graph A if every* full *edge-coloring of* (G, E^r, E^g) *with colors red and green contains a monochromatic subgraph isomorphic to A.*

The game $G_{\text{Avoid-Ramsey}}$ is the generalization of Sim to graph Ramsey theory. Note that in the following definitions, the precolorings (E^r, E^g) are part of the input, allowing us to analyze arbitrary game positions.

Definition 3. *The graph Ramsey avoidance game* $G_{\text{Avoid-Ramsey}}(G, A, E^r, E^g)$ *is played on a graph* $G = (V, E)$, *another graph A, and two nonintersecting sets* $E^r \cup E^g \subseteq E$ *that contain edges initially colored in red and green, respectively. Two players, Red and Green, take turns in selecting at each move one so far uncolored edge from E and color it in red for player Red respectively in green for player Green. However, both players are forbidden to choose an edge such that a subgraph of the red or the green part of G becomes isomorphic to A. Red moves first. The first player that is unable to move loses.*

Definition 4. $G_{\text{Avoid'-Ramsey}}$ *is a variant of* $G_{\text{Avoid-Ramsey}}$ *in which completing a monochromatic subgraph isomorphic to graph A is possible but the first player who does so loses the game. In contrast to* $G_{\text{Avoid-Ramsey}}$, *this game ends in a tie if no edges are left to color.*

Clearly, these two avoidance variants coincide whenever $(G, E^r, E^g) \to A$ (the proof is straightforward):

Lemma 1. *If* $(G, E^r, E^g) \to A$, *then player Red has a winning strategy in* $G_{\text{Avoid-Ramsey}}$ (G, A, E^r, E^g) *iff player Red has a winning strategy in* $G_{\text{Avoid'-Ramsey}}(G, A, E^r, E^g)$.

We call $G_{\text{Avoid-Ramsey+}}$ the game variant where each player selects *at least one* so far uncolored edge per move (the "+" here is borrowed from Unix regular expression syntax where it also means "at least one"). This variant intuitively corresponds even closer to the spirit of Ramsey theory because any combination in the number of red and green edges is possible. In the other graph Ramsey games, red and green edges are added at the same rate, and thus the colorings are more restricted than required by Ramsey theory.

Definition 5. $G_{\text{Avoid-Ramsey}+}(G, A, E^r, E^g)$: *Everything is as in Definition 3, except that each player selects* at least one *so far uncolored edge from E during one move.*

In the case of graph Ramsey achievement games, three major variants can be distinguished, as follows:

Definition 6. $G_{\text{Achieve-Ramsey}}(G, A, E^r, E^g)$: *everything is as in Definition 4, except that the first player who builds a monochromatic subgraph isomorphic to A wins.*

Definition 7. *A simple strategy-stealing argument tells us that with optimal play on an uncolored board, $G_{\text{Achieve-Ramsey}}$ must be either a first-player win or a tie, so it is only fair to count a tie as a second-player win. Let us call this variant $G_{\text{Achieve'-Ramsey}}$, even when playing on already precolored boards, that is, the graphs in the game $G_{\text{Achieve'-Ramsey}}(G, A, E^r, E^g)$ are to be interpreted as in the other games.*

We know from the fundamental theorem of combinatorial game theory (see e.g. [17]) that there exists a winning strategy for this game. It is straightforward that when $(G, E^r, E^g) \to A$, $G_{\text{Achieve-Ramsey}}$ and $G_{\text{Achieve'-Ramsey}}$ are in fact the same game.

Definition 8. *Following the terminology of [1], let us call the variant of $G_{\text{Achieve-Ramsey}}$ where all the second player does is to try to prevent the first player to build A (without winning by building it himself), the "weak" graph Ramsey achievement game $G_{\text{Achieve''-Ramsey}}$.*

Again, it is straightforward that when the first player has a winning strategy or when there is no possibility for the second player to build a green subgraph isomorphic to A, $G_{\text{Achieve'-Ramsey}}$ and $G_{\text{Achieve''-Ramsey}}$ are the same game.

[26] studied both $G_{\text{Achieve-Ramsey}}$ and $G_{\text{Avoid-Ramsey}}$ where G is restricted to complete finite graphs of small size, A being an arbitrary graph. For graph Ramsey achievement games, several tractable subcases are known:

Theorem 2 ([13]). *The first player has a winning strategy in $G_{\text{Achieve-Ramsey}}(K_n, K_k, \{\}, \{\})$ if*

$$k \leq \frac{1}{2} \log_2 n$$

and the game ends in a tie if

$$2^l > \frac{n}{k}, \quad \text{where} \quad l = \frac{k}{2} - 1,$$

i.e., it is a tie if

$$k \geq 2\,(1 + o(1))\,\log_2 n.$$

While these results do not cover all cases with complete graphs such as for example $\text{Sim}_a = G_{\text{Achieve-Ramsey}}(K_6, K_3, \{\}, \{\})$, small instances of $G_{\text{Achieve-Ramsey}}$ generally seem to be very easy to analyze. Figure 3 shows, for instance, a complete winning strategy for the first player in Sim_a. [2] and [1] have generalized these results to games where the players alternate in choosing among previously unchosen elements of the

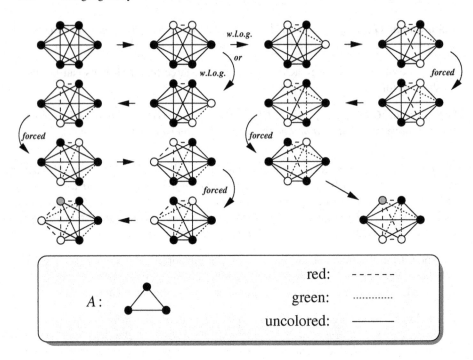

Fig. 3. Trivial winning strategy for player Red in $\mathrm{Sim}_a = G_{\text{Achieve-Ramsey}}(K_6, K_3, \{\}, \{\})$.

complete k-uniform hypergraph of N vertices K_N^k, and the first player wins if he has selected all k-tuples of an n-set. For the case $k = 2$, their results subsumes Theorem 2. They also study infinite Ramsey games where the edges of the hypergraphs are required to be infinite but countable, for which they show that there always exist winning strategies for the first player. Several games of this kind are analyzed, all featuring simple winning strategies that imply their tractability.

In avoidance games, it is intuitively clear that for many graphs G and A, making the first move is an essential disadvantage. [27] analyze situations when the second player really benefits from this and has a winning (or at least non-losing) strategy. They identify a large class of graphs G for which the second player always has a so-called "symmetric" winning strategy, that is, the second player can ensure that the subgraphs of both colors are isomorphic after every round of the game. Nevertheless, they also prove that for a large subclass of the former, actually playing the winning strategy requires **NP**-powerfulness of the second player. They also show that when $A = K_{1,2}$ the second player has a symmetric winning strategy on complete graphs $G = K_n$ for $n > 2$ even though they could show that in general there is no symmetric strategy on complete graphs.

We also restate the definitions of two known **PSPACE**-complete problems that we will use in the proofs:

Definition 9 ([44]). $G_{\text{Achieve-POS-CNF}}(F)$: *We are given a positive CNF formula F, that is, a conjunctive normal form formula in which $\neg$ does not occur. A move consists of* *choosing some variable of F which has not yet been chosen. Player I starts the game. The game ends after all variables of F have been chosen. Player I wins iff F is* true *when all variables chosen by player I are set to* true *and all variables chosen by player II are set to* false.

For example, on input $x_1 \wedge (x_2 \vee x_3) \wedge (x_2 \vee x_4)$ player II has a winning strategy, whereas on input $(x_1 \vee x_4) \wedge (x_2 \vee x_3) \wedge (x_2 \vee x_4)$ player I has a winning strategy.

Definition 10 ([44]). $G_{\text{Achieve-POS-DNF}}(F)$ *We are given a positive disjunctive normal form formula F. A move consists of choosing some variable of F which has not yet been chosen. Player I starts the game. The game ends after all variables of F have been chosen. Player I wins iff F is* true *when all variables chosen by player I are set to* true *and all variables chosen by player II are set to* false. *In other words, player I wins iff he succeeds in playing all variables in at least one clause.*

3   Complexity Results

In the following, we always mean the complexity of deciding whether the first player has a winning strategy when we speak of the complexity of a game's decision problem. Our proofs imply that, under the condition that $\mathbf{P} \neq \mathbf{PSPACE}$, there exists no efficient algorithm to decide whether certain endgame positions feature a winning strategy for the first player. Our proofs say nothing about the uncolored graph, besides that the game played on it is in **PSPACE**. This is typical for hardness results of well-known games, e.g., the game Othello [28] or Go [31], and actually poses no difficulties when modeling real world problems, where initial situations tend to be far from being "uncolored" anyway.

The intricate parts of the proofs of Theorems 3–8 will be found in their hardness parts. The following lemma establishes the membership parts of all proofs:

Lemma 2. *All game decision problems corresponding to the graph Ramsey games defined in Sect. 2 are in* **PSPACE**.

Proof. Let $n \stackrel{\text{def}}{=} |(G, A, E^r, E^g)|$ denote the size of the input. The number of moves in any graph Ramsey game is limited by the number of initially uncolored edges in the graph G, so any game will end after at most $|E| - |E^r| - |E^g| < n$ edge colorings. If everything is know about a given game position, any immediately following game position is fully characterized by that previous game position plus the edge that was just colored. The memory requirement for that additional edge is $O(\log n)$. It is easy to enumerate in some lexicographic order all game positions that can originate from a particular game position through the coloring of one edge. Altogether, this implies membership in **PSPACE** by the following argument: Given an initial game position, a depth-first algorithm that checks all possible game sequences but keeps in memory only one branch of the game tree at a time, backtracking to unexplored branching points in order to scan through the whole game tree, can decide whether there is a winning strategy for player Red using memory bounded by the maximum stack size, which is $O(n^2)$ and thus polynomial in the size of the input.   □

Theorem 3. *The $G_{\text{Avoid-Ramsey}}$ problem is* **PSPACE**-*complete*.

Proof sketch. Because of Lemma 2 it only remains to show that there exists a logarithmic space reduction from the known **PSPACE**-complete problem $G_{\text{Achieve-POS-CNF}}$ to the $G_{\text{Avoid-Ramsey}}$ problem.

We first define the **LOGSPACE** reduction from the game problem for game $G_{\text{Achieve-POS-CNF}}(F)$ to the game problem for game $G_{\text{Avoid-Ramsey}}(G, A, E^r, E^g)$: Let a positive CNF formula F be given. Assume without loss of generality that $F = C_1 \wedge \ldots \wedge C_m$ where each conjunct C_j is a disjunction of n_j distinct positive literals, that is, $C_j = l_{j,1} \vee \ldots \vee l_{j,n_j}$ where $l_{j,k} \in \{x_1, \ldots, x_n\}$ and all n variables appear at least once in F. We then define the graphs $G \stackrel{\text{def}}{=} (V, E)$, $A \stackrel{\text{def}}{=} (V^A, E^A)$ and the edge-sets E^r, E^g, by

$$\triangle(\alpha, \beta, \gamma) \stackrel{\text{def}}{=} \{\alpha, \beta\}, \{\alpha, \gamma\}, \{\beta, \gamma\} \quad ,$$

$$V \stackrel{\text{def}}{=} \bigcup_{0 \leq i \leq n} X_i \, ,$$

$$X_0 \stackrel{\text{def}}{=} \bigcup_{0 \leq j \leq m} B_j \, ,$$

$$B_0 \stackrel{\text{def}}{=} \{u_{0,0}, u_{0,1}, u_{0,2}, r_{0,\text{t}}, r_{0,\text{b}}\} \, ,$$

$$B_j \stackrel{\text{def}}{=} \{u_{j,0}, u_{j,1}, u_{j,2}, d_{j,\text{t}}, d_{j,\text{b}}\} \cup \bigcup_{1 \leq p < j} \{w_{j,p}\} \cup \bigcup_{1 \leq k \leq n_j} \{f_{j,k}\} \quad \text{for} \quad 1 \leq j \leq m \, ,$$

$$X_i \stackrel{\text{def}}{=} \{v_{i,0}, v_{i,1}, v_{i,2}, r_{i,\text{t}}, r_{i,\text{b}}, v_{i,3}, y_{i,\text{t}}, y_{i,\text{b}}, \\ v_{i,4}, g_{i,\text{t}}, g_{i,\text{b}}, v_{i,5}, v_{i,6}, v_{i,7}\} \quad \text{for} \quad 1 \leq i \leq n \, ,$$

$$E \stackrel{\text{def}}{=} \bigcup_{0 \leq i \leq n} P_i \, ,$$

$$P_0 \stackrel{\text{def}}{=} \bigcup_{0 \leq j \leq m} D_j \, ,$$

$$D_0 \stackrel{\text{def}}{=} \triangle(u_{0,0}, u_{0,1}, u_{0,2}) \cup \triangle(u_{0,2}, r_{0,\text{t}}, r_{0,\text{b}}) \, ,$$

$$D_j \stackrel{\text{def}}{=} \triangle(u_{j,0}, u_{j,1}, u_{j,2}) \cup \triangle(u_{j,2}, d_{j,\text{t}}, d_{j,\text{b}}) \cup \\ \bigcup_{1 \leq p < j} \{w_{j,p}, d_{p,\text{t}}\}, \{w_{j,p}, d_{p,\text{b}}\}, \{w_{j,p}, d_{j,\text{t}}\}, \{w_{j,p}, d_{j,\text{b}}\} \cup \\ \bigcup_{1 \leq k \leq n_j} \{f_{j,k}, d_{j,\text{t}}\}, \{f_{j,k}, d_{j,\text{b}}\}, \{f_{j,k}, g_{h,\text{t}}\}, \{f_{j,k}, g_{h,\text{b}}\} \mid l_{j,k} = x_h \\ \text{for} \quad 1 \leq j \leq m \, ,$$

$$P_i \stackrel{\text{def}}{=} \triangle(v_{i,0}, v_{i,1}, v_{i,2}) \cup \triangle(v_{i,2}, r_{i,\text{t}}, r_{i,\text{b}}) \cup \triangle(v_{i,3}, r_{i,\text{t}}, r_{i,\text{b}}) \cup \\ \triangle(v_{i,3}, y_{i,\text{t}}, y_{i,\text{b}}) \cup \triangle(v_{i,4}, y_{i,\text{t}}, y_{i,\text{b}}) \cup \triangle(v_{i,4}, g_{i,\text{t}}, g_{i,\text{b}}) \cup \\ \triangle(v_{i,5}, g_{i,\text{t}}, g_{i,\text{b}}) \cup \triangle(v_{i,5}, v_{i,6}, v_{i,7}) \quad \text{for} \quad 1 \leq i \leq n \, ,$$

$$V^A \stackrel{\text{def}}{=} \{a_0, a_1, a_2, a_3, a_4\} \, ,$$

$$E^A \stackrel{\text{def}}{=} \triangle(a_0, a_1, a_2) \cup \triangle(a_2, a_3, a_4),$$

$$E^r \stackrel{\text{def}}{=} \bigcup_{0 \le i \le n} P_i^r,$$

$$P_0^r \stackrel{\text{def}}{=} \bigcup_{1 \le j \le m} \left(\triangle(u_{j,0}, u_{j,1}, u_{j,2}) \cup \{u_{j,2}, d_{j,t}\}, \{u_{j,2}, d_{j,b}\} \right),$$

$$P_i^r \stackrel{\text{def}}{=} \{v_{i,3}, r_{i,t}\}, \{v_{i,3}, r_{i,b}\}, \{v_{i,3}, y_{i,t}\}, \{v_{i,3}, y_{i,b}\},$$
$$\{v_{i,5}, g_{i,t}\}, \{v_{i,5}, g_{i,b}\} \cup \triangle(v_{i,5}, v_{i,6}, v_{i,7}) \quad \text{for} \quad 1 \le i \le n,$$

$$E^g \stackrel{\text{def}}{=} \bigcup_{0 \le i \le n} P_i^g,$$

$$P_0^g \stackrel{\text{def}}{=} \bigcup_{0 \le j \le m} D_j^g,$$

$$D_0^g \stackrel{\text{def}}{=} \triangle(u_{0,0}, u_{0,1}, u_{0,2}) \cup \{u_{0,2}, r_{0,t}\}, \{u_{0,2}, r_{0,b}\},$$

$$D_j^g \stackrel{\text{def}}{=} \bigcup_{1 \le p < j} \{w_{j,p}, d_{p,t}\}, \{w_{j,p}, d_{p,b}\}, \{w_{j,p}, d_{j,t}\}, \{w_{j,p}, d_{j,b}\} \cup$$
$$\bigcup_{1 \le k \le n_j} \{f_{j,k}, d_{j,t}\}, \{f_{j,k}, d_{j,b}\}, \{f_{j,k}, g_{h,t}\}, \{f_{j,k}, g_{h,b}\} \mid l_{j,k} = x_h$$
$$\text{for} \quad 1 \le j \le m,$$

$$P_i^g \stackrel{\text{def}}{=} \triangle(v_{i,0}, v_{i,1}, v_{i,2}) \cup \{v_{i,2}, r_{i,t}\}, \{v_{i,2}, r_{i,b}\},$$
$$\{v_{i,4}, y_{i,t}\}, \{v_{i,4}, y_{i,b}\}, \{v_{i,4}, g_{i,t}\}, \{v_{i,4}, g_{i,b}\} \quad \text{for} \quad 1 \le i \le n.$$

It immediately follows from the construction that there is a simple **LOGSPACE** transducer that computes (G, A, E^r, E^g) from input F.

Let us sketch here the idea of the proof using Fig. 4. Since printing and copying in color was not universally available when this paper was written, and to avoid confusion resulting from the large number of vertices and edges, the graph in Fig. 4 uses certain conventions to represent colors, vertices and edges as indicated on its right-hand side. For instance, we use r_3 as a shortcut for the edge $\{r_{3,t}, r_{3,b}\}$, where "t" marks the vertex at the "top" of the edge and "b" the one at the "bottom" in G's drawing in Fig. 4 (it is understood that these are only, albeit convenient, names of vertices since there is no inherent top and bottom in the graph). Each set of three edges r_i, y_i, g_i corresponds to the boolean variable x_i of F. Each edge d_j corresponds to the jth clause of F. The links between edges g_i and d_j correspond to the occurrence of the variables in the conjuncts. Player Red can only color edges r_i and y_i, whereas player Green can only color edges g_i, y_i, and possibly one of the edges d_j if the g_i's connected to it are uncolored. By counting the number of possible moves, one sees that Green has a winning strategy if he succeeds in coloring one edge d_j at move $2n + 2$. Coloring edge y_i means that the other player can only color one edge in that set (r_i in case of Red, g_i in case of Green). Thus,

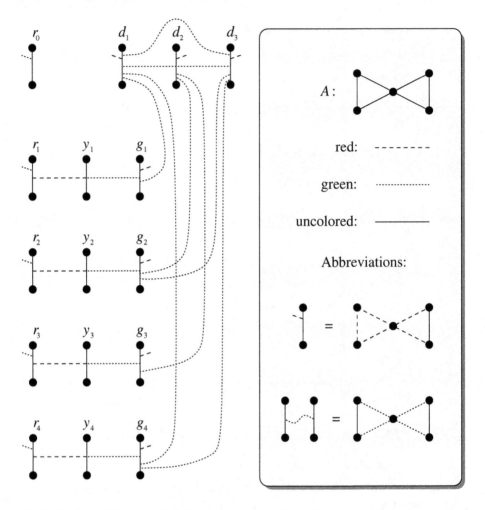

Fig. 4. Example of the reduction $G_{\text{Achieve-POS-CNF}} \leq_{\log} G_{\text{Avoid-Ramsey}}$ from the proof of Theorem 3. The graph G is shown on the left and corresponds to the input formula $F = (x_1 \vee x_4) \wedge (x_2 \vee x_3) \wedge (x_2 \vee x_4)$, featuring a winning strategy for player I.

the players first will race to color all edges y_i, since by doing so, Red could possibly hinder Green from coloring any edge d_j at the end, whereas Green could possibly leave enough edges g_i uncolored so that he can color one edge d_j at the end.

From the definition of $G_{\text{Achieve-POS-CNF}}$, we easily see that player I wins iff he succeeds in playing some variable in each conjunct. This is mirrored in $G_{\text{Avoid-Ramsey}}$ as follows: Player Red can win iff he succeeds in coloring some edge y_i so that player Green later on can only choose edges g_i in these particular triples, making it impossible for Green to color any edge d_j at the end.

The rest of the proof consists in a lengthy but straightforward analysis of several cases showing that there is a winning strategy for player I in $G_{\text{Achieve-POS-CNF}}(F)$ iff there is a winning strategy for player Red in $G_{\text{Avoid-Ramsey}}(G, A, E^r, E^g)$, independently of what the opponent of the player with a winning strategy does (a detailed proof is contained in [51]). ◻

Theorem 4. *The $G_{\text{Avoid'-Ramsey}}$ problem is* **PSPACE**-*complete*.

Proof. Hardness follows from Definition 4 as well as Theorem 3 and the construction in its proof, which makes sure that $G_{\text{Avoid'-Ramsey}}(G, A, E^r, E^g)$ never ends in a tie, by forcing $(G, E^r, E^g) \rightarrow A$. Indeed, the construction features, among others, n triples r_i, y_i, g_i for $i = 1, \ldots, n$, such that coloring more than *one* edge in any triple would end a $G_{\text{Avoid'-Ramsey}}$ game for that player. Since each triple contains three edges but there are only two player, no $G_{\text{Avoid'-Ramsey}}$ game will ever end in a tie because all edges have been occupied. Therefore, Lemma 1 ensures that $G_{\text{Avoid'-Ramsey}}(G, A, E^r, E^g)$ will have the same winning strategy as $G_{\text{Avoid-Ramsey}}(G, A, E^r, E^g)$ for one of its players, and the proof of Theorem 3 carries over. ◻

Theorem 5. *The $G_{\text{Avoid-Ramsey}+}$ problem is* **PSPACE**-*complete*.

Proof sketch. A careful analysis of the proof of Theorem 3 reveals that we can reuse the reduction of that proof to show the **PSPACE**-completeness of the $G_{\text{Avoid-Ramsey}+}$ problem. Indeed, all arguments go through even when both players are allowed to color more than one edge per move. The difficulty here lies in the analysis of the cases when the opponent plays worse than optimally:

Let us assume that player I has a winning strategy and player Red has so far played according to it, as explained in the proof of Theorem 3. We notice that if player Green colors at least two edges in his current move, the best he can hope for is that he will have colored *one* edge d_j at the end of the game. At any rate, the number of edges Green has left to color after his move will decrease by at least two. Red is none the worse off by Green's move and actually just needs to continue to choose *one* uncolored edge r_i after the other per move to win, since there is no urge now to force Green to color edges g_i. Conversely, let us assume player II has a winning strategy and player Green so far followed it. If player Red colors at least two edges in some move, the best he can hope for is that this will disable Green to color an edge d_j as his last move, so Green has only one edge less left to color during the rest of the game. However, the number of edges Red has left to color decreases by the number of edges he colored, so Green is none the worse off and still wins, now without having to worry to color one additional edge d_j at the end of the game. Thus, Green just needs to choose one uncolored edge g_i after the other per move to win, since he is the second player and he has now at least as many edges left to color as Red. ◻

Theorem 5 facilitates the matching between abstract problems and real life applications as it allows to drop the artificial requirement that players must color exactly one edge per move. Let us observe, however, that **PSPACE**-completeness of avoidance games such as the avoidance games played on propositional formulas and on sets described by [44] do not automatically imply the **PSPACE**-completeness of their avoidance$^+$ variants: Most of these **PSPACE**-complete single-choice-per-move avoidance games have trivially decidable, and thus tractable, avoidance$^+$ variants. We also note that even in case both the avoidance$^+$ and the single-choice-per-move avoidance variant are **PSPACE**-complete, it is easy to see that the players having winning strategies can be different for the two games, and that even if in both games the first player has a winning strategy, completely new game situations requiring different playing behavior may arise in an avoidance$^+$ variant.

Corollary 1. *The $G_{\text{Avoid-Ramsey}}$, $G_{\text{Avoid'-Ramsey}}$, and $G_{\text{Avoid-Ramsey}^+}$ problems remain* **PSPACE***-complete even if the avoidance graph A is restricted to a specific fixed graph (the bow-tie graph A depicted in Fig. 4).*

Proof. Follows directly from the bow-tie construction of A in the proof of Theorem 3.
□

For achievement games, the situation is similar:

Theorem 6. *The $G_{\text{Achieve''-Ramsey}}$ problem is* **PSPACE***-complete.*

Proof sketch. We show that there is a logarithmic space reduction from problem $G_{\text{Achieve-POS-DNF}}$ to problem $G_{\text{Achieve''-Ramsey}}$. By a result of [44] showing that $G_{\text{Achieve-POS-DNF}}$ is **PSPACE**-complete and Lemma 2, the result follows. Figure 5 shows an example of the reduction. It is relatively straightforward to see that a winning strategy for $G_{\text{Achieve-POS-DNF}}$ is directly translated into a winning strategy for the achievement games by coloring edge X_i whenever variable x_i of F needs to be chosen (the reduction's detailed description and all eventualities of the opponent's play can be found in [51]).
□

Theorem 7. *The $G_{\text{Achieve'-Ramsey}}$ problem is* **PSPACE***-complete.*

Proof. Since the construction used in the proof of Theorem 6 leaves no possibility open for Green to construct a green subgraph isomorphic to A, we can reinterpret the whole proof according to the rules of $G_{\text{Achieve'-Ramsey}}$. It is easy to see that $G_{\text{Achieve''-Ramsey}}$ and $G_{\text{Achieve'-Ramsey}}$ are in fact the same game when the second player cannot build a green subgraph isomorphic to A, and so the **PSPACE**-completeness proof remains true if we replace every occurrence of $G_{\text{Achieve''-Ramsey}}$ by an occurrence of $G_{\text{Achieve'-Ramsey}}$. Therefore, the **PSPACE**-completeness of $G_{\text{Achieve''-Ramsey}}$ carries over to $G_{\text{Achieve'-Ramsey}}$.
□

Theorem 8. *The $G_{\text{Achieve-Ramsey}}$ problem is* **PSPACE***-complete.*

Proof sketch. We will adapt the reduction from the proof of Theorem 6 for the present proof. Indeed, all we have to make sure is that player Red has a winning strategy in $G_{\text{Achieve''-Ramsey}}$ iff Red also wins the corresponding $G_{\text{Achieve-Ramsey}}$ game. The modifications we describe below ensure that, on the one hand, if player Red has a winning strategy in $G_{\text{Achieve''-Ramsey}}$ and thus can construct a red subgraph isomorphic to graph A, this winning strategy will carry over to $G_{\text{Achieve-Ramsey}}$ without change. On the other hand, if player Green can prevent Red from constructing such a red subgraph, player Green has a winning strategy in $G_{\text{Achieve''-Ramsey}}$ but not (yet) in $G_{\text{Achieve-Ramsey}}$, so for the (modified) latter game we need to add a gadget that makes sure Green can build a green subgraph isomorphic to A, of course only in case Red cannot build one earlier in red.

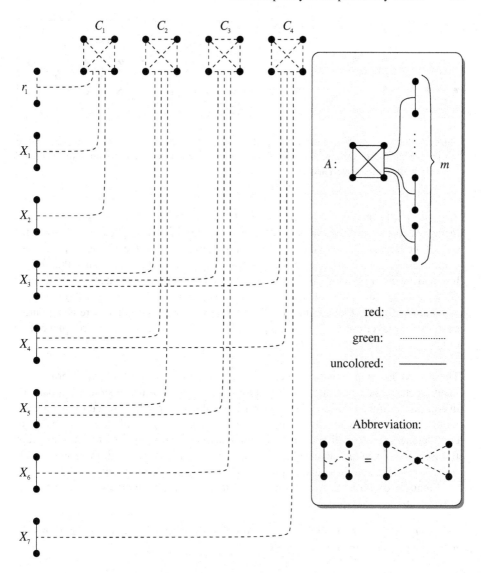

Fig. 5. Instance of the $G_{\text{Achieve''-Ramsey}}$ game corresponding to the $G_{\text{Achieve-POS-DNF}}$ input formula $F = (x_1 \wedge x_2) \vee (x_3 \wedge x_4 \wedge x_5) \vee (x_3 \wedge x_5 \wedge x_6) \vee (x_3 \wedge x_4 \wedge x_7)$, which features a winning strategy for player II. The number m is the size of the largest clause in F. The graph A that player Red has to complete in his color looks like an "m-legged octopus", we therefore call it an "m-topus".

This would make sure that Green would be given a winning strategy in $G_{\text{Achieve-Ramsey}}$ in case Green had a winning strategy in $G_{\text{Achieve''-Ramsey}}$.

Almost everything is defined as in the proof of Theorem 6 besides the redefinition of m, of G, and of E^g. The changes add the new graph $H \overset{\text{def}}{=} (V^H, E^H)$, namely a

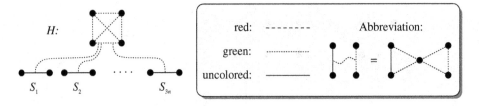

Fig. 6. $3n$-topus graph H that must be added to the construction in Fig. 5 besides the redefinition of m to the number of variables in order to transform the reduction used in the proof of Theorem 6 to one that allows to prove Theorem 8.

$3n$-topus partly precolored in green with all $3n$ feet-edges still uncolored, as depicted in Fig. 6, to the initial game situation, increase the number of "legs" in A from the size of the largest clause to the number of variables n, and add legs already precolored completely in red to each gadget C_i corresponding to the clauses of F such that in case Red had a winning strategy in $G_{\text{Achieve"-Ramsey}}$, it can be reused without change in the new $G_{\text{Achieve-Ramsey}}$ game. We note that the reduction can still be done in logarithmic space.

It easily follows that this modified reduction will make sure that there is a winning strategy for player I of $G_{\text{Achieve-POS-DNF}}(F)$ iff there is a winning strategy for player Red of $G_{\text{Achieve-Ramsey}}(G, A, E^r, E^g)$. □

The **PSPACE**-completeness of the achievement games came a bit as a surprise since many tractable subcases are known and Sim_a has a trivial winning strategy, apparently in contrast to Sim and Sim^+. This illustrates that the asymptotic intractability of a game family does not imply much for particular game instances. Nevertheless, our **PSPACE**-completeness results still allow us to claim that when increasingly more complex graph Ramsey avoidance and achievement game instances are considered, there is a certain point from which on these games plausibly will become very difficult.

Finally, we would like to mention two obvious open problems whose solutions would at least satisfy our academic curiosity.

Problem 1. Are the graph Ramsey game problems **PSPACE**-complete when we restrict the game graphs G to complete graphs?

Problem 2. Are the graph Ramsey game problems **PSPACE**-complete without precolored edges?

However, both problems seem rather difficult: To solve Problem 1, we most likely need to precolor almost all edges in G, leaving out only a few edges that constitute some constrained gadget so that a reduction from another **PSPACE**-complete problem can be done. Since G can be arbitrarily large, graph A must also grow: Otherwise we would not be able to make sure that the precolored part of G contains no monochromatic subgraph isomorphic to A. However, finding such a precoloring with a logarithmic space or even with a polynomial time restricted transducer as would be necessary for the reduction seems rather unlikely.

The difficulty in Problem 2 comes from the fact that without precolored edges, we cannot constrain the games in any way: According to the rules, players can choose arbitrary edges, and so the existence of gadget type reductions that force a certain pattern seems also rather unlikely.

4 Concluding Remarks

Sim and Sim$^+$ are very easy to learn and can be played on a small piece of paper, a typical game taking only a few minutes. Nevertheless, they are fascinating to play because they are much more difficult than they first appear while at the same time being simple and elegant. In this paper, we proved that these games belong to a family of graph Ramsey games that are **PSPACE**-complete, implying that they are the most difficult problems in a class of problems generally *believed* to be intractable, though a formal proof that **P** $\neq$ **PSPACE** is lacking. At the very least, our results imply that the studied games are equivalent from the point of view of computational complexity theory to a large number of well-known games (e.g., Othello [28]) and problems of industrial relevance (e.g., decision-making under uncertainty such as stochastic scheduling [37]) generally recognized as very difficult. The new characterization of **PSPACE**-complete problems as graph Ramsey games might help in studying competitive situations from industry, economics or politics where opposing parties try to achieve or to avoid a certain pattern in the structure of their commitments, in particular situations that may arise in distributed networks, maybe in a future not too far away, for instance the mobile Internet agent warfare scenarios described in [52].

Acknowledgments

This research was partially supported by Austrian Science Fund Project N Z29-INF. I am grateful to the anonymous reviewers for their valuable comments.

References

1. J. Beck and L. Csirmaz. Variations on a game. *Journal of Combinatorial Theory, Series A*, 33:297–315, 1982.
2. József Beck. Van der Waerden and Ramsey type games. *Combinatorica*, 1(2):103–116, 1981.
3. József Beck. Biased Ramsey type games. *Studia Scientiarum Mathematicarum Hungarica*, 18:287–292, 1983.
4. József Beck. Graph games. In *Proceedings of the International Colloquium on Extremal Graph Theory*, 1997.
5. Peter J. Cameron. *Combinatorics: topics, techniques, algorithms*. Cambridge University Press, 1994.
6. V. Chvátal and Frank Harary. Generalized Ramsey theory for graphs. *Bull. Amer. Math. Soc.*, 78:423–426, 1972.
7. Michael L. Cook and Leslie E. Shader. A strategy for the Ramsey game "Tritip". In *Proc. 10th southeast. Conf. Combinatorics, Graph Theory and Computing, Boca Raton*, volume 1 of *Congr. Numerantium 23*, pages 315–324, 1979.

8. Michael Cornelius and Alan Parr. *What's your game?: A resource book for mathematical activities*. Cambridge University Press, 1991.

9. Carl Darby and Richard Laver. Countable length Ramsey games. In Carlos Augusto Di Prisco et al., editor, *Set theory: techniques and applications. Proceedings of the conferences, Curacao, Netherlands Antilles, June 26–30, 1995 and Barcelona, Spain, June 10–14, 1996*, pages 41–46. Dordrecht: Kluwer Academic Publishers, 1998.

10. A. P. DeLoach. Some investigations into the game of SIM. *J. Recreational Mathematics*, 4(1):36–41, January 1971.

11. W. Deuber. A generalization of Ramsey's theorem. In A. Hajnal, R. Radó, and V.T. Sós, editors, *Infinite and finite sets*, volume 10 of *Colloq. Math. Soc. János Bolyai*, pages 323–332. North-Holland, 1975.

12. Douglas Engel. DIM: three-dimensional Sim. *J. Recreational Mathematics*, 5:274–275, 1972.

13. P. Erdős and J. L. Selfridge. On a combinatorial game. *Journal of Combinatorial Theory, Series A*, 14:298–301, 1973.

14. Paul Erdős, András Hajnal, and Latjos Pósa. Strong embeddings of graphs into colored graphs. In A. Hajnal, R. Radó, and V.T. Sós, editors, *Infinite and finite sets*, volume 10 of *Colloq. Math. Soc. János Bolyai*, pages 585–595. North-Holland, 1975.

15. Geoffrey Exoo. A new way to play Ramsey games. *J. Recreational Mathematics*, 13(2):111–113, 1980-81.

16. A. S. Fraenkel and D. Lichtenstein. Computing a perfect strategy for $n \times n$ Chess requires time exponential in n. *Journal of Combinatorial Theory, Series A*, 31:199–214, 1981. Preliminary version in Proc. 8th Internat. Colloq. Automata, Languages and Programming (S. Even and O. Kariv, eds.), Acre, Israel, 1981, *Lecture Notes in Computer Science* **115**, 278–293, Springer Verlag, Berlin.

17. Aviezri S. Fraenkel. Complexity of games. In Richard K. Guy, editor, *Combinatorial Games*, volume 43 of *Proceedings of Symposia in Applied Mathematics*, pages 111–153. American Mathematical Society, 1991.

18. Aviezri S. Fraenkel. Dynamic surveys in combinatorics: Combinatorial games: Selected bibliography with a succinct gourmet introduction. *Electronic Journal of Combinatorics*, 1994 (revised January 1, 2000).

19. W. W. Funkenbusch. SIM as a game of chance. *J. Recreational Mathematics*, 4(4):297–298, October 1971.

20. Martin Gardner. Mathematical games. *Scientific American*, 228(1):108–115, January 1973.

21. Martin Gardner. Mathematical games. *Scientific American*, 228(5):102–107, May 1973.

22. Martin Gardner. *Knotted doughnuts and other mathematical entertainments*, chapter 9, pages 109–122. W. H. Freeman and Company, 1986.

23. Michael R. Garey and David S. Johnson. *Computers and intractability: A guide to the theory of NP-completeness*. Freeman and Co., 1979.

24. Ronald L. Graham, Bruce L. Rothschild, and Joel H. Spencer. *Ramsey theory*. Wiley, 2nd edition, 1990.

25. A. Hajnal and Zs. Nagy. Ramsey games. *Transactions of the American Mathematical Society*, 284(2):815–827, 1984.

26. Frank Harary. Achievement and avoidance games for graphs. In Béla Bollobás, editor, *Proceedings of the Conference on Graph Theory*, volume 13 of *Ann. Discrete Math.*, pages 111–119, Cambridge, 1982. North-Holland, mathematics studies 62.

27. Frank Harary, Wolfgang Slany, and Oleg Verbitsky. A symmetric strategy in graph avoidance games. Manuscript presented at the Combinatorial Game Theory Research Workshop held at the Mathematical Sciences Research Institute in Berkeley, California, July 2000.

28. S. Iwata and T. Kasai. The Othello game on an $n \times n$ board is PSPACE-complete. *Theoret. Comput. Sci. (Math Games)*, 123:329–340, 1994.

29. Martin Knor. On Ramsey-type games for graphs. *Australasian J. Comb.*, 14:199–206, 1996.
30. Péter Komjáth. A simple strategy for the Ramsey-game. *Studia Scientiarum Mathematicarum Hungarica*, 19:231–232, 1984.
31. D. Lichtenstein and M. Sipser. Go is polynomial-space hard. *J. Assoc. Comput. Mach.*, 27:393–401, 1980. Earlier draft appeared in Proc. 19th Ann. Symp. Foundations of Computer Science (Ann Arbor, MI, Oct. 1978), IEEE Computer Soc., Long Beach, CA, 1978, 48–54.
32. Ernest Mead, Alexander Rosa, and Charlotte Huang. The game of SIM: A winning strategy for the second player. *Mathematics Magazine*, 47(5):243–247, November 1974.
33. Thomas E. Moore. SIM on a microcomputer. *J. Recreational Mathematics*, 19(1):25–29, 1987.
34. John H. Nairn and A. B. Sperry. SIM on a desktop calculator. *J. Recreational Mathematics*, 6(4):243–251, Fall 1973.
35. Jaroslav Nešetřil. Ramsey theory. In R.L. Graham, M. Grötschel, and L. Lovász, editors, *Handbook of Combinatorics*, chapter 25, pages 1331–1403. Elsevier, 1995.
36. G. L. O'Brian. The graph of positions in the game of SIM. *J. Recreational Mathematics*, 11(1):3–9, 1978-79.
37. Christos H. Papadimitriou. Games against Nature. *Journal of Computer and System Sciences*, 31:288–301, 1985.
38. Christos H. Papadimitriou. *Computational complexity*. Addison-Wesley, 1994.
39. Aleksandar Pekeč. A winning strategy for the Ramsey graph game. *Combinatorics, Probability & Computing*, 5(3):267–276, September 1996.
40. Frank P. Ramsey. On a problem of formal logic. *Proc. London Math. Soc.*, 30(2):264–286, 1930. Reprinted in I. Gessel and G.-C. Rota, editors. Classic Papers in Combinatorics. Pages 2–24. Birkhäuser Boston 1987.
41. Duncan C. Richer. Maker-breaker games. Poster-presentation at Erdős'99 July 4–11, 1999, in Budapest, Hungary, 1999.
42. Vojtech Rödl. A generalization of Ramsey theorem and dimension of graphs. Msc thesis, Charles University, Prague, 1973.
43. E. M. Rounds and S. S. Yau. A winning strategy for SIM. *J. Recreational Mathematics*, 7(3):193–202, Summer 1974.
44. Thomas J. Schaefer. On the complexity of some two-person perfect-information games. *Journal of Computer and System Sciences*, 16(2):185–225, April 1978.
45. Benjamin L. Schwartz, editor. *Mathematical solitaires and games*, pages 37–81. Baywood Publishing Company, Farmingdale, NY, 1979.
46. Benjamin L. Schwartz. SIM with non-perfect players. *J. Recreational Mathematics*, 14(4):261–265, 1981-82.
47. Leslie E. Shader. Another strategy for SIM. *Mathematics Magazine*, 51(1):60–64, January 1978.
48. Leslie E. Shader and Michael L. Cook. A winning strategy for the second player in the game Tri-tip. In *Proc. Tenth S.E. Conference on Computing, Combinatorics and Graph Theory*, Utilitas, Winnipeg, 1980.
49. Gustavus J. Simmons. The game of SIM. *J. Recreational Mathematics*, 2(2):66, April 1969.
50. Wolfgang Slany. HEXI: A happy perfect hexagone player. Technical Report 31, Institut für Statistik und Informatik der Universität Wien, January 1988.
51. Wolfgang Slany. Graph Ramsey games. Technical Report DBAI-TR-99-34, Institut für Informationssysteme der Technischen Universität Wien, 1999.
52. Bent Thomsen and Lone Leth Thomsen. Towards global computations guided by concurrency theory. *Bulletin of the European Association for Theoretical Computer Science*, (66):92–98, October 1998.
53. Bidan Zhu. Ramsey theory illustrated through a Java based game that plays heuristically and can learn in a client-server style or optimally if possible. Master's thesis, Technische Universität Wien, 1997.

Virus Versus Mankind

Aviezri S. Fraenkel

Department of Computer Science and Applied Mathematics
The Weizmann Institute of Science
Rehovot 76100, Israel
fraenkel@wisdom.weizmann.ac.il
http://www.wisdom.weizmann.ac.il/~fraenkel

Humanity is but a passing episode in the eternal life of the virus

Abstract. We define a two-player virus game played on a finite cyclic digraph $G = (V, E)$. Each vertex is either occupied by a single virus, or is unoccupied. A move consists of transplanting a virus from some u into a selected neighborhood $N(u)$ of u, while devouring every virus in $N(u)$, and replicating in $N(u)$, i.e., placing a virus on all vertices of $N(u)$ where there wasn't any virus. The player first killing all the virus wins, and the opponent loses. If there is no last move, the outcome is a draw. Giving a minimum of the underlying theory, we exhibit the nature of the games on hand of examples. The 3-fold motivation for exploring these games stems from complexity considerations in combinatorial game theory, extending the hitherto 0-player and solitaire cellular automata games to two-player games, and the theory of linear error correcting codes.

Keywords: two-player cellular automata games, generalized Sprague-Grundy function

1 Introduction

The virus is engaged mainly in the following two activities:

- feed: devour and get devoured, or
- replicate[1].

Man tries to employ genetic engineering to destroy the virus totally, winning if he succeeds. The two activities of the virus, however, have a single mission in mind: bring about the speedy demise of mankind. But if it realizes that it fails in this, it will strive to bring about its own death, making itself the last move, rather than give man the satisfaction of destroying the virus. If the virus succeeds in this, it won, having successfully committed suicide. Both agree that if there is no last move, i.e., there is always a virus, then the outcome is a draw.

These activities can be modeled as follows. A finite cyclic digraph $G = (V, E)$ represents the universe, where V denotes the location of the virus: $w(u) = 0$ if u isn't occupied by a virus, $w(u) = 1$ if a virus is at u. There is at most one virus at u for

---

[1] Actually, mankind is similarly active. The first of these activities, however, is then termed more explicitly as warfare: kill and get killed.

T.A. Marsland and I. Frank (Eds.): CG 2000, LNCS 2063, pp. 204–213, 2001.
© Springer-Verlag Berlin Heidelberg 2001

all $u \in V$. Both players alternate in selecting a virus, i.e., a vertex u with $w(u) = 1$ and *firing* it, that is, "complementing" it together with a selected neighborhood $N(u)$ of vertices. By complementing we mean that $w(u)$ switches to 0, and $w(v)$ reverses its parity for every vertex $v \in N(u)$. In biological language, the virus at u moves away from u, devouring every virus in $N(u)$, and replicating in v for all $v \in N(u)$ for which $w(v) = 0$ prior to the move. However, if u is a leaf, the virus at u can neither devour nor replicate, so it dies of frustration: it's dead wood, though $w(u)$ remains 1. The player making the last move wins (after which all vertices, except possibly leaves, have *weight* $w = 0$), and the opponent loses. If there is no last move, i.e., there is always a vertex with weight $w = 1$, the outcome is a draw. A precise definition of the games follows.

Given a finite digraph $G = (V, E)$, also called *groundgraph*. Order V in some way, say

$$V = \{z_0, \ldots, z_{n-1}\}.$$

This ordering, with $|V| = n$, is assumed throughout. The set of *followers* of a vertex z_i is denoted by $F(z_i) = \{z_j \in V : (z_i, z_j) \in E\}$. If $F(z_i) = \emptyset$, then u is a *leaf*.

Let $\mathbf{G} = (\mathbf{V}, \mathbf{E})$ denote the following game graph of the virus game played on $G = (V, E)$. The digraph $\mathbf{G}$ is also called the *virus graph* or *game graph* of G. Any vertex in $\mathbf{G}$ can be described in the form of a binary vector $\mathbf{u} = (u^0, \ldots, u^{n-1})$, where $u^k = 1$ if $w(z_k) = 1$, $u^k = 0$ if $w(z_k) = 0$. In particular, $\Phi = (0, \ldots, 0)$ is a leaf of $\mathbf{V}$, and $|\mathbf{V}| = 2^n$.

The field consisting of two elements $\{0, 1\}$ under Nim-addition $\oplus$ (also called exclusive or — XOR) and and ordinary multiplication is known as GF(2), the Galois field of two elements. Thus $\mathbf{V}$ is an abelian group under $\oplus$ with identity Φ. Every nonzero element has order 2. Moreover, $\mathbf{V}$ is a vector space over GF(2) satisfying $1\mathbf{u} = \mathbf{u}$ for all $\mathbf{u} \in \mathbf{V}$. For $i \in \{0, \ldots, n-1\}$, define unit vectors $\mathbf{z}_i = (z_i^0, \ldots, z_i^{n-1})$ with $z_i^j = 1$ if $i = j$; $z_i^j = 0$ otherwise. They span the vector space.

For defining $\mathbf{E}$, let $\mathbf{u} \in \mathbf{V}$ and let $0 \leq k \leq n - 1$. For $0 \leq q = q(k) \leq |F(z_k)|$, let $F^q(z_k) \subseteq F(z_k)$ be any subset of $F(z_k)$ satisfying

$$|F^q(z_k)| = q. \tag{1}$$

Define,

$$(\mathbf{u}, \mathbf{v}) \in \mathbf{E} \text{ if there exists } k \text{ such that } u^k = 1 \text{ and}$$
$$\mathbf{v} = \mathbf{u} \oplus \mathbf{z}_k \oplus \{\mathbf{z}_\ell : z_\ell \in F^q(z_k)\},$$

for some $F^q(z_k)$ satisfying (1).

Informally, an edge $(\mathbf{u}, \mathbf{v})$ reflects the firing of u^k in $\mathbf{u}$ (with $u^k = 1$), i.e., the complementing of the weights of z_k and $F^q(z_k)$. Such an edge exists for every $F^q(z_k)$ satisfying (1). Note that if $z_k \in G$ is a leaf, then there is no move from $\mathbf{z}_k$, since then $\{z_\ell \in F^q(z_k)\} = \emptyset$.

Remark The strategy of a cellular automata game on a finite *acyclic* digraph is the same as that of a sum-game on $F^q(z_k)$, i.e., a game without complementation of the 1s of $F^q(z_k)$. This follows from the fact that $a \oplus a = 0$ for any nonnegative integer a.

For $s \in \mathbb{Z}^+$, an *s-game* on a digraph $G = (V, E)$ is a virus game on G satisfying $q \leq s$ for all $k \in \{0, \ldots, n-1\}$ (q as in (1)). An *s-regular game* is an *s*-game such that for

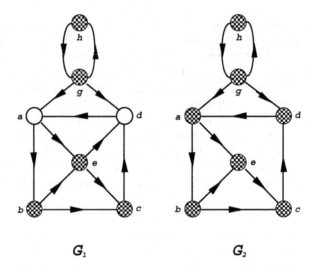

Fig. 1. A two-player virus game. © 2000

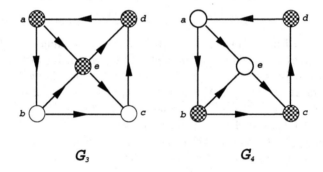

Fig. 2. Adding two more components. © 2000

all k we have $q = s$ if $|F(z_k)| \geq s$, otherwise $q = |F(z_k)|$. Thus if $s \geq \max_{u \in V} |F(u)|$, then firing u in an s-regular game entails complementing all of $F(u)$, for all $u \in V$, in addition to complementing u, unless u is a leaf.

The emphasis here is on exhibiting examples of two-player virus games. Their theory and complexity are explored in [11], where also formal proofs are given. Motivation for defining this particular family of games and how they fit into the general theory and complexity of combinatorial games are given in the epilogue (§7).

2   Initial Examples

In Figures 1 and 2, the shaded vertices have weight 1, the rest have weight 0.

Example 1 We play a 3-regular game on the *sum* of the two components G_1, G_2 of Figure 1, for which we compute the so-called γ-*values* 3, 2 respectively (more about the

γ-function follows in the next sections). A sum of games is a finite collection of disjoint component games, where a move consists of selecting any component and making a move in it. Thus a move in the sum of two-player cellular automata games consists of selecting a component, and firing a vertex in it. Then the γ-value of the sum is given by $3 \oplus 2 = 1$. Since this value is finite but > 0, the first player can win, namely by moving to a position with γ-value 0. This is can be affected in one of three ways, all involving moves in G_1:

- Fire vertex b, which complements the weights of b, c, e, namely it makes $w(b) = w(c) = w(e) = 0$, leaving the other weights unchanged. Then the γ-value of the resulting position of G_1 changes to 2, leading to $2 \oplus 2 = 0$.
- Fire c, complementing the weights of c and d, again transforming the γ-value of the position of G_1 to 2.
- Fire e, complementing the weights of c, d, e, resulting in a new position of γ-value 2.

Example 2 Play a 2-regular game on the two components of Figure 2. The γ-value of the position of each is 1, so the γ-value of the sum is $1 \oplus 1 = 0$. This implies that whatever move the first player will do, the second player can win.

Example 3 We now wish to play a 3-regular game on all four components G_1, G_2, G_3, G_4. The value of the sum is $3 \oplus 2 \oplus 1 \oplus 1 = 1$. Thus the first player can win, either by selecting one of the above three moves, or by moving to a position with γ-value 0 in one of G_3 or G_4. This can be affected by firing a or d or e in G_3, or by firing b or c or d in G_4.

An elaboration on these three examples follows in §4.

3 Background

For explaining the above examples, we need some basic facts. Let $G = (V, E)$ be any finite cyclic digraph. A digraph is cyclic if it *may* contain cycles. The *Generalized Sprague-Grundy function* γ is a mapping $\gamma: V \to \mathbb{Z}^0 \cup \{\infty\}$ which satisfies certain conditions, where the symbol ∞ indicates a value larger than any natural number. If $\gamma(u) = \infty$, we also say that $\gamma(u)$ is infinite. We wish to define γ also on certain subsets of vertices. Specifically: $\gamma\ F(u) = \gamma(v) < \infty: v \in F(u)$. If $\gamma(u) = \infty$ and if we denote the set $\gamma\ F(u)$ by K for brevity, then we also write $\gamma(u) = \infty(K)$. Next we define equality of $\gamma(u)$ and $\gamma(v)$: if $\gamma(u) = k$ and $\gamma(v) = \ell$ then $\gamma(u) = \gamma(v)$ if one of the following holds: (a) $k = \ell < \infty$; (b) $k = \infty(K)$, $\ell = \infty(L)$ and $K = L$. We also use the notations

$$V_i = \{u \in V : \gamma(u) = i\}\ (i \in \mathbb{Z}^0),\quad V^f = \ u \in V: \gamma(u) < \infty\ ,\quad V^\infty = V \setminus V^f.$$

We next formulate an algorithm for computing γ. Initially a special symbol ν is attached to the label $\ell(u)$ of every vertex u, where $\ell(u) = \nu$ means that u has no label. We also introduce the notation $V_\nu = \{u \in V : \ell(u) = \nu\}$.

Algorithm GSG for computing the Generalized Sprague-Grundy function for a given finite cyclic digraph $G = (V, E)$.

1. (Initialize labels.) Put $i \leftarrow 0$, $\ell(u) \leftarrow \nu$ for all $u \in V$.
2. (Label.) As long as there exists $u \in V_\nu$ such that no follower of u is labeled i and every follower of u which is either unlabeled or labeled ∞ has a follower labeled i, put $\ell(u) \leftarrow i$.
3. (∞-label.) For every $u \in V_\nu$ which has no follower labeled i, put $\ell(u) \leftarrow \infty$.
4. (Increase label.) If $V_\nu \neq \emptyset$, put $i \leftarrow i + 1$ and return to 2; otherwise end.

We then have $\gamma(u) = \ell(u)$. The γ function has been defined in [22]. See also [4], Ch. 11. The above algorithm is a simplified version of Algorithm G in [16].

Informally, a *P-position* is any position u from which the *P*revious player can force a win, that is, the opponent of the player moving from u. An *N-position* is any position v from which the *N*ext player can force a win, that is, the player who moves from v. The next player can win by moving to a *P*-position. A *D-position* is any position from which neither player can win, but both have a nonlosing next move, namely, moving to some *D*-position. Denote the set of all *P*-positions by $\mathcal{P}$, all *N*-positions by $\mathcal{N}$ and all *D*-positions by $\mathcal{D}$.

It turns out that the γ-function on the *sum* of the exponentially large virus graphs enables us to compute a strategy: We first compute the γ-values of each of the components V_i of sizes $O(|V_i|)$. Their *generalized Nim sum* is then the γ-function of the sum [16], [22]. The generalized Nim sum of nonnegative integers is their XOR as for the common Nim sum. For any finite subsets K, L of nonnegative integers and any nonnegative integer a, $\infty(K) \oplus a = a \oplus \infty(K) = \infty(K \oplus a)$, and $\infty(K) \oplus \infty(L) = \infty(\emptyset)$.

The connection between γ on the game graph $\mathbf{G} = (\mathbf{V}, \mathbf{E})$ of the sum of one or more disjoint games and P, N, D is given by:

$$\mathcal{P} = \{\mathbf{u} \in \mathbf{V} : \gamma(\mathbf{u}) = 0\}, \quad \mathcal{D} = \{\mathbf{u} \in \mathbf{V} : \gamma(\mathbf{u}) = \infty(L), \ 0 \notin L\}, \quad (2)$$

$$\mathcal{N} = \{\mathbf{u} \in \mathbf{V} : 0 < \gamma(\mathbf{u}) < \infty\} \cup \{\mathbf{u} \in \mathbf{V} : \gamma(\mathbf{u}) = \infty(L), \ 0 \in L\}, \quad (3)$$

see [16], [22].

The γ-function of the sum of the virus graphs $\mathbf{G}$ induced by the digraphs G on which the game is played, enables us to play optimally. We compute it in the next section.

4   The First Three Examples Revisited

If $\{u_i, u_j, \ldots, u_k\}$ is a collection of vertices of G with $w(u_i) = w(u_j) = \ldots = w(u_k) = 1$ and all other vertices have weight 0, we denote the γ-value of this position of $\mathbf{G}$ by $\gamma(u_i u_j \ldots u_k)$.

Consider G_3 for ≥ 2-regular play. We use Algorithm GSG and inspection to note that $\gamma(e) = 0$: e has only the as yet unlabeled follower cd, which has the follower $\Phi \in \mathbf{V}_0$. Inspection further shows that also $\gamma(ac) = \gamma(bd) = 0$. It turns out that γ is a homomorphism from $\mathbf{V}^f$ onto $GF(2)^t$ for some nonnegative integer t with kernel $\mathbf{V}_0$ and quotient space $\mathbf{V}^f / \mathbf{V}_0 = \{\mathbf{V}_i : 0 \leq i < 2^t\}$ (collection of cosets), and $\dim(\mathbf{V}^f) = t + \dim(\mathbf{V}_0)$.

In particular, $\mathbf{V}_0$ is a linear subspace of $\mathbf{V}$, so it contains all Nim sums of Φ, e, ac, bd. Inspection shows that it contains nothing else. Hence

$$\mathbf{V}_0 = \{\Phi, e, ac, bd, ace, bde, abcd, abcde\}. \tag{4}$$

Inspection further shows that $\gamma(cd) = 1$. Thus we get the coset

$$\mathbf{V}_1 = cd \oplus \mathbf{V}_0 = \{cd, cde, ad, bc, ade, bce, ab, abe\}.$$

We see that $ade \in \mathbf{V}_1$, as claimed in Example 2. Also $\dim(\mathbf{V}_0) = 3$, $\dim(\mathbf{V}^f) = 4$, $t = 1$. (We also have $\gamma(a) = \infty$, but $\mathbf{V}^\infty$ is clearly not a subspace. However, $\mathbf{W} = \{a, \Phi\}$ is a subspace such that every vector in $\mathbf{V}$ can be written uniquely as a sum of two vectors: one in $\mathbf{V}^f$ and one in $\mathbf{W}$.)

The same considerations show that for G_4,

$$\mathbf{V}_0 = \{\Phi, b, ac, de, abc, bde, acde, abcde\}, \tag{5}$$

$$\mathbf{V}_1 = cd \oplus \mathbf{V}_0 = \{cd, bcd, ad, ce, abd, bce, ae, abe\}.$$

The γ-function of the sum of G_3 and G_4 is thus $1 \oplus 1 = 0$. By (2) this is a P-position, substantiating the claim of Example 2.

Note that the lower part of G_1 is congruent to G_3, and the lower part of G_2 is congruent to G_4. Inspection shows that for both G_1 and G_2, $\gamma(gh) = 2$. Therefore $t = 2$ for both components. Further, $\mathbf{V}_0$ and $\mathbf{V}_1$ of G_1 (G_2) are the same as $\mathbf{V}_0$ and $\mathbf{V}_1$ of G_3 (G_4) respectively. For G_1 we thus get the cosets $\mathbf{V}_2 = gh \oplus \mathbf{V}_0$, $\mathbf{V}_3 = adgh \oplus \mathbf{V}_0$ where $\mathbf{V}_0$ is given by (4), $\dim \mathbf{V}_0 = 3$, $\dim \mathbf{V}^f = 5$. The same holds for G_2, but with $\mathbf{V}_0$ given by (5).

We thus conclude that for G_1, $\gamma(bcegh) = 3$, and for G_2, $\gamma(abcdegh) = 2$, as claimed in Example 1. The winning moves suggested in Examples 1–3 all yield positions with γ-value 0, so they are P-positions by (2).

5   Further Examples

Example 4 We now wish to play a 2-regular game on the sum of the two components of Figure 1, rather than a 3-regular game. This affects only vertex g, whose outdegree is 3 in both components. The outcome is now a draw by (2), since $\gamma(gh) = \infty(0)$ for both components. Note that not every sequence of moves will maintain a draw: thoughtless moves might give the opponent the opportunity to win.

Example 5 Play a 2-regular game on the sum of the two components G_2 with position g, and G_4 with the position shown in Figure 2. We have $\gamma(g) = \infty(1)$, so the sum has γ-value $\infty(1) \oplus 1 = \infty(1 \oplus 1) = \infty(0)$. By (3) this is an N-position. The first player has a unique winning move by firing g and complementing a and d. Any other move leads to either a draw; or to an N-position, from which the opponent can win.

The next examples involve play on the digraph $G(p)$ (Figure 3) which depends on a parameter $p \in \mathbb{Z}^+$. It has vertex set $\{x_1, \ldots, x_s, y_1, \ldots, y_s\}$, and edges:

$$F(x_i) = y_i \quad \text{for } i = 1, \ldots, s,$$
$$F(y_k) = \{y_i : 1 \le i < k\} \cup \{x_j : 1 \le j \le s \text{ and } j \ne k\} \text{ for } k = 1, \ldots, s.$$

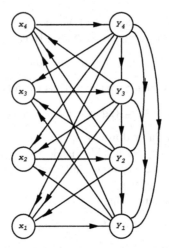

Fig. 3. Virus games on a parametrized digraph. © 2000

Example 6 Play a 3-regular game on $G(7)$, where $w(x_7) = w(y_1) = \ldots = w(y_7) = 1$, and all the other vertices have weight 0. Inspection shows that all collections of an even number of x_i are in $\mathbf{V}_0$ (i.e., an even number of x_i having weight 1, all other vertices weight 0); and $\mathbf{V}^f$ consists precisely of all collections of an even number of 1s. Further, $\gamma(x_j y_j)$ = smallest nonnegative integer not the Nim sum of at most three $\gamma(x_i y_i)$ for $i < j$. Thus $\gamma(x_7 y_1 \ldots y_7) = 0 \oplus 1 \oplus 2 \oplus 4 \oplus 8 \oplus 15 \oplus 16 \oplus 32 = 48$. So firing y_7 and complementing y_6 and any two of the x_i ($i < 7$) is a winning move. Incidentally, the sequence $\{1, 2, 4, 8, 15, 16, 32, 51, \ldots\}$ has been used in [1] for a special case of the game "Turning Turtles".

Example 7 Play a 4-regular game on $G(7)$, on the same initial position as in the previous example. Here any collection of x_i is in $\mathbf{V}_0$. (More generally, in an s-regular game on $G(p)$ any collection of x_i is in $\mathbf{V}_0$ if s is even, otherwise any collection of an even number of x_i is in $\mathbf{V}_0$.) Moreover, $\gamma(y_i)_{i \geq 1} = (1, 2, 4, 8, 16, 31, 32, 64, 103, \ldots)$. It is the sequence defined as the smallest nonnegative integer not the Nim sum of at most four earlier terms. This sequence appears in Table 3, Ch. 14 of [1]. Note that the strategy of our two-player virus games on just Fig. 3 alone subsumes and unifies that of a battery of games there. Thus $\gamma(x_7 y_1 \ldots y_7) = 0 \oplus 1 \oplus 2 \oplus 4 \oplus 8 \oplus 16 \oplus 31 \oplus 32 = 32$. Firing y_7 and complementing any four x_i ($i < 7$) results in a P-position.

Example 8 Play the sum of the games in the two previous examples. In the 3-regular games, fire y_6 and complement any two of the x_i ($i \neq 6$). This results in γ-value 32 for this game, so its sum with the 4-regular game constitutes a P-position.

Example 9 Now consider a play on $G(7)$ which is not regular: play a 2-regular game on the subset of all the y_{2i-1} and on all the x_i, and a 3-regular game on the subset of all the y_{2i} ($i \geq 1$). Clearly any collection of x_{2i-1} is in $\mathbf{V}_0$, and any collection of even size of x_{2i} is in $\mathbf{V}_0$. We can now use induction on the number of even indexed x_i to show that actually any collection of x_i is in $\mathbf{V}_0$. Inspection shows

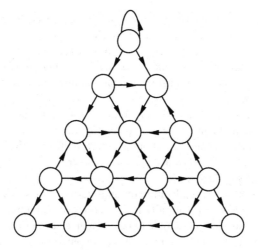

Fig. 4. Triple Strike game. © 2000

that $\gamma(y_i)_{i\geq 1} = (1, 2, 4, 8, 7, 16, 11, 29, 13\ldots)$. This strange looking sequence is not (yet) in the useful encyclopædia of integers of Neil Sloane available on-line at http://www.research.att.com/~njas/sequences/

Example 10 Play a 1-regular game on the subset of all the y_{2i-1} and on all the x_i, and a 2-regular game on the subset of all the y_{2i} $(i \geq 1)$. We can show again that any collection of x_i is in $\mathbf{V}_0$. Inspection shows further that $\gamma(y_i)_{i\geq 1} = \{1, 2, 3, 4, 5, 8, 6, 15, 7, 16, 9, 26, 10, 32, 11, 44, 12, 49, 13\ldots\}$. This is also a sequence that is not yet in the encyclopædia of integers.

6 Homework

Consider the digraph depicted in Fig. 4. It is suggested to play three types of games on it.

- A weight distribution over $\mathbb{Z}^0$ on the vertices is given initially. A move consists of selecting a vertex u with weight > 0 and shifting a weight of 1 to any follower v. After the move, $w(u)$ has diminished by 1, and $w(v)$ has increased by 1.
- Play a 2-regular game on the digraph.
- Play a 3-regular game on the digraph.
- Play a sum of three instances of the previous games.

Compute an optimal strategy for each of the four games.

7 Epilogue

The special case of 1-regular games are the so-called *annihilation games*, analyzed in [8], [13], [15]. For annihilation games it is convenient to consider a vertex with weight 1 to be occupied by a token, and one with weight 0 to be unoccupied. Two tokens are then mutually annihilated on impact, hence the name. Misère play (in which the player making the last move loses, and the opponent wins) of annihilation games was investigated in [7]. Our motivation for examining annihilation games, suggested to us by John Conway, was to create games which exhibit some interaction between tokens and yet still have a polynomial strategy.

Annihilation games are "barely" polynomial, in several senses. Their complexity is $O(|V|^6)$, the polynomial computation of a winning move may require a "strategy in the broad sense" (see [10] §4) and just about any perturbation of them yields Pspace-hard games (see [14], [12], [17]). They also lead to linear error correcting codes [9], but their Hamming distance is ≤ 4.

The current work is a generalization of annihilation games. It was motivated by the desire to create games which naturally induce codes of Hamming distance > 4. Virus games may provide such codes, a topic to be taken up elsewhere. In practice, $\mathbf{V}_0$, which is all that's needed for the codes, can often be computed by inspection. The best codes may be derived from a digraph which is a simplification of that of Figure 3, where $\mathbf{V}_0$ can also be computed by direct inspection. The lexicodes method [6], [5], [21], related to games, is clearly exponential.

Another motivation was to further explore polynomial games with "token interactions". Last but not least was the desire to create two-player virus games, which traditionally have been solitaire only.

Cellular Automata Games have traditionally either been 0-player games such as Conway's "Life", or solitaire games played on a grid or digraph $G = (V, E)$. Two commercial manifestations are *Lights Out* manufactured by Tiger Electronics, and *Merlin Magic Square* by Parker Brothers (but Arthur-Merlin games are something else again). Quite a bit is known about such solitaire games. Background and theory can be found e.g., in [18], [20], [23], [24], [25], [26], [27]. Incidentally, related but different solitaires are *chip firing games*, see e.g., [3], [19], [2]. In the present paper we have extended cellular automata games to two-player games in a natural way, and indicated a strategy for them, which is based on three key ideas: I. Showing that γ is essentially additive: $\gamma(\mathbf{u}_1 \oplus \mathbf{u}_2) = \gamma(\mathbf{u}_1) \oplus \gamma(\mathbf{u}_2)$ if $\mathbf{u}_1 \in \mathbf{V}^f$ or $\mathbf{u}_2 \in \mathbf{V}^f$. II. Showing that $\mathbf{V}_0$, $\mathbf{V}^f$ and γ can be computed by restricting attention to the linear span of "sparse" vectors. III. Computing a winning move.

References

1. E. R. Berlekamp, J. H. Conway and R. K. Guy. *Winning Ways for your mathematical plays.* (Two volumes), Academic Press, London, 1982.
2. N. L. Biggs. Chip-firing and the critical group of a graph. *J. Algebr. Comb.* **9**, 25–45, 1999.
3. A. Björner and L. Lovász. Chip-firing on directed graphs. *J. Algebr. Comb.* **1**, 305–328, 1992.
4. J. H. Conway. *On Numbers and Games.* Academic Press, London, 1976.
5. J. H. Conway. Integral lexicographic codes. *Discrete Math.* **83**, 219–235, 1990.
6. J. H. Conway and N. J. A. Sloane. Lexicographic codes: error-correcting codes from game theory. *IEEE Trans. Inform. Theory* **IT-32**, 337–348, 1986.
7. T. S. Ferguson. Misère annihilation games. *J. Combin. Theory* (Ser. A) **37**, 205–230, 1984.

8. A. S. Fraenkel. Combinatorial games with an annihilation rule. In The Influence of Computing on Mathematical Research and Education, *Proc. Symp. Appl. Math.* (J. P. LaSalle, ed.), Vol. 20, Amer. Math. Soc., Providence, RI, pp. 87–91, 1974.

9. A. S. Fraenkel. Error-correcting codes derived from combinatorial games. In *Games of No Chance,* Proc. MSRI Workshop on Combinatorial Games, July, 1994, Berkeley, CA, MSRI Publ. (R. J. Nowakowski, ed.), Vol. 29, Cambridge University Press, Cambridge, pp. 417–431, 1996.

10. A. S. Fraenkel. Scenic trails ascending from sea-level Nim to alpine chess. In *Games of No Chance,* Proc. MSRI Workshop on Combinatorial Games, July, 1994, Berkeley, CA, MSRI Publ. (R. J. Nowakowski, ed.), Vol. 29, Cambridge University Press, Cambridge, pp. 13–42, 1996.

11. A. S. Fraenkel. Two-player games on cellular automata. To appear in: *More Games of No Chance,* Proc. MSRI Workshop on Combinatorial Games, July, 2000, Berkeley, CA, MSRI Publ. (R. J. Nowakowski, ed.), Cambridge University Press, Cambridge, 2001.

12. A. S. Fraenkel and E. Goldschmidt. Pspace-hardness of some combinatorial games. *J. Combin. Theory* (Ser. A) **46**, 21–38, 1987.

13. A. S. Fraenkel and Y. Yesha. Theory of annihilation games. *Bull. Amer. Math. Soc.* **82**, 775–777, 1976.

14. A. S. Fraenkel and Y. Yesha. Complexity of problems in games, graphs and algebraic equations. *Discrete Appl. Math.* **1**, 15–30, 1979.

15. A. S. Fraenkel and Y. Yesha. Theory of annihilation games — I. *J. Combin. Theory* (Ser. B) **33**, 60–86, 1982.

16. A. S. Fraenkel and Y. Yesha. The generalized Sprague–Grundy function and its invariance under certain mappings. *J. Combin. Theory* (Ser. A) **43**, 165–177, 1986.

17. A. S. Goldstein and E. M. Reingold. The complexity of pursuit on a graph. *Theoret. Comput. Sci.* (Math Games) **143**, 93–112, 1995.

18. E. Goles. Sand piles, combinatorial games and cellular automata. *Math. Appl.* **64**, 101–121, 1991.

19. C. M. López. Chip firing and the Tutte polynomial. *Ann. of Comb.* **1**, 253–259, 1997.

20. D. H. Pelletier. Merlin's magic square. *Amer. Math. Monthly* **94**, 143–150, 1987.

21. V. Pless. Games and codes. In *Combinatorial Games,* Proc. Symp. Appl. Math. (R. K. Guy, ed.), Vol. 43, Amer. Math. Soc., Providence, RI, pp. 101–110, 1991.

22. C. A. B. Smith. Graphs and composite games. *J. Combin. Theory* **1**, 51–81, 1966. Reprinted in slightly modified form in: *A Seminar on Graph Theory* (F. Harary, ed.), Holt, Rinehart and Winston, New York, NY, 1967.

23. D. L. Stock. Merlin's magic square revisited. *Amer. Math. Monthly* **96**, 608–610, 1989.

24. K. Sutner. On σ-automata. *Complex Systems* **2**, 1–28, 1988.

25. K. Sutner. Linear cellular automata and the Garden-of-Eden. *Math. Intelligencer* **11**, 49–53, 1989.

26. K. Sutner. The σ-game and cellular automata. *Amer. Math. Monthly* **97**, 24–34, 1990.

27. K. Sutner. On the computational complexity of finite cellular automata. *J. Comput. System Sci.* **50**, 87–97, 1995.

Creating Difficult Instances
of the Post Correspondence Problem

Richard J. Lorentz

Department of Computer Science
California State University
Northridge, CA 91330-8281 USA
lorentz@csun.edu

Abstract. Computational aspects of the Post Correspondence Problem (PCP) are studied. Specifically, we describe our efforts to find difficult instances of the PCP, where a "difficult" instance is defined to mean an instance whose shortest solution is long. As a result, we attempt to quantify the difficulty of the PCP in the same way the Busy Beaver Problem does for the Turing Halting Problem. We find instances of the PCP that have quite long solutions even when the number of pairs and the length of the strings is small, e.g., four and three, respectively. We discuss algorithms for solving the PCP and for generating difficult PCP instances. This problem poses unique difficulties because the size of the search space is unbounded.

Keywords: single-agent search, PCP, Post Correspondence Problem, hash table, transposition table

1   Introduction

Most games and puzzles that are interesting to programmers are interesting because they are difficult and this difficulty can often be formalized for generalized versions of these games and puzzles. For example, go [8], checkers [4], and sokoban [1] are all either PSPACE hard or complete. Nevertheless, now standard techniques in AI allow programmers to develop programs that play these games with considerable skill. We have decided to try to tackle a theoretically more difficult problem using these same techniques. We have developed programs that both solve and generate instances of an undecidable problem: the Post Correspondence Problem.

The Post Correspondence Problem was first introduced by Post in 1946 [13] where he also showed that the problem is undecidable. It has since become a classic undecidable problem. For example, it can be used to easily show many problems about context-free languages and grammars are undecidable. (See [2] for a number of such proofs.) An instance of a Post Correspondence Problem (we often just say "a PCP") is as follows: Given an alphabet Σ and a set of pairs of strings over the alphabet, $\{(w_i, x_i) : w_i, x_i \in \Sigma^*, i = 1, \ldots, n\}$, does there exist a sequence of integers

T.A. Marsland and I. Frank (Eds.): CG 2000, LNCS 2063, pp. 214–228, 2001.

$j_1,\dots,j_l$ such that $w_{j_1}w_{j_2}\dots w_{j_l} = x_{j_1} x_{j_2}\dots x_{j_l}$? We call n the *size* of the PCP and if l is the smallest integer providing a solution to the PCP we say that l is the *length* of the PCP solution. Of course, it is possible that there is no such l, that is, the PCP has no solution. When $w_{j_1}w_{j_2}\dots w_{j_k}$ is a prefix of $x_{j_1}x_{j_2}\dots x_{j_k}$ (or conversely) we say that they are *solution strings*, even if there is no solution that begins with these strings. We are especially interested in PCP's where all of the strings are small, so we define m to be the size of the longest string in the PCP and we call m the *width* of the PCP. Since we are interested in structural properties of the Post Correspondence Problem we will restrict our alphabet so that $\Sigma = \{0, 1\}$. Larger alphabets can easily be coded using the binary alphabet.

We use the standard two-dimensional representation to display an instance. For example, we denote the PCP $\{(01, 0), (01, 110), (0, 100), (100, 10)\}$ which has size 4 and width 3 as:

$$\begin{pmatrix} 01 & 01 & 0 & 100 \\ 0 & 110 & 100 & 10 \end{pmatrix} \tag{1}$$

It turns out that this PCP does have solutions and the minimum solution is of length 76.

Here is an even simpler instance:

$$\begin{pmatrix} 10 & 0 & 00 \\ 0 & 000 & 01 \end{pmatrix} \tag{2}$$

Let us solve this size 3, width 3 example. Clearly, we must begin with pair 2 since this is the only pair where one string is a prefix of the other. We display this by showing the two solution strings and the pair(s) involved as:

0
000
(2)

How might a solution continue? There are now two choices. We could concatenate another pair 2 giving us:

0<u>0</u>
000<u>000</u>
(2) (2)

where we underline the most recent concatenation on the solution strings for emphasis. In fact, we see that we can continue concatenating an arbitrary number of pair 2 strings and as the number of such concatenations increases, the likelihood of this leading to a minimal solution decreases, but we are obliged to search this path. As it turns out this first choice does not give us the shortest solution so we select the other choice and concatenate pair 3 to our solution strings giving the following:

0<u>00</u>
000<u>01</u>
(2) (3)

Now there is only one way to continue. We must choose pair 2, so we have:

0000
00001000
(2) (3) (2)

Again, there is only one choice (pair 1), giving us:

000010
000010000
(2) (3) (2)(1)

There are now two choices, but as it turns out, the shortest solution never uses pair 2 again, so the shortest solution plays out as follows:

00001000	0000100000	000010000010
00001000001	0000100000101	00001000001010
(2)(3) (2)(1)(3)	(2) (3) (2)(1)(3)(3)	(2) (3)(2)(1)(3)(3)(1)
00001000001010	0000100000101000	000010000010100010
000010000010100	00001000001010001	000010000010100010
(2)(3)(2)(1)(3)(3)(1)(1)	(2)(3)(2)(1)(3)(3)(1)(1)(3)	(2)(3)(2)(1)(3)(3)(1)(1)(3)(1)

One of our motivations for looking at this problem is the fact that the literature shows very few interesting examples of the PCP. The following is typical of examples given in textbooks [5]:

$$\begin{pmatrix} 01 & 110010 & 1 & 11 \\ 0 & 0 & 1111 & 01 \end{pmatrix} \qquad (3)$$

It has the same size as PCP (1) above, it has double the width, yet the length of the solution is only 6. The most interesting example we have found in the literature is the following [9]:

$$\begin{pmatrix} 001 & 1 & 011 & 010 \\ 01 & 011 & 10 & 01 \end{pmatrix} \qquad (4)$$

It has the same size and width as PCP (1) and has the reasonably large solution length of 66 but even this instance has some undesirable features that will be discussed below.

Since the PCP is undecidable we know that the solution length viewed as a function of the size and width cannot be computable, and so must eventually grow extremely quickly. The purpose of this work is to show just how large solutions can become, even for small sizes and widths, to show how standard techniques in AI can be used to establish this, and to demonstrate how PCP instances with long solutions can be created. Large solutions of PCP's that have small size and width demonstrate in a way how difficult the PCP is as well as providing an aesthetic measure. It is interesting to compare our efforts at automatically finding interesting PCP problems with attempts to automatically find interesting sokoban instances [12] and go problems [17].

The idea of looking at small instances of an undecidable problem is not new. In 1962 the "Busy Beaver Problem" [14] was first proposed and studied. The original statement of the problem asks: Among all Turing Machines with n states, if the Turing Machine starts with a blank tape and eventually halts, how many 1's can be placed on the tape? Another version of the problem asks how many state transitions can a halting n-state Turing Machine make. It was shown that even for a very small number of states a Turing Machine can be constructed to do an enormous amount of computation as measured by either the number of steps the machine makes or the number of symbols it places on its output tape. Since then the numbers have steadily increased, with the current published records displaying small Turing Machines that place millions of symbols on the tape [10]. Even more recently, the same two authors and a third collaborator claim to have a 6-state Turing Machine that deposits on the order of 10^{21} 1's on an initially blank tape [11].

In Section 2 we describe various algorithms for solving specific instances of the PCP. In Section 3 we describe how we use our solving algorithms to attempt to find the most difficult PCP instances for some fixed sizes and widths. Section 4 explains how we modify the solving algorithms to help generate interesting PCP's for some other sizes and widths. The last section deals with conclusions, open problems, and suggestions for future work.

2   Solving PCP's

Of course, because of the undecidability of the Post Correspondence Problem, we will never be able to write a program that can solve PCP's in general. That said, the obvious approach to solving PCP's is quite straightforward. It is just the standard single-agent search that is used to solve puzzles and is typically implemented as a depth-first tree search through all the possible legal configurations. However, as simple as this idea is, and as easy as it is to implement, it is typically insufficient for solving any problems of practical interest because the size of the tree is often enormous, as typically happens with PCP's. Witness the effort that has gone into the problem of solving sokoban problems [7]. Many man-years have been invested, developing heuristics specific to the problem so that a program could be written that can solve them. Yet this program is still not up to the level of the best human solvers. Hence, we cannot be too optimistic about being able to solve PCP's, especially given the "difficulty" of this problem.

The first decision to be made when solving a PCP is how deep are we willing to look, that is, how long do we think the solution might be? Immediately we run into the undecidability bottleneck since an algorithm that could determine an upper bound on the length of a solution would imply an algorithm for solving the PCP. This is so because once provided with an upper bound k we would know that if we searched the space completely to depth k we would be guaranteed either to find a solution or know there is none. But since there is no algorithm we cannot expect to be able to do this. Hence, what we have done is to use an idea similar to that of iterative deepening from the world of game programming. We start by setting the bound to a fairly low value, and then if no solution is found we repeatedly increase the bound by a fixed amount until we find a solution or we run out of time and give up. And, as in iterative deepening, we can use the information obtained from one level to help speed up the computation at the next deeper level. We will explain how when we describe our use of hashing. This is the basic algorithm and serves as the major engine for all of our other solving and instance generating algorithms.

Many PCP's can be proven to have no solution based strictly on syntactic concerns and so do not need to be searched at all. All of our algorithms do this filtering before actually calling the search engine. There are three filters we consider.

The first is called the *prefix filter*. (There is a symmetric *postfix filter* as well.) In order for a solution to be possible a solution must begin with one of the pairs and so one of the pairs must be such that one of the strings is a prefix of the other. (We assume no trivial solutions of length 1 where both strings of the pair are the same.) Put another way, one of the PCP pairs must be a solution string. Consider the following example:

$$\begin{pmatrix} 001 & 00 & 01 & 10 \\ 01 & 011 & 101 & 001 \end{pmatrix} \tag{5}$$

A quick observation reveals that (5) cannot have a solution since there is no way for it to begin. No pair is such that one is a prefix of the other. So, the prefix filter will eliminate this instance. Note, however, that this instance does pass the postfix filter, that is, it is possible for this PCP to end. It can end with pair number 1 or pair number 3.

The second filter is called the *length balance filter*. Again, assuming no trivial solutions of length 1, since the first pair of a solution makes one of the solution strings longer than the other at least one of the other pairs must balance this by allowing the longer solution string to become smaller. For example, in the following:

$$\begin{pmatrix} 001 & 00 & 1 & 101 \\ 00 & 01 & 0 & 10 \end{pmatrix} \tag{6}$$

A solution might begin with pair 1 or pair 4, but at this point the top solution string is longer than the bottom solution string and can never become shorter, so no solution is possible.

The final filter we employ is called the *element balance filter*. It is similar to the length balance filter only it is concerned with the count of the number of 1's and 0's in a solution string. If one of the two solution strings has an excess number of 1's say, and there is no pair in the PCP instance that will allow the other solution string to gain 1's then this filter will reject the instance as having no solution. For example, given this PCP:

$$\begin{pmatrix} 001 & 11 & 01 & 010 \\ 00 & 011 & 000 & 10 \end{pmatrix} \tag{7}$$

If there is a solution it would have to begin with the first pair, but at this point the top solution string has more 1's than the bottom string and no other pair in the instance will allow the bottom string to catch up on the 1's. This completes the description of the filters.

The major execution time enhancement we employ is *hashing* to allow tree *cutoffs* at nodes that have already been visited, which directly corresponds to the idea of transposition tables used in game playing. The idea is that if a configuration has already appeared and has already been evaluated then if the same configuration appears later in the search for a solution, it may not have to be evaluated again, so we might be able to cut off this part of the tree. The idea of using hashing for single-agent searches has already been studied and applied to the 15-puzzle and to the Traveling Salesman Problem [15] where it is shown to be quite powerful. Though not so consistently helpful in this setting, it nevertheless allows us to solve many PCP's we would not be able to solve otherwise. Another attempt to prune duplicate nodes involves automatically creating a finite automaton that is then traversed as the search tree is descended [16]. This technique is especially useful when there are many duplicate nodes in the search tree and these duplicates tend to be near each other. Since this is usually not the case with PCP's the cost of the additional overhead would seldom pay off.

When solving a PCP, a configuration corresponds to the part of the longer solution string that remains to be matched, that is, the suffix of the longer solution string beyond the prefix that matches the shorter solution string. For example, in PCP (1), one might begin trying to solve it by first concatenating pairs 4, 1, 1, 4, 2, and 1. After doing this the top solution string is 10001011000101, the bottom solution string is 1000101100, and the configuration is such that the top string is longer and 0101 remains to be matched.

Since we are searching depth-first, at any given configuration we will determine either that a solution is possible from here, no solution is possible, or that up to our depth limit neither is known. If a solution is found, certainly no further searching need be done. On the other hand, for the other two cases, if in later branches of the search we come upon this same configuration (e.g., 0101 is again the extra part of the longer top string) this is referred to as a move *transposition* [15] and if this configuration appears at a depth equal to or deeper than where it was first found there is no need to search again, as we already know it will not find a solution. There is actually one other case called a move *cycle* [15]. Interesting, for the problems examined in [15] move cycles were either trivially eliminated or simply did not pay off whereas for the PCP detecting cycles turns out to be extremely useful.

A cycle occurs when a configuration reappears further down the same search branch. It means that we have entered a loop, that is, if a solution is found from this point a shorter solution must exist further up the tree, where the first occurrence of the configuration appeared. This has the potential of pruning huge portions of the tree. In most games this kind of loop has a different meaning. For example, in go it is referred to as *super-ko* and a move that would create such a position is considered illegal. In chess it would correspond to a repeated position which, should it occur yet again, would indicate that the game is drawn.

Before discussing implementation details, let us describe a couple of specific examples. As mentioned above, PCP (1) has a solution of length 76. It turns out that in solving this problem, 3,803 nodes are visited. However, if we use hashing we find that we get a total of 9 cutoffs and only 3,600 nodes need to be visited. It should be pointed out that since a solution was found the number of nodes visited depends on the actual order the nodes are visited (recall that searching ceases as soon as a solution is found) so a truer measure of the cutoffs due to hashing might be obtained by seeing how many nodes are visited when proving that there is no solution of length 75 or less. In this case 10,308 nodes are visited without hashing and 8,573 nodes are visited with hashing with 19 cutoffs. Again, the number of nodes and cutoffs with hashing depends on the order the nodes are visited, but without hashing the number of nodes will not change. It might be prudent to try to order the nodes at each level based on some heuristic(s) to try to increase the number of cutoffs (as is done to improve alpha-beta cutoffs in game look-ahead trees) but we have chosen not to do this because the most useful cutoffs seem to be move cycles, while move transpositions do not seem to occur so often.

An extreme example of the power of hashing is demonstrated in the following:

$$\begin{pmatrix} 101 & 0 & 10 & 11 \\ 1 & 01 & 110 & 10 \end{pmatrix} \tag{8}$$

This PCP has a solution of length 84, and to solve it with hashing 927 nodes are visited but without hashing it requires more than 20 billion nodes. Yet there are only 3 cutoffs. This is possible because the cutoffs occur at low levels of the tree (depths 4, 6, and 7) and the branches pruned are on the same branch, effectively pruning the bulk of the tree.

Hashing was implemented two different ways. The first method tries to store every configuration in a tree structure. There are actually two trees corresponding to whether the top or the bottom string is longer. A configuration is stored in the tree as a path starting from the root and moving either down the '0' edge or the '1' edge according to what the string values are. Though this representation has the feature of being general and allowing arbitrarily large configurations, it was undesirable in practice. The major problem was that it took too long to traverse the trees looking for matches. Instances with few or no cutoffs run at least 10 times slower using hashing than without. This is unacceptably slow. Also, many PCP's create so many configurations that memory ended up becoming an issue and so the length of the configurations had to be limited, defeating one of the major features of this approach.

The more successful implementation used two arrays, again corresponding to whether the top or bottom string was longer. Each array is a hash table of boolean values indexed by the configuration string, where the configuration string is hashed to an integer value by directly converting it to an integer. The values in the hash table indicate whether this configuration has appeared or not and if so, how deep in the tree. Since arrays cannot be arbitrarily large, this again puts a limit on the number of configurations that can be stored, but since longer configurations are less likely to be matched the penalty does not seem to be too great. Also, we could choose to use a more sophisticated hash function, allowing longer strings to hash into the table, but in general this does not seem to be worth the extra execution time.

This approach to hashing also lent itself to a more efficient hashing algorithm. Rather than recalculate the hash from scratch for each configuration, we are able to use the previous configuration and simply change the hash based on the bits that are removed from one end and added to the other end. The result is that the penalty for hashing is about a 5% decrease in running speed verses no hashing at all in the cases where no matches are found. We feel this is a reasonable price to pay given the fact that huge improvements are possible if matches are found.

We did one other optimization. Having a purely recursive search routine provides the simplest and clearest code, however, because of limitations on the compiler we were using we found that we occasionally ran out of stack space when trying to solve some of the more difficult PCP instances. So, in an effort to deal with this and to squeeze a little more efficiency out of the program we simulated a recursive call to the searching routine with a *goto* whenever there was exactly one possible pair available to continue the search, or whenever we were at the last of multiple options. Since our recursive calls appear at the end of the search routine we don't have to worry about returning any information from the recursive calls except the return value, and this value must be the same for the calling and the called searches, so this technique is indeed safe. We found that on average our maximum stack size would be about half of what it was without the *goto* so we no longer run out of stack space for even the most difficult problems we have worked on. Also, the program does run marginally faster, but not enough to make the technique worthwhile if this were its only benefit.

3   Solving the PCP for Small Sizes and Widths

Our main interest is in creating interesting instances of the PCP. One way to do this for very small sizes and widths is to try to find the instance with the longest solution among all instances for some fixed size and width. Our techniques almost allow us to do this with all sizes and widths up to 3. The one exception is the size 3, width 3 case where we believe we know the answer, but it is not yet proven. Let us look at the other cases first.

The case where the size is 1 is trivial since there is a solution if and only if both strings of the pair are the same. The case where the width is 1 is similarly trivial. Where the size and width are both 2 there are very few distinct PCP's that satisfy the filters so it is easy to analytically determine that PCP (9) below, which obviously has a solution of length 2, is the best. Naturally the program discovers this too.

$$
\begin{pmatrix} 10 & 1 \\ 1 & 01 \end{pmatrix} \tag{9}
$$

The size 2, width 3 case is a little more interesting. There are now enough different PCP's that it is difficult to manually determine the longest one. The program, however, again quickly finds it, but there is a complication. There are 36 instances the program can't resolve, so we must deal with those manually. I will explain more about these instances the program is unable to resolve when we discuss the more difficult case where the size and width are both 3. PCP (10), with solution length 4, has the longest solution among the size 2, width 3 PCP's.

$$
\begin{pmatrix} 100 & 0 \\ 0 & 001 \end{pmatrix}
\tag{10}
$$

The size 3, width 2 case is also easy for the program to deal with, leaving us with only 15 unresolved instances that are all easy to deal with manually. PCP (11) has a solution of length 5, longest among the size 3, width 2 PCP's.

$$
\begin{pmatrix} 11 & 0 & 00 \\ 1 & 00 & 11 \end{pmatrix}
\tag{11}
$$

The size 3, width 3 situation is considerably more difficult than the previous cases. We calculate the total number of possible PCP's for this case by noting that there are 14 different strings possible in each of the six positions, for a total of $14^6 = 7,529,536$ different instances. The filters, however, cut this number down considerably. The length balance filter immediately reduces it to 307,328 PCP's. The element balance filter further reduces it to 99,814. Finally, applying the prefix and postfix filters leaves us with 42,388 PCP's. When we apply the solver to these instances, searching for solutions up to length 10, we find that 15,099 do have solutions, 11,396 do not have solutions, and the remaining 15,893 cannot be resolved by the solver. The unresolved instances correspond to cases where either there is a solution but it is longer than 10, or there is no solution. In the case where there is no solution it might be that all branches terminate, but at a depth greater than 10, or it might be that some branches simply continue on indefinitely, never forming a solution. By increasing our search depth from 10 to 20 we don't gain much more information: 15,115 have solutions, 11,480 do not, and 15,793 are still unresolved. Continued increases in the depth reduces the number of unresolved instances by just a few, indicating that most of the unresolved instances are probably due to nonterminating branches. Also, time constraints prevent us from running the solver on all instances for values much higher than 20. We must find other ways to deal with the unresolved instances.

So far we have been ignoring the considerable symmetry that exists among PCP instances. Actually, we have been implicitly removing some symmetries already, because the length balance filter assumes both that the first pair must have a longer string on top and that longer string must begin with a 1. Also, it assumes that the second pair must have a shorter string on top. The numbers reported above already take these symmetries into account, and help account for what might otherwise seem like surprisingly small values.

But there are other symmetries. We may exchange 0's and 1's to form an equivalent PCP. We may reverse all the strings in an instance. And we may rearrange the pairs into any order and still have an equivalent PCP. By checking for all the symmetries and searching to depth 20 we find that there are 8,225 unique PCP's that survive the filters where 2,989 have solutions, 2,361 do not and 2,875 are unresolved.

This is considerable progress, and demonstrates that our techniques are producing good results. But the number of unresolved instances is still too large to deal with manually and so we are still unable to completely solve the size 3, width 3 case.

Included in this list of unresolved instances is:

$$\begin{pmatrix} 100 & 0 & 1 \\ 1 & 100 & 0 \end{pmatrix} \tag{12}$$

PCP (12) has solution length 75 and we conjecture it is the maximum among all the size 3, width 3 PCP's. One reason we believe this is that when generating good random PCP's (as discussed in the next section) this instance comes up often and no better one has appeared. Perhaps even more compelling, however, is that when examining the unresolved instances we find that almost all of them have one of two forms that makes it unlikely they have solutions at all.

The first of these includes situations where a certain pair is made up entirely of a single character. Consider the following example from the list of unresolved instances:

$$\begin{pmatrix} 11 & 01 & 011 \\ 1 & 110 & 0 \end{pmatrix} \tag{13}$$

By simply enumerating cases it is not hard to see that PCP (13) does not have a solution. In particular, we notice that if a solution begins with 3 or more occurrences of pair 1 it will be impossible to ever use either of the other two pairs, so no solution can begin in this way. Given this observation, there are just a couple of other cases that are easily dismissed. However, this is difficult for the program to prove.

The other case includes situations like PCP (14), again from the unresolved instances list:

$$\begin{pmatrix} 11 & 01 & 101 \\ 0 & 011 & 1 \end{pmatrix} \tag{14}$$

In this case it is again not too difficult for a human to determine that there is no solution. But we have a situation such that when the top solution string is longer than the bottom one (which can occur immediately if we begin with pair number 3) we are always able to either continue with pair 1 or pair 3 since one of those will be able to match its single character with whatever remains in the top string. Of course doing this will simply continue to make the top string longer, never leading to a solution, but again, this is difficult for the program to detect. So, to be able to prove that PCP (12) is the best we will need to find ways to automatically prove that instances like PCP (13) and (14) do not have solutions so that the collection of unresolved instances can be made sufficiently small.

It is interesting to note that when studying the list of unresolved instances, the vast majority quickly lend themselves to simple proofs that there is no solution. In particular, instances that have a pair made up entirely of a single character fall especially easily, and usually along the lines of the argument used on PCP (13). So, in an effort to concentrate on instances that are more likely to yield long solutions, we eliminated from the unresolved list any instance that had a pair made up of a single character. This left us with a list of only 99 PCP's. From this list four were found to have solutions (including PCP (12), of course) and most of the rest again fell reasonably easily to analysis. A few, however, had very bushy search trees and so we were not able to prove they have no solution, though we strongly suspect this is the case. For example, we suspect the following PCP's have no solutions, but we are currently unable to confirm this.

$$\begin{pmatrix} 100 & 0 & 10 \\ 0 & 100 & 100 \end{pmatrix} \quad \begin{pmatrix} 101 & 1 & 01 \\ 1 & 101 & 101 \end{pmatrix} \quad \begin{pmatrix} 100 & 1 & 00 \\ 0 & 100 & 100 \end{pmatrix} \tag{15}$$

4   Generating PCP's

Having an efficient solver makes it at least feasible to construct a program to generate PCP's. Certainly generating PCP's is a simple matter. However recognizing interesting instances is another matter that depends more on modifying the solver than on improvements in generating.

After generating an instance we need to be able to quickly determine if it is unlikely to have a solution and if it does have one if it is likely to be a long solution. We pass all generated PCP's through the filters described above, of course, but after that we do not want to spend too much time on each instance since we want to look at as many instances as we can. We must balance the evaluation so that it does not waste too much time trying to solve PCP's that do not have solutions but also doesn't readily eliminate PCP's that do have interesting solutions.

The first step is being willing to give up on a search for a solution if we do not seem to be finding one. We do this by counting the number of nodes visited and terminating the search after a certain number have been visited. This number is selected as a multiple of the maximum depth of the search since the deeper we are looking the more nodes we should be willing to examine. The multiple is chosen large enough so that for shallow searches this number is likely to include all nodes, so we can be reasonably effective in detecting instances with short solutions. But since the search tree grows exponentially with depth while the number of nodes examined grows linearly, the deeper the search, the ratio of nodes visited decreases and so the less likely we are to find a solution if there is one. We attempt to counter this problem by doing a kind of iterative deepening. The program attempts to solve the instance for a sequence of depths, usually starting at 10 and progressing by increments of 10 up to the length where we are hoping to find solutions. Increments of size 10 seem to be just about right for the kinds of problems we are looking at. Smaller increment sizes would not give enough new information for the time it would have to spend; the new search is too similar to the search that was just completed. We would be better off increasing the multiplier instead and thus searching more completely at the previous level. On the other hand, increasing the increment would give too coarse of a search and would make it harder to find solutions, especially those of intermediate length. With the current values, if there is a very short solution it is likely to be found immediately, if there is a relatively short solution the repeated searches to deeper levels are likely to find it, and since the number of nodes increases as the depth increases we still frequently find reasonably long solutions, too.

There is a delicate balance between time spent and useful information being obtained. If we spend too much time with shallow searches we will never examine deeper ones and so we will never find those instances that we are looking for – the ones with long solutions. However, if we move too quickly to deep searches we will incorrectly report a long solution when there might also be quite short solutions that were skipped due to our haste to move on to search for longer ones. Equally bad is spending a lot of time on instances that have no solution at all since the goal is to find PCP's with solutions. It seems that the scheme described above balances these factors quite well.

There is still one problem with this approach. The solver always searches the tree in the same order, so if there is a solution just out of reach of our node count we will never find it. To deal with this we have the solver traverse subtrees in random order. So, for example, if a solution is missed at a level 10 search because it is beyond the node count, that same solution might be found at the level 20 search since the nodes are visited in a different order. For higher depths we also search more than once with the same node count cutoff. Since the searches are random we examine many more paths this way. This increases our chances of finding solutions and has proven very effective. Of course interesting solutions can still slip by and certainly many have. But by using this technique we have found many interesting PCP's. In fact, every PCP in this paper, except for those cited as being from other authors, were found using this algorithm.

A few other heuristics are used with mixed success. One has to do with the format of instances. Besides needing to pass through the filters, we find that most interesting PCP's do not have strings of length 3 or greater composed entirely of the same symbol. So when generating instances we discourage this. As with any heuristic, care must be taken as one of the most interesting PCP's below contains both the strings 000 and 111.

Also, we find that most interesting PCP's do not contain solutions where the same pair is repeated more than 4 or 5 times in succession. So, when attempting to solve instances and choosing one of the possible paths to take, even though the paths are chosen randomly, we bias the randomness to discourage further repetitions of a pair if that pair appears too many times in succession. In particular, as we saw with PCP (2) and the discussion about PCP (13), if a pair like (000, 0) occurs in the instance this pair can be chosen an arbitrary number of times as the start of a possible solution, but it is unlikely that there will be a solution where a configuration has too many consecutive 0's.

Finally, configurations tend not to get too large, so we have added another feature to the solver that discourages configurations from getting too large. The discussion concerning PCP (14) explains one situation where this would be important. If there is more than one choice at a particular node, and the configuration is beyond some cutoff value (usually set around 20) then we encourage paths that will decrease the size of the configuration. After considerable testing, it is not all clear, however, that this heuristic really provides any benefits.

We now mention a few of the more interesting PCP's we have found using these programs. In terms of confirmed difficulty the best we have is:

$$\begin{pmatrix} 100 & 00 & 1 & 1 \\ 1 & 110 & 00 & 011 \end{pmatrix} \tag{16}$$

In this case we know there is a solution of length 160 (83,027 nodes visited with 13 hash cutoffs), but there is no solution of length 159 or less (77,624 nodes visited with 13 cutoffs).

On the other hand we have another instance:

$$\begin{pmatrix} 01 & 00 & 1 & 001 \\ 0 & 011 & 101 & 1 \end{pmatrix} \tag{17}$$

Here we have found a solution of length 160 by repeatedly checking random paths. In fact, every time we run the program that searches for solutions by choosing random paths it will find a solution of length 160 within seconds, typically after starting no more than 20 searches. On the other hand if we try using the same program to find a solution of length 159 or less it does not find one, even if run for hours with hundreds of thousands of different starts. Hence we strongly suspect that the length of PCP (17) is 160 but we are currently unable to verify this. We have, of course, tried to prove there is no solution of length 159 or less by running the solver over an extended period of time, but 10 billion nodes was not enough to prove there is no solution. Checking the number of nodes visited after trying to solve it at different depths we found that every increase in 10 in length of solution increases the number of nodes visited by a factor of at least 5.5. Noting that a length 20 search visits about 100 nodes, we conclude that evaluating the total length 159 tree would require examining at least 10^{12} nodes. This is 100 times more nodes than we have examined so far, and on our current hardware would require about a month of execution time.

Our most interesting PCP to date is this one:

$$\begin{pmatrix} 000 & 0 & 11 & 10 \\ 0 & 111 & 0 & 100 \end{pmatrix} \tag{18}$$

The random solver will always eventually find a solution of length 204, but not always quickly – sometimes requiring thousands of starts. So far it has not been able to find a solution of length 203 or less despite running hundreds of millions of starts. Also, an attempt to prove there is no solution of length 203 or less was abandoned after more than 200 billion nodes were examined. So, the empirical evidence strongly suggests the length of this PCP is 204, but it remains unproven for now. Also, doing an analysis as we did above for PCP (17), we found that every depth increase of 10 multiplies the number of nodes visited by at least 20 and the search to depth 20 visits about 1000 nodes so to prove that there is no solution of length 203 or less would require us to visit approximately 10^{30} nodes. Clearly, to prove this, some new techniques need to be developed.

Notice the aesthetic simplicity of this instance. Six of the strings are composed of a single character (contrary to one of our heuristics) and these six are paired up with each other while the other two strings are simple and similar. This instance is difficult to forget.

5   Conclusions, Open Problems, and Future Work

By using standard techniques and employing some simple heuristics we are able to generate interesting PCP instances. In particular, PCP (18) has a disarmingly simple looking structure, yet apparently a very difficult solution. This is in stark contrast to PCP's one finds in the textbooks where solutions tend to either be very short as is the case for PCP (3) or if not short then not very branching, as is the case of PCP (4). Even though PCP (4) has a solution of considerable length, 66, the number of nodes visited to determine this is exactly 66.

We have found a number of interesting PCP instances where the size, i.e., the number of pairs, is 3. If the maximum string length, the width, is also 3 we have PCP (12), which has a solution of length 75. We conjecture that this pretty PCP has the longest solution among all PCP's whose length and width is 3. However, to prove this new techniques need to be developed that can more readily detect PCP's that do not have solutions, especially those that satisfy the patterns described in Section 3.

If we allow a width of 4 we have found the following instance that has solution length 119.

$$\begin{pmatrix} 1 & 10 & 1101 \\ 0 & 1011 & 1 \end{pmatrix} \qquad (19)$$

We also conjecture that this PCP has the longest solution among the size 3, width 4 instances, but with less confidence. We could increase our confidence if we were to perform a systematic attack similar to the one we performed on the size 3, width 3 case, but techniques that cull more unresolved instances need to be developed first.

Finally, for the size 4, width 3 case we have PCP (18) that we conjecture has a solution length of 204. We are not, however, ready to conjecture that it is the most difficult instance.

We do suspect, though, that for size 2, the most difficult instances have length two less than double the width and these maximal instances all have the same form. For example, for width 6 and size 2 the following PCP has solution length 10.

$$\begin{pmatrix} 100000 & 0 \\ 0 & 000001 \end{pmatrix} \qquad (20)$$

For larger widths, the pattern is simply to extend the number of 0's in the two longer strings. For small widths it is easy to prove that these are indeed the most difficult

PCP's. If this can be proven in general, it would provide a simple proof that the PCP problem with size 2 is decidable. Compare with the difficult proof of [3].

Besides hoping to eventually prove that PCP (18) has a solution of length 204, we would like to try to find PCP's of this same size that have longer solutions and to eventually find larger PCP's with much longer lengths.

References

1. Culberson, J.: Sokoban is PSPACE-complete. In: Int. Conf. Fun with Algorithms, Elba, June 1998
2. Denning, P.J., Dennis, J.B., and Qualitz, J.E.: Machines, Languages and Computation. Prentice-Hall Inc, 1978
3. Ehrenfeucht, A, Karhumaki, J., and Rozenberg, G.: The (generalized) post correspondence problem with lists consisting of two words is decidable. Theoret. Comput. Sci., 21, 2, 119-144, 1982
4. Fraenkel, A.S., Garey, M. R., Johnson, D. S., Schäfer, T., and Yesha, Y.: The complexity of checkers on an N × N board – preliminary report. In: Proc. 19th IEEE Symp. On the Foundations of Computer Science, pp. 55–64, 1978
5. Gurari, E.: An Introduction to the Theory of Computation. Computer Science Press, 1989
6. Hopcroft, J.E., and Ullman, J.D.: Introduction to Automata Theory, Languages and Computation. Addison-Wesley Publ., 1979
7. Junghanns, A. and Schaeffer, J: Sokoban: Improving the Search with Relevance Cuts. To appear in: Journal of Theoretical Computing Science, 1999
8. Lichtenstein, D., and Sipser, M.: Go is polynomial-space hard. J.ACM 23, pp. 710–719, 1976
9. Manna, Zohar: Mathematical Theory of Computation. McGraw Hill Inc, 1974
10. Marxen, H., and Buntrock, J.: Attacking the busy beaver. Bul. EATCS, 40, 247–251, 1990
11. Marxen, H, Buntrock, J, and Thompson, C: http://www.drb.insel.de/~heiner/BB/index.html
12. Murase, Y., Matsubaara, H., and Hiraga, Y.: Automatic Making of Sokoban Problems. In: Fourth Pacific Rim International Conference in Artificial Intelligence, Cairns, Australia, Aug 26 – 30, 1996
13. Post, E.L.,: A variant of a recursively unsolvable problem. Bull. of the Am. Math. Soc., 52, 264–268, 1946
14. Rado, T.: On non-computable functions. Bell Sys. Tech. J., 41, 3, 877–884, 1962
15. Reinefeld, A, and Marsland, T.A.: Enhanced Iterative-Deepening Search. In: IEEE Trans. Pattern Anal. Machine Intell., Vol. 16, No. 7, 701–710, July, 1994
16. Taylor, L, and Korf, R.: Pruning Duplicate Nodes in Depth-First Search. In: Proc. 11th National Conf. on AI, AAAI-93, 756–761, July 11 – 15, 1993
17. Wolf, T.: Generating tsume go problems with GoTools. In: Proceedings of the Fourth International Computer Olympidade, London, 1992

Integer Programming Based Algorithms
for Peg Solitaire Problems

Masashi Kiyomi and Tomomi Matsui

Department of Mathematical Engineering and Information Physics,
Graduate School of Engineering, University of Tokyo,
7-3-1, Hongo, Bunkyo-ku, Tokyo 113-8656, Japan
masashi,tomomi@misojiro.t.u-tokyo.ac.jp
Tel: +81-3-5841-6921

Abstract. Peg solitaire is a one player game using pegs and a board with some
holes. The game is classical, and nowadays sold in many parts of the world under
the trade name of Hi-Q.

In this paper, we dealt with the peg solitaire problem as an integer programming
problem. We proposed algorithms based on the backtrack search method and
relaxation methods for integer programming problem.

The algorithms first solve relaxed problems and get an upper bound of the number
of jumps for each jump position. This upper bound saves much time at the next
stage of backtrack searching. While solving the relaxed problems, we can prove
many peg solitaire problems are infeasible. We proposed two types of backtrack
searching, forward-only searching and forward-backward searching. The performance of these two methods highly depends on the symmetricity and the length
of the sequence of required jumps. Our algorithm can solve all the peg solitaire
problem instances we tried and the total computational time is less than 20 minutes
on an ordinary notebook personal computer.

Keywords: peg solitaire, integer programming, backtrack searching

1   Introduction

Peg solitaire is a one player game using pegs and a board with some holes. In each
configuration, each hole contains at most one peg (see Figures 1, 2). A peg solitaire game
has two special configurations, the starting configuration and the finishing configuration.
The aim of this game is to get the finishing configuration from the starting configuration
by moving and removing pegs as follows.

> When there are (vertically or horizontally) consecutive three holes satisfying
> that first and second holes contain pegs and third hole is empty, the player can
> remove two pegs from the consecutive holes and place a peg in the empty hole
> (see Figure 3).

The move obeying the above rule is called a *jump*.

In this paper, we propose an algorithm for peg solitaire games based on integer programming problems. In Sections 2, 3, and 4, we consider a peg solitaire problem defined

T.A. Marsland and I. Frank (Eds.): CG 2000, LNCS 2063, pp. 229–240, 2001.

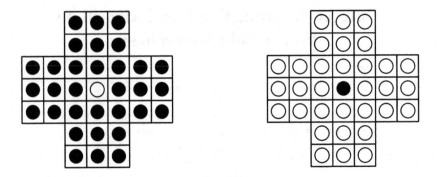

Fig. 1. Starting configuration example   **Fig. 2.** Finishing configuration example

● implies a hole with a peg, and ○ implies a hole with no peg.

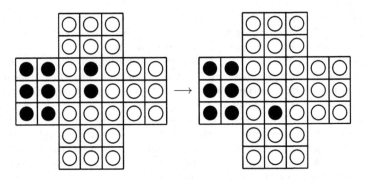

Fig. 3. An example of a jump

below, which is a natural extension of the ordinary peg solitaire game. In Section 2, we formulate the peg solitaire problem as an integer programming problem. In Section 3, we show a relation between the pagoda function defined by Berlekamp, Conway and Guy [3] and our integer programming problem. Section 4 proposes an integer programming problem which is useful for pruning the backtrack search described in Section 6. We report computational results of our algorithm in Section 5.

There are many types of peg solitaire boards and various pairs of starting and finishing configurations. The most famous peg solitaire board is that of Figure 1, which is called the English board.

A peg solitaire problem is defined by a peg solitaire board and a pair of starting and finishing configurations. If there exists a sequence of jumps which transforms the starting configuration into the finishing configuration, we say that the given peg solitaire problem is *feasible*, and the sequence of jumps is a *feasible sequence*. The peg solitaire problem finds a feasible sequence of jumps if it is feasible; and answers "infeasible", if the problem is not feasible.

In [8], Uehara and Iwata dealt with the generalized Hi-Q problems which are equivalent to the above peg solitaire problems and showed the NP-completeness. In the well-known book "Winning ways for Mathematical Plays [3]", Berlekamp, Conway and Guy discussed variations of problems related to peg solitaire problems. They showed the infeasibility of the peg solitaire problem "sending scout 5 paces out into desert" by using the pagoda function approach. In [7], Kanno proposed a linear programming based algorithm for finding a pagoda function which guarantees the infeasibility of a given peg solitaire problem, if it exists. Recently, Avis and Deza [1] formulated a peg solitaire problem as a combinatorial optimization problem and discussed the properties of the feasible region called "a solitaire cone".

2   Integer Programming

In this section, we formulate the peg solitaire problem as an integer programming problem.

We assume that all the holes on a given board are indexed by integer numbers $\{1, 2, \ldots, n\}$. The board of Figure 1 has 33 holes and so $n = 33$. We describe a state of certain configuration (pegs in the holes) by the n-dimensional 0-1 vector p satisfying that the ith element of p is 1 if and only if the hole i contains a peg. In the rest of this paper, we denote the starting configuration by p_s and the finishing configuration by p_f.

Let J be the family of all the sequences of consecutive three holes on a given board. Each element in J corresponds to a certain jump and so we can denote a jump by a unit vector indexed by J. In the rest of this paper, we assume that all the elements in J are indexed by $\{1, 2, \ldots, m\}$. For example, the board of Figure 1 contains 76 sequences of consecutive three holes and so $m = 76$. Given a peg solitaire board, we define $n \times m$ matrix $A = (a_{ij})$, whose rows and columns are indexed by holes and jumps respectively, by

$$a_{ij} = \begin{array}{l} 1 \text{ (a peg on the hole } i \text{ is removed by the jump } j), \\ -1 \text{ (a peg is placed on the hole } i \text{ by the jump } j), \\ 0 \text{ (otherwise).} \end{array}$$

For any 0-1 vector p, $\#p$ denotes the number of 1s in the elements of p. We denote $\#p_s - \#p_f$ by l. If the given peg solitaire problem is feasible, any feasible sequence consists of l jumps. For example, the peg solitaire problem defined by Figures 1 and 2 is feasible and there exists a feasible sequence whose length $l = 32$. Since each jump corresponds to an m-dimensional unit vector, a feasible sequence corresponds to a sequence of l unit vectors $(x^1, x^2, \ldots, x^l)$ such that $x^k = (x_1^k, x_2^k, \ldots, x_m^k)^\top$ for all $k \in \{1, 2, \ldots, l\}$ and

$$x_j^k = \begin{array}{l} 1 \text{ (the } k\text{th move is the jump } j), \\ 0 \text{ (the } k\text{th move is not the jump } j). \end{array}$$

If a configuration p' is obtained by applying the jump j to a configuration p, then $p' = p - Au$ where u is the jth unit vector in $\{0, 1\}^m$. From the above discussion, we can formulate the peg solitaire problem as the following integer programming problem;

IP1:   find $(\boldsymbol{x}^1, \boldsymbol{x}^2, \ldots, \boldsymbol{x}^l)$

s. t. $A(\boldsymbol{x}^1 + \boldsymbol{x}^2 + \cdots + \boldsymbol{x}^l) = \boldsymbol{p}_s - \boldsymbol{p}_f,$

$0 \leq \boldsymbol{p}_s - A(\boldsymbol{x}^1 + \boldsymbol{x}^2 + \cdots + \boldsymbol{x}^k) \leq 1 \;\; (\forall k \in \{1, 2, \ldots, l\}),$

$x_1^k + x_2^k + \cdots + x_m^k = 1 \qquad\qquad (\forall k \in \{1, 2, \ldots, l\}),$

$\boldsymbol{x}^k \in \{0, 1\}^m \qquad\qquad\qquad\quad (\forall k \in \{1, 2, \ldots, l\}).$

The problem IP1 has a solution if and only if the given peg solitaire problem is feasible. Clearly, any solution $(\boldsymbol{x}^1, \ldots, \boldsymbol{x}^l)$ of IP1 corresponds to a feasible sequence of jumps.

If we formulate the peg solitaire problem defined by Figures 1 and 2, then the number of variables is $m \times l = 76 \times 32 = 2,432$, the number of equality constraints is $n + l = 32 + 33 = 65$, and the number of inequality constraints is $2 \times n \times l = 2 \times 33 \times 32 = 2,112$. Thus, the size of the integer programming problem is huge and so it is hard to solve the problem by commercial integer programming software.

In Section 5, we propose an algorithm for peg solitaire problem which is a combination of backtrack search and pruning technique based on the above integer programming problem.

3   Linear Relaxation and Pagoda Function

In [3], Berlekamp, Conway and Guy proposed the pagoda function approach for showing the infeasibility of some peg solitaire problems including the well-known problem "sending scout 5 paces out into desert". In this paper, we show that the pagoda function approach is equivalent to the relaxation approach for the integer programming problem.

A real valued function pag : $\{1, 2, \ldots, n\} \to R$ defined on the set of holes is called a *pagoda function* when $\mathrm{pag}(\cdot)$ satisfies the properties that for every (vertically or horizontally) consecutive three holes (i_1, i_2, i_3), the pagoda function values $\{\mathrm{pag}(i_1), \mathrm{pag}(i_2), \mathrm{pag}(i_3)\}$ satisfies $\mathrm{pag}(i_1) + \mathrm{pag}(i_2) \geq \mathrm{pag}(i_3)$. (Clearly, the sequence (i_3, i_2, i_1) is also a consecutive three holes, and so the inequality $\mathrm{pag}(i_3) + \mathrm{pag}(i_2) \geq \mathrm{pag}(i_1)$ also holds.) A pagoda function corresponds to an assignment of real values to holes on the board satisfying the above properties. Figure 4 is an example of pagoda function defined on English board.

For any configuration $\boldsymbol{p} \in \{0, 1\}^n$, we denote the sum total $\sum_{i=1}^{n} \mathrm{pag}(i) \times p_i$ by $\mathrm{pag}(\boldsymbol{p})$. The definition of the pagoda functions implies that if a configuration $\boldsymbol{p}'$ is obtained by applying a jump to a configuration $\boldsymbol{p}$, then $\mathrm{pag}(\boldsymbol{p}) \geq \mathrm{pag}(\boldsymbol{p}')$. Thus, if a given peg solitaire problem is feasible, then the inequality $\mathrm{pag}(\boldsymbol{p}_s) \geq \mathrm{pag}(\boldsymbol{p}_f)$ holds for any pagoda function $\mathrm{pag}(\cdot)$. So, the existence of a pagoda function $\mathrm{pag}(\cdot)$ satisfying $\mathrm{pag}(\boldsymbol{p}_s) < \mathrm{pag}(\boldsymbol{p}_f)$ shows that the given peg solitaire problem is infeasible.

In [7], Kanno showed that there exists a pagoda function which guarantees the infeasibility of the given peg solitaire problem if and only if the optimal value of the following linear programming problem is negative.

PAG-D: min. $(\boldsymbol{p}_s - \boldsymbol{p}_f)^\top \boldsymbol{y}$

s. t. $A^\top \boldsymbol{y} \geq \boldsymbol{0}.$

It is easy to see that for any feasible solution $\boldsymbol{y}$ of PAG-D, the function $\mathrm{pag}(\cdot)$ defined by $\mathrm{pag}(i) = y_i$ for all $i \in \{1, 2, \ldots, n\}$ is a pagoda function. Thus, it is clear that if the

		-0.3	0.4	0		
		1.0	0	1.0		
0.5	0	0.5	0.4	0.1	0.3	-0.1
0	0.9	0.7	0.3	0.9	1.1	0.4
0.5	0.6	0.1	0.5	0.2	0.6	0.2
		0.8	0	0.8		
		0	0.5	-0.2		

Fig. 4. An example of assignment of Pagoda functions

optimal value of PAG-D is negative, then the given peg solitaire problem is infeasible. Unfortunately, the inverse implication does not hold; that is, there exists an infeasible peg solitaire problem instance such that the optimal value of the corresponding linear programming problem (PAG-D) is equal to 0 (see Kanno [7] for example).

The dual of the above linear programming problem is

$$\max\{\mathbf{0}^\top \boldsymbol{x} \mid A\boldsymbol{x} = \boldsymbol{p}_s - \boldsymbol{p}_f, \; \boldsymbol{x} \geq \mathbf{0}\}.$$

Since the objective function $\mathbf{0}^\top \boldsymbol{x}$ is always 0, the above problem is equivalent to the following problem;

PAG-P: find $\boldsymbol{x}$

s. t. $A\boldsymbol{x} = \boldsymbol{p}_s - \boldsymbol{p}_f, \quad \boldsymbol{x} \geq \mathbf{0}.$

Thus, there exists a pagoda function which shows the infeasibility of the given peg solitaire problem if and only if the above linear inequality system PAG-P is infeasible.

In the rest of this section, we show that the problem PAG-P is obtained by relaxing the integer programming problem IP1. First, we introduce a new variable $\boldsymbol{x}$ satisfying $\boldsymbol{x} = \boldsymbol{x}^1 + \cdots + \boldsymbol{x}^l$. Next, we relax the first constraint in IP1 by $A\boldsymbol{x} = \boldsymbol{p}_s - \boldsymbol{p}_f$, remove second and third constraints, and relax the last 0-1 constraints of original variables by non-negativity constraints of artificial variables $\boldsymbol{x}$. Then the problem IP1 is transformed into PAG-P.

We applied pagoda function approach to problems defined on English board and found many infeasible problem instances which do not have any pagoda function showing the infeasibility. Although, the pagoda function approach was a powerful tool for proving the infeasibility of the problem "sending scout 5 paces out into desert", it is not so useful for peg solitaire problems defined on English board.

4   Upper Bound of the Number of Jumps

In this section, we propose a method for finding an upper bound of the number of jumps for each (fixed) jump j contained in a feasible sequence. And additionally, this method is proved to be a very strong tool to check the feasibilities of the given problems. In the next section, We propose a pruning technique for backtrack search using the upper bound described below.

We consider the following integer programming problem for each jump j;

$$
\begin{aligned}
\text{UB}j: \quad \text{max. } & x_j^1 + x_j^2 + \cdots + x_j^l \\
\text{s. t. } & A(x^1 + x^2 + \cdots + x^l) = p_s - p_f, \\
& 0 \le p_s - A(x^1 + x^2 + \cdots + x^k) \le 1 \quad (\forall k \in \{1, 2, \ldots, l\}), \\
& x_1^k + x_2^k + \cdots + x_m^k = 1 \qquad\qquad (\forall k \in \{1, 2, \ldots, l\}), \\
& x^k \in \{0, 1\}^m \qquad\qquad\qquad\quad (\forall k \in \{1, 2, \ldots, l\}).
\end{aligned}
$$

Since the set of constraints of UBj is equivalent to that of IP1, the given peg solitaire problem is feasible, if and only if UBj has an optimal solution. We denote the optimal value of UBj by z_j^*, if it exists. It is clear that any feasible sequence of the given problem contains the jump j at most z_j^* times. However, the size of the above problem is equivalent to the original problem IP1 and so, it is hard to solve. In the following, we relax the above problem to a well-solvable problem.

To decrease the number of variables, we consider only the pair of starting and finishing configurations and ignore intermediate configurations. The above relaxation corresponds to the replacement of the variables $x^1 + \cdots + x^l$ by x. We decrease the number of constraints by dropping second and third constraints of UBj. Then we have the following relaxed problem of UBj;

$$
\begin{aligned}
\text{RUB}j: \quad \text{max. } & x_j \\
\text{s. t. } & Ax = p_s - p_f, \\
& x_1, x_2, \ldots, x_m \text{ are non-negative integers.}
\end{aligned}
$$

If a problem RUBj is infeasible for an index $j \in \{1, 2, \ldots, m\}$, the original peg solitaire problem is infeasible and the problem RUBj is also infeasible for each index j. Since the above problem is a relaxed problem of UBj, the optimal value is an upper bound of the optimal value of UBj. If we deal with the problem defined by Figures 1 and 2, the problem RUBj has $m = 76$ integer variables and $n = 33$ equality constraints. The relaxed problems RUBj of the problem defined by Figure 7 are infeasible and so the original problem is also infeasible.

Figure 6 shows the optimal values of RUBj for each jump $j \in \{1, 2, \ldots, m\}$ of the peg solitaire problem defined by Figure 5. Since RUBj is a relaxation of UBj, feasibility of RUBj does not guarantee the feasibility of the given peg solitaire problem. For example, all the relaxed problems RUBj ($j \in \{1, 2, \ldots, m\}$) have optimal solutions as shown in Figure 6 and the given peg solitaire problem defined by Figure 5 is infeasible.

If the problem PAG-P is infeasible, then the problem RUBj is also infeasible for each $j \in \{1, 2, \ldots, m\}$. However, the inverse implication does not hold. For example, PAG-P as defined by Figure 7 is feasible, and RUBj is infeasible for each $j \in \{1, 2, ..., m\}$.

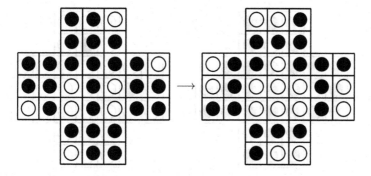

Fig. 5. An example of peg solitaire problem

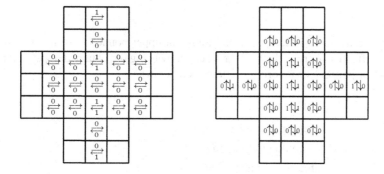

Fig. 6. Maximal numbers of jumps of Figure 5

Kanno [7] gave 10 infeasible peg solitaire problem instances such that the pagoda function approach failed to show the infeasibility. Our infeasibility check method can show infeasibility of all 10 examples given by Kanno. From the above, our infeasibility check method compares severely with the pagoda function approach.

Here we note an important comment related to the next section. See Figure 6, we can see that there are many jumps not to be done at all and not to be done twice and so on. So, we can prune many and many branches during the backtrack searching which we deal with in the next section.

Clearly, there are variations of relaxation problems of UBj. We have examined many relaxation problems and chose the above problem RUBj by considering the trade-off between the required computational efforts and the tightness of the obtained upper bound. However, the choice depends on the available softwares and hardwares for solving integer programming problems. In Section 6, we will describe the environment and results of our computational experiences in detail.

5 Backtrack Search

First, we propose forward-only backtrack search algorithm for solving peg solitaire problems. Before executing the backtrack search, we solved the problem RUBj for each

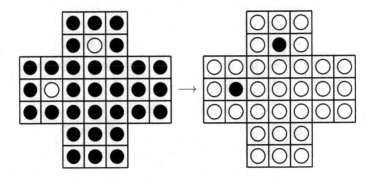

Fig. 7. A problem which can be proved to be infeasible

jump position $j \in \{1, 2, \ldots, m\}$ and found an upper bound of the number of jumps. We can effectively prune the backtrack search by using the obtained upper bound of the number of jumps. The algorithm is described below:

```
void
solve(configuration start,
        configuration end) {
   table_of_all_jumps  upper_bound;
   int              rest;

   for (each jump j) {
     solve RUBj;
     if (RUBj is infeasible) {
       print "This problem is infeasible." and exit;
     }
     set the optimal value of RUBj upper_bound[j];
   }

   rest =  the number of pegs of the configuration start
            − the number of pegs of the configuration end;

   if (search(start, end, upper_bound, rest) ≠ true) {
     print "This problem is infeasible." and exit;
   }
}
bool
```

```
search(configuration    start,
       configuration    end,
       table_of_all_jumps upper_bound,
       int              rest) {

    if (rest ≤ 0) {
      if (start = end)
        return true;
      else
        return false;
    }

for (all possible jumps) {
  if (upper_bound of the jump ≤ 0)
    continue;   /* It is no use searching about this jump. */

  upper_bound of the jump = upper_bound of the jump − 1;
  update the configuration start by applying the jump operation.

  if (search(start, end, upper_bound, rest − 1) = true) {

    display the configuration start;
    restore the configuration start by applying the reverse jump operation;
    return true;

  } else {
    upper_bound of the jump = upper_bound of the jump + 1;
    restore the configuration start by applying the reverse jump operation;
    }
  }

  return false;
}
```

To avoid searching the same configurations more than twice, we used a hash table with 2,097,169 entries for maintaining all the scanned configurations. If the backtrack search algorithm finds that the present configuration is contained in the hash table, we

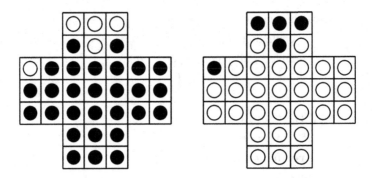

Fig. 8. An example problem to which hash is efficient

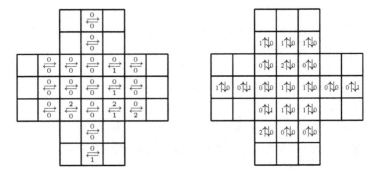

Fig. 9. Jump upper bounds of the problem defined by Fig. 8

can stop searching the present configuration. We used the hash table whose size is the maximum prime number currently available at our computer.

Here we point out the importance of the hash technique for solving peg solitaire problem. For example, if we use both IP and hash method, we can solve the problem defined by Figure 8 in 2 seconds on the notebook personal computer we will use in the next section. But if we use only IP upper bound or use only hash technique, the program doesn't stop in ten minutes.

Second, we propose forward-backward backtrack search algorithm. The second algorithm uses the properties of a peg solitaire problem that the number of jumps required to solve the problem is known and solving the problem in forward direction is essentially equivalent to solving the problem in backward direction. Here, solving problem in backward direction means that we start from the finishing configuration, repeat the 'reverse jump' operation and aim to get the starting configuration. Our second algorithm executes backtrack searching from the finishing configuration to the half depth of the search tree and maintains all the obtained configurations by the hash table. Next, the algorithm begins backtrack searching from the starting configuration to the half depth of the search tree. When an obtained configuration is in the hash table, the original solitaire problem is feasible. If all the scanned configurations are not in the hash table, the original problem is infeasible. The idea of the above second algorithm is very simple and

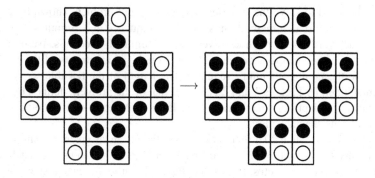

Fig. 10. A peg solitaire problem fit for backward and forward searching

effective for some problems but not all. When we applied the second algorithm, some problems require a very large hash table whose size is greater than 2,097,169.

6   Computational Results

In this section, we deal with peg solitaire problems defined on the English board. We used a notebook personal computer with MMX-Pentium 233MHz CPU, 64MB memory, and Linux OS. We solved the relaxed problem RUBj for each j, by the software *lp_solve 3.0*. This *lp_solve* version is released under the LGPL license. One can find the latest version of *lp_solve* at the following ftp site.

ftp://ftp.ics.ele.tue.nl/pub/lp_solve/

In the following, we discuss the problems RUBj ($j \in \{1, 2, \ldots, m\}$), which are called *relaxed problems* in the following. It took about 16 minutes to solve all the 76 relaxed problems defined by Figures 1 and 2. (Since the pair of starting and finishing configurations are symmetrical, we actually need to solve only 12 relaxed problems. However, our computer program does not use the information depending on the symmetricity.) We tried more than 20 peg solitaire problem instances, and the computational time required for solving 76 relaxed problems for each instance is less than 16 minutes. Here we note that there are some problems which took only 10 seconds to solve whole 76 relaxed problems.

We made a comparison of the forward-only backtrack searching and the forward-backward search method. The forward-only backtrack search method solves the peg solitaire problem defined by Figures 1 and 2 in 1 second. However, if we apply the forward-backward search method, the hash overflows after 3.5minutes. (Here we note that the program concludes that the hash has overflowed when the 80 percent of the hash is used). Since this problem is symmetric, the forward-only search method finds a feasible sequence easily. However, the symmetricity increases the upper-bound of the number of jumps and so the backward search generates many configurations and the hash overflows.

We also tried to solve the symmetrical peg solitaire problem shown by Figure 10. The forward-only search method finds a feasible sequence in 37 minutes. However, the forward-backward search method solves the problem in 1.4 seconds. We think that the performance of two methods depends not only on the symmetricity of the problem but also on the the number of jumps required. The length of the feasible sequence of the problem defined by Figures 1 and 2 is 31 and that of the problem defined by Figure 10 is 13. Since the depth of search tree of the latter problem is small, the hash does not overflow during the algorithm.

We solved more than 20 problems including the instances given in Kanno [7]. Either forward-only search method or forward-backward search method solves each instances we tried in at most 20 minutes. If we select faster algorithm for each instance we solved, the total computational time of our algorithm is less than 20 minutes.

References

1. Avis, D., and Deza, A.: Solitaire Cones, *Technical Report No. SOCS-96.8*, 1996.
2. Beasley, J. D.: Some notes on Solitaire, *Eureka*, 25 (1962), 13–18.
3. Berlekamp, E. R., Conway, J. H. , and Guy, R. K.: *Winning Ways for Mathematical Plays*. Academic Press, London, 1982.
4. de Bruijn, N. G.: A Solitaire Game and Its Relation to a Finite Field. *Journal of Recreational Mathematics*, 5 (1972), 133–137.
5. Cross, D. C.: Square Solitaire and variations. *Journal of Recreational Mathematics*, 1 (1968), 121–123.
6. Gardner, M.: *Scientific American*, 206 #6(June 1962), 156–166; 214 #2(Feb. 1966), 112–113; 214 #5(May 1966), 127.
7. Kanno, E.: Linear Programming Algorithm for Peg Solitaire Problems, Bachelor thesis, Department of Mathematical Engineering, Faculty of Engineering, University of Tokyo, 1997 (in Japanese).
8. Uehara, R., Iwata, S.: Generalized Hi-Q is NP-complete, *Trans. IEICE*, 73 (1990), 270-273.

Ladders Are PSPACE-Complete

Marcel Crâşmaru[1] and John Tromp[2]

[1] Department of Mathematical and Computing Science,
Tokyo Institute of Technology,
2-12-1 Oo-okayama, Meguro-ku, Tokyo, Japan, 152,
`marcel@is.titech.ac.jp`
[2] CWI
Kruislaan 413, 1098 SJ Amsterdam, The Netherlands
`tromp@cwi.nl`

Abstract. In the game of Go, the question of whether a ladder—a method of capturing stones—works, is shown to be PSPACE-complete. Our reduction closely follows that of Lichtenstein and Sipser [2], who first showed PSPACE-hardness of Go by letting the outcome of a game depend on the capture of a large group of stones. A greater simplicity is achieved by avoiding the need for pipes and crossovers.

1 Introduction

Consider the following Go[1] problem: Black to capture the marked white stone.

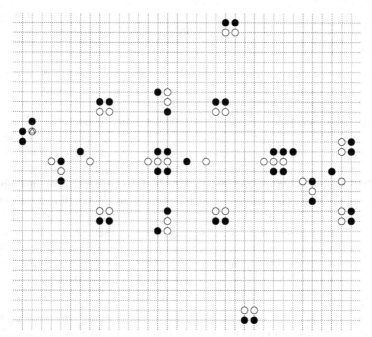

---

[1] details of the rules may be found at `http://www.cwi.nl/~{}tromp/go.html`.

T.A. Marsland and I. Frank (Eds.): CG 2000, LNCS 2063, pp. 241–249, 2001.
© Springer-Verlag Berlin Heidelberg 2001

We will show how this position encodes the Quantified Boolean Formula (**QBF**) $\forall x \exists y (x \vee y) \wedge (\neg x \vee \neg y)$. Since this formula is true, the ladder should work, as the reader may verify.

The problem of deciding the truth of **QBF** is complete for the class PSPACE of all problems that can be decided using an amount of space that is polynomial in the length of the input (see Theorem 7.10 of [1]). Completeness means that not only is **QBF** in the class PSPACE, but that every other problem in this class can be efficiently (in polynomial time) reduced to **QBF**, so that **QBF** is, essentially, a hardest problem in PSPACE. For many games, we can consider the problem of whether a position on an arbitrarily large board (say, n by n) is a win for the player to move. This can usually be determined by a recursive search, which uses space proportional to the product of board description size and the maximum length of the game. The former is certainly polynomial and the latter quite often is too. Those games are therefore in PSPACE, and showing completeness of such a game establishes that there is some intrinsic hardness to the game. E.g. there can be no 'shortcuts' by which the results of a position can be computed efficiently, in polynomial time, if we accept the widely held belief that the class P of polynomial time solvable problems is a strict subset of PSPACE.

Our main result is

Theorem 1 LADDERS *is PSPACE-complete.*

We formalize the game of Go and the ladder problem as follows:

GO: Given a position on an arbitrarily-sized Go board, does Black have a winning strategy?

LADDERS : Given a position on an arbitrarily-sized Go board, and a white group with 2 liberties, can Black keep putting white in atari—that is, reduce white to 1 liberty—until capture?

As shown by Lichtenstein and Sipser [2], one can construct positions in which black victory hinges upon the survival of a very large eyeless black group, that White has almost entirely surrounded. To survive, it needs to connect to a 2-eyed group through a structure of pipes and junctions that can be modeled after a Quantified Boolean Formula. This proved **GO** to be PSPACE hard. Robson [3] used the same idea but introduced a collection of *ko*'s into the structure, so that the large group could connect out only if its owner held an appropriate subset of all the *ko*'s. Such ko-games were shown to be EXPTIME-complete. But even though the owner of the large group might not be able to obtain an appropriate subset of ko's, he might be able to keep cycling through the ko's, so that the outcome of the game depends on the exact rule dealing with whole-board-repetition. Robson assumed a null-ruling[2], so he established EXPTIME-completeness of the question whether an arbitrary position is a forced win for Black or a null result.

Both constructions employ *pipes*, a pipe being a line of white stones sandwiched between 2 lines of black stones. Pipes are essential in containing the flow of play between the other gadgets (similar to the ones we will introduce) used in the constructions.

---

[2] With the superko rule that simply forbids the whole-board position from repeating, Black can win trivially In Robson's construction. We speculate that superko does not make Go harder than PSPACE, a topic of future research.

A disadvantage of pipes is that they take up space, and thus cannot simply cross on a Go board. Both Robson (directly), and Lichtenstein and Sipser (indirectly; at the conceptually higher level of graphs), constructed ingenious but somewhat complicated pipe-intersections. How much easier it would be to model play not as a flow to be contained but as light that travels unaided through empty space; bent by mirrors where need be.

2   Enter the Ladder

One of the first aspects of the game that beginners familiarize themselves with, the ladder (Figure 1) is a straightforward method of capturing stones by repeated atari on alternate sides. As shown in diagram L2, the ladder travels diagonally across the board and its fate will depend on what meets its path. A ladder will *work*, i.e. result in capture, if it either hits the edge of the board, or an existing solitary black stone, as in diagram W2. It will fail if it hits or borders on a solitary white stone, as in diagram F2. In that case White's move at 12 puts the black stone at 9 in atari, and if Black persists at 13, White captures her way to freedom. There are of course many more complicated situations where the ladder approaches both black and white stones in each others vicinity, or where these stones are short on liberties. There we cannot easily determine whether the ladder works. In fact we will exploit these possibilities in our own construction.

Ladders are forced sequences that can run all across the board, causing plays in one area of the board to affect other, remote areas. Ladders are also ubiquitous in Go; they come up many times per game, if not in actual play then at least in the variations that a player considers to decide on his next move.

We show how ladders can take the place of pipes in constructing hard capture problems.

3   Of Forks, Joins, and Mirrors

Our introductory ladder problem features the four different *gadgets* listed in Figure 2: the black choice (B), the white choice (W), the join (J) and the mirror (M) (for conciseness a black choice is partially merged with the join to its right in the center of the problem).

In diagram B1, we see a Black choice gadget with a projected ladder approaching from the top left. When Black plays the ladder, he'll have a choice of playing move 9 on the right or the left of White, leading respectively to diagram B2 or B3. From Black's viewpoint, the top-left ladder works if either the bottom-left ladder in B2, or bottom-right ladder in B3, work.

In diagram W1, we see a White choice gadget with a projected ladder approaching from the top left. When Black plays the ladder, White's move 10 puts the marked black stone in atari, and Black must play above it to prevent White from getting too many liberties. Now White can choose to either capture the marked stone, or extend to the right, leading respectively to diagram W2 or W3. Black's moves 15 and 17 in diagram W2 are needed to route the ladder around the rightmost white stone, which would otherwise interfere. From Black's viewpoint, the top-left ladder works if both the right-down ladder in W2, and right-up ladder in W3, work.

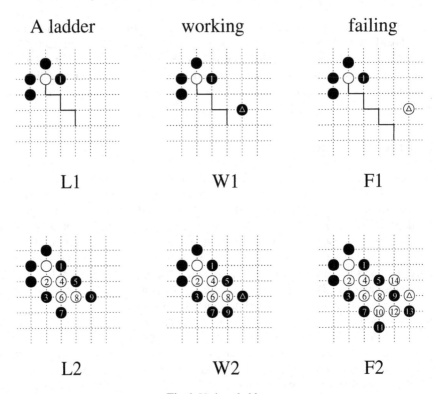

Fig. 1. Various ladders

In diagram J1, we see a Join gadget with projected ladders approaching from the top left and top right. Diagram J2 shows what happens with a ladder from the top left. The forced sequence ends with the ladder continuing to the bottom left. Diagram J3 shows the symmetrical case of a ladder from the top right. From Black's viewpoint, either top ladder works if the bottom-left ladder does.

Finally, in diagram M1, we see a Mirror gadget with a projected ladder approaching from the top left. When Black plays the ladder, he is forced to send it back up with move 11. Mirrors allow us to direct ladders from one gadget to the next.

3.1 Problem Analysis

Figure 3 shows a line of play in our original problem.

This line of play is entirely forced except for White's choice of playing 'a' and Black's choice of playing 'b'. If we let boolean variable x represent whether White chose to send the ladder up, and let y represent whether Black chose to send the ladder up, then the current line of play corresponds to setting $(x, y) = ($true$,$ false$)$. In general we have for each variable a choice gadget, an upper and lower mirror, and a join gadget, positioned at the corners of an imaginary diamond shape. The setting of the variable determines which (upper or lower) edges of the diamond get covered and which get exposed. All gadgets are placed sufficiently far apart to ensure their correct operation. (Recall that this

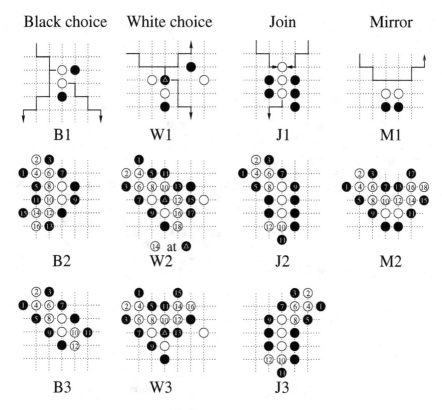

Fig. 2. Ladder gadgets

specific instance differs from the general one in that we saved some space by merging a Join and Black choice gadget in the center.)

Now consider the top black choice gadget. If play arrives here and the ladder leaves to the bottom-left (T), then it works if and only if x is true. If it leaves to the bottom-right, then it works if and only if y is true. It follows that the ladder going up from White's choice at 'u'—which after bouncing off 2 mirrors enters the top black choice gadget—works if $x \vee y$ holds. Similarly, the ladder going down from White's choice at 'd' works if $\neg x \vee \neg y$ holds. Hence, after both variables have been set, the ladder works if $(x \vee y) \wedge (\neg x \vee \neg y)$. This shows that our original problem indeed encodes the truth of the formula $\forall x \exists y (x \vee y) \wedge (\neg x \vee \neg y)$,

To prove Theorem 1, i.e. PSPACE-completeness, we must show two things: first, that **LADDERS** belongs to PSPACE, and second, that **QBF** (known to be PSPACE-complete) reduces to **LADDERS**.

3.2 LADDERS ∈ PSPACE

Membership in PSPACE follows if capturability can be determined by a polynomial-depth-limited search. As long as white keeps adding stones to his group, the search must reach an end before the group becomes bigger than the whole board. Consider then a

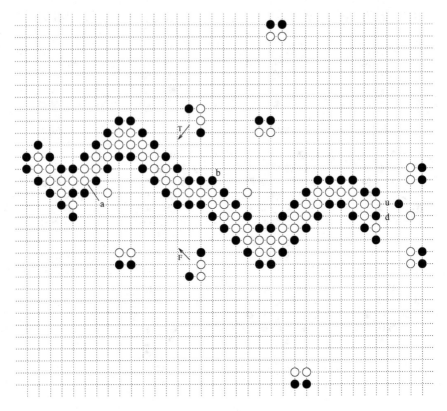

Fig. 3. A forced line apart from choices 'a' and 'b'

line of play where instead, white on his move always captures some black stones to gain
extra liberties. Since the search ends in failure for Black when white gains 2 liberties (for
a total of 3), we may assume that only one stone of each captured black group is adjacent
to White's group. Let us analyze how many times black can replay on that point.

If White captured 2 or more stones, and Black replays on the liberty, then White can
recapture and either Black's 2nd replay is suicide, or it captures white in a 'snapback'
(Figure 4, diagram A), both settling the situation.

If White captured 1 stone, then Black can only replay there by capturing White's
stone back. If the latter captures multiple White stones, then White can recapture too
and settle the situation (Figure 4, diagram B). If instead Black captures just the one
White stone, then the recaptures can continue back and forth, a situation known as 'ko'
(Figure 4, diagram C).

The rules of go forbid taking back immediately in a ko, since this recreates the
position of 2 moves back. The stronger "superko" rule forbids repetition of *any* earlier
position, but this rule is not universally accepted as opposed to the *basic* ko rule above.

Now, if there are at least 4 kos adjacent to the ladder, then White, in atari, has at
least 3 choices of where to capture, while Black has only 2 choices of capture. Under

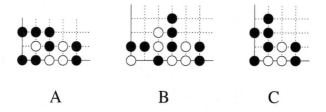

Fig. 4. 3 types of recapturing

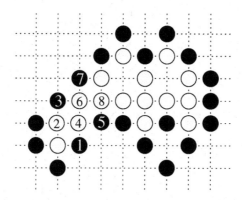

Fig. 5. A ladder depending on a triple ko

the basic ko rule, this allows both players to cycle forever, while the superko rule forbids Black first. In both cases the search ends in failure for Black.

With at most 3 kos, there are at most 6 configurations (001, 010, 011, 100, 101, 110, according to what player holds which kos). Figure 5 shows an example where the ladder runs into a such a "triple ko". With superko, White will be forbidden to cycle in this case (examples of superko forbidding Black are equally well possible) and the ladder works, but without superko, it will cycle forever and Black fails.

In conclusion, White can temporarily avoid extending his group by captures but once all non-ko situations are settled, then this is only possible by starting multiple kos. If White can start enough then he prevails, else the result is determined by the exact ko rules. Altogether, White needs to extend his group at least once every 6 times board-size moves, so the search may be limited to a depth of 6 times board-size squared, showing that **LADDERS** is in PSPACE.

3.3   QBF Reduces to LADDERS

Consider the standard PSPACE-complete problem

QBF: Given a quantified boolean formula $F = Q_1 x_1 Q_2 x_2 \ldots Q_n x_n E$, where E is a Boolean expression involving $x_1, \ldots, x_n$ and each Q_i is either "$\forall$" or "$\exists$", determine if F is true.

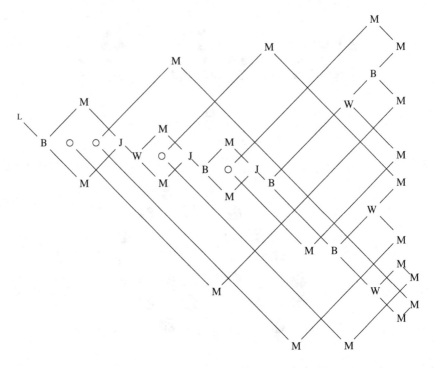

Fig. 6. Schematic of ladder instance $\exists x \forall y \exists z (x \wedge \neg y) \vee (\neg x \wedge y) \vee (\neg z \wedge (\neg x \vee z))$

We show how to reduce **QBF** to **LADDERS** by way of the example

$$\exists x \forall y \exists z (x \wedge \neg y) \vee (\neg x \wedge y) \vee (\neg z \wedge (\neg x \vee z)).$$

The construction for this example is illustrated in Figure 6. The sequence of diamonds is similar to that in the opening problem, with a Black choice gadget for each $\exists$ quantifier, and a White choice gadget for each $\forall$. The size of each diamond is made proportional to the maximum of the number of positive and the number of negative occurrences of the corresponding variable, so that we can have a disjoint incoming ladder for each occurrence. Inside each diamond we place extra white stones to act as ladder-breakers. This ensures the failing of any incoming ladder at the uncovered side of the diamond. Next, to the right of the last diamond, the boolean expression is laid out. Each $\vee$ is mapped to a Black choice gadget, and each $\wedge$ to a White choice gadget. The two sub-expressions are then recursively laid out to the upper right, and lower right, spaced sufficiently apart to allow for disjoint ladders. At the leaves we place mirrors directing the ladder to the appropriate diamond. For leaves with opposite incoming and outgoing direction, such as used for the $(x \wedge \neg y)$ sub-expression in the example, we use 2 mirrors to ensure disjointness of incoming and outgoing ladder paths. Ladder paths are free to intersect since the actual line of play can only follow one path back to a diamond. It should be obvious how to apply this method to any formula in **QBF**.

As explained in section 3, the ladder thus constructed works if and only if the formula is true.

4   Conclusions

For the first time, we have identified a natural aspect of the game of Go—the ladder—which is not only PSPACE hard, but PSPACE-complete. This may surprise many Go players who think reading out ladders is an elementary exercise in visualization.

Our reduction improves on that of Lichtenstein and Sipser [2] in simplicity (by avoiding the need for intersection gadgets), economy (using a number of stones only linear in formula size), and aesthetic appeal (the opening problem would not look out of place in a go magazine).

References

1. Garey, M., R., Johnson, D., S., Computers and Intractability, Bell Telephone Laboratories, (1979)
2. Lichtenstein, D. and Sipser, M., GO is Polynomial-Space Hard, Journal of the ACM, Vol. **27**, No. 2, (April 1980) 393–401.
3. Robson, J., The Complexity of Go, Proc. IFIP (International Federation of Information Processing), (1983) 413–417.
4. Robson, J., Combinatorial games with exponential space complete decision problems, Proc. 11th Symposium on Mathematical Foundations of Computer Science, (1984) 498–506.
5. Robson, J., Alternation with Restrictions on Looping, Information and Control, Vol. **67**, 2-11 (1985).
6. Papadimitriou, H., Computational complexity, Addison-Wesley, (1994)

Simple Amazons Endgames and Their Connection to Hamilton Circuits in Cubic Subgrid Graphs

Michael Buro

NEC Research Institute, Princeton NJ 08540, USA
mic@research.nj.nec.com

Abstract. Amazons is a young board game with simple rules and a high branching factor, which makes it a suitable test-bed for planning research. This paper considers the computational complexity of Amazons puzzles and restricted Amazons endgames. We first prove the NP-completeness of the Hamilton circuit problem for cubic subgraphs of the integer grid. This result is then used to show that solving Amazons puzzles is an NP-complete task and determining the winner of simple Amazons endgames is NP-equivalent.

Keywords: Amazons endgame, puzzle, NP-complete, planning

1 Introduction

The success of full–width search and total enumeration in certain combinatorial problems – such as Rubik's cube [8], Othello [1], checkers [11], and chess [3] – masks the lack of progress in the planning and reasoning departments. The consequences are apparent: in spite of vast hardware speed-ups, hardly any AI system can pass the Turing test except for very specialized tasks. In the domain of games the problem becomes evident if we increase the number of move choices from dozens to thousands. If a system uses sophisticated pruning techniques it may still find reasonable moves. However, we can easily turn up the heat by decreasing the impact of single moves (which increases the length of move sequences) or replacing slow turn based play by fast real-time action. At this point even the greatest systems using traditional approaches look pathetic compared to human abilities. Prominent examples are real-time war simulation games – such as Starcraft[1] – in which the computer AIs desperately try to coordinate combat units. Currently, their only way of winning against humans is by starting with a considerable material advantage or simply by cheating.

In order to push planning and reasoning research, we need to focus on tasks that require goal directed search in order to cope with vast state spaces. Moreover, the major goals should be simple enough to be in reach of current machine learning techniques. Finally, the tasks should be suited to human mental abilities because this is the current AI benchmark per se.

Amazons is a young board game that is beginning to attract researchers for these reasons. It is played on an $n \times n$ board (usually $n = 10$). Both players have four amazons. A move consists of picking an amazon to move like a chess queen and shooting an arrow

---

[1] Starcraft is a trademark of Blizzard Entertainment

T.A. Marsland and I. Frank (Eds.): CG 2000, LNCS 2063, pp. 250–261, 2001.

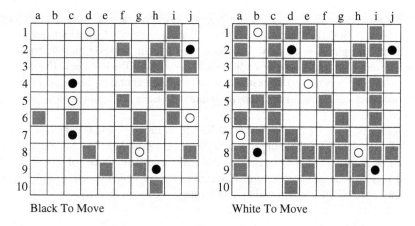

Fig. 1. Typical middle and endgame positions

in queue direction from the amazon's destination to an empty square. This square gets blocked for the remainder of the game and no amazon or arrow can pass it. Arrows are not allowed to pass amazons either and amazons can not be captured. Blocking squares is mandatory. The game proceeds in turns and the first player without any legal move loses. Figure 1 shows two typical Amazons positions. In the standard starting position all amazons are evenly distributed along the four edges. To make the game more interesting one can also place amazons randomly after blocking a small number of squares. Both players then estimate the final game outcome expressed as a move surplus of the player to move. The average estimate is used to assign colors and to determine the winner when the game is finished.

Amazons strategy is based on mobility and territory. Its high branching factor (more than 2000 opening moves on a 10x10 board!) limits the scope of full-width search and known forward pruning techniques considerably. State space sizes of PSPACE-complete puzzles like Sokoban [2] or RushHour [4] are also huge. The difference, however, is that the length of Amazons games is limited by the board size whereas move sequences in the mentioned puzzles can have exponential length. This makes solving hard instances of those puzzles less attractive for human players. Compared with the Asian board game Go, Amazons shares the property that in endgames the position gets split into separate subgames. This allows combinatorial game theory to step in and provide means of finding optimal moves faster than traditional approaches. On the other hand, the notorious problem of evaluating Go positions statically [9] does not seem to have an Amazons counterpart. As shown in past computer Amazons tournaments and in computer games against advanced human players, evaluations based on square-access-distance lead to reasonable (but still far from perfect) play.

In this paper we consider the computational complexity of solving *simple* Amazons puzzles and endgames. In these games amazons of equal color are located in their own, entirely sealed off territories. Thus, both opponents are separated and the winner is determined by the total number of moves each player can make in her own territories and whose turn it is. Because this scenario often occurs in actual games it would be helpful to incorporate automatic endgame scorers into Amazons game servers (e.g. the Generic Game Server (GGS) at telnet://ftp.nj.nec.com:5000), which quickly short-cut boring straight-forward move sequences. It turns out, however, that determining the winner even of simple Amazons endgames in general is NP-equivalent. This means that most likely there is no fast algorithm for solving the general problem, and we have to rely on clever heuristics to find (approximate) solutions to small problem instances in limited time.

In what follows, we first show that the Hamilton circuit problem and related problems are NP-complete for cubic subgraphs of the integer grid. We then use these results to prove that deciding whether an amazon can make a certain number of moves in a given board region is an NP-complete task, too. Finally, we conclude that simple Amazons endgames are NP-equivalent and motivate future Amazons research.

2   Hamilton Circuits in Cubic Subgrid Graphs

Definition 1. *Let G^∞ be the infinite graph consisting of all points of the plane with integer coordinates and edges connecting points with Euclidean distance one. Finite subgraphs of G^∞ are called subgrid graphs. Subgrid graphs with nodes of degree at most three are called cubic subgrid graphs. Grid graphs are finite node induced subgraphs of G^∞.*

In [6] it is shown that the Hamilton circuit problem for grid graphs is NP-complete. Nodes in grid graphs can have degree four, which makes this result impractical for proving the hardness of Amazons problems. This is because there is no easy way of modeling 4-way intersections that can be traversed only once – as we will see later. However, the proof ideas in [6] can be refined such that the reduction leads to cubic subgrid graphs which can be modeled by Amazons positions without much difficulty.

Theorem 1. *The set HC3G of all cubic subgrid graphs with a Hamilton circuit is NP-complete.*

Proof. Guessing a potential Hamilton circuit in a given cubic subgrid graph and verifying it in polynomial time shows that HC3G belongs to NP. In what follows we show that the set HCB3P of bipartite cubic planar graphs with a Hamilton circuit can be polynomial time reduced to HC3G. This concludes the proof because HCB3P is known to be NP-complete [10,6].

Mapping a given bipartite cubic planar graph G into a cubic subgrid graph G_3 while preserving the Hamilton circuit property is a three step process illustrated in Figure 2: M_1 transforms G into a cubic orthogonal drawing. This task can be accomplished in linear time and space as shown in [7]. Cubic orthogonal drawings are not necessarily cubic subgrid graphs because node connections may be longer than the unit grid length. Adding the missing intermediate nodes solves this problem. In general, however, the resulting graph does not preserve the Hamilton circuit property. To save this property, a second mapping, M_2, scales up the augmented orthogonal drawing by a factor of four first. Then – if necessary – it moves images of G nodes one grid position to the left or right and reconnects the edges to adjust the parity $(x(v) + y(v) \bmod 2)$ with respect

G (bipartite cubic planar graph)

$\underset{M_1}{\longmapsto}$ G_1 (cubic orthogonal drawing + intermediate nodes)

$\underset{M_2}{\longmapsto}$ G_2 (parity preserving cubic subgrid graph)

$\underset{M_3}{\longmapsto}$ G_3 (Hamilton circuit preserving cubic subgrid graph)

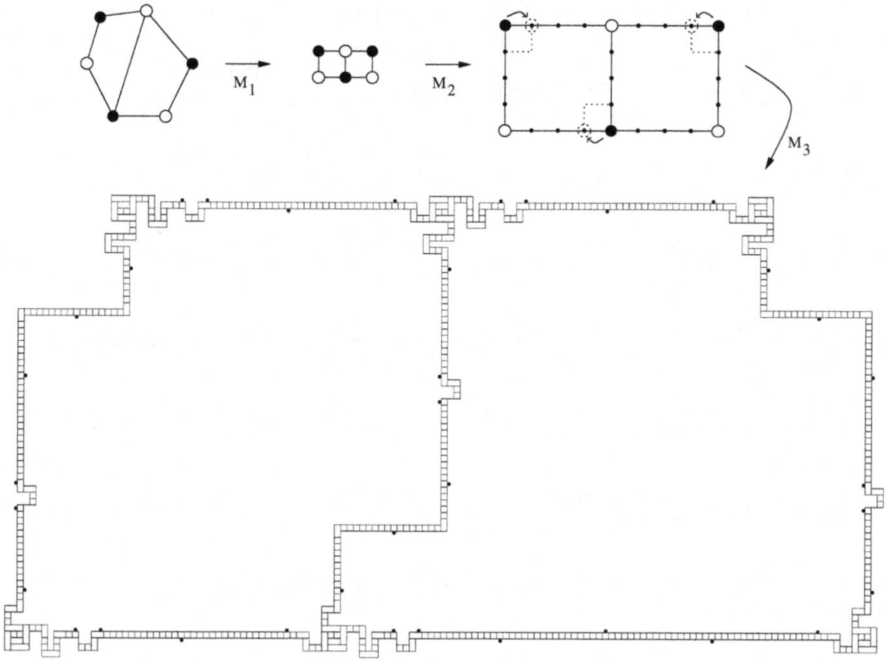

Fig. 2. Transformation example

to the original node partition (G is bipartite). Thus, in G_2 the images of the original nodes are connected by simple paths of odd length. This is necessary for applying the last transformation, M_3, which replaces all nodes of G_2 by (adjusted) copies of the 17x17 cluster and strips shown in Figure 3. Original nodes of degree two are replaced by clusters from which one tentacle has been removed (w.l.o.g. there are no nodes of degree one). Each component has some outgoing edges marked with black dots. When connecting components the respective markers have to match. The odd distance of original node images in G_2 ensures a unique matching. Finally, one reflector gadget (shown in Figure 3) is placed in each component connecting strip. The resulting graph G_3 is a cubic subgrid graph because all nodes in the clusters and strips have degree at most three and connecting the components does not increase degrees. Since the entire graph transformation obviously can be computed in polynomial time, the proof rests on showing

regular nodes:

intermediate nodes:

reflector gadget:

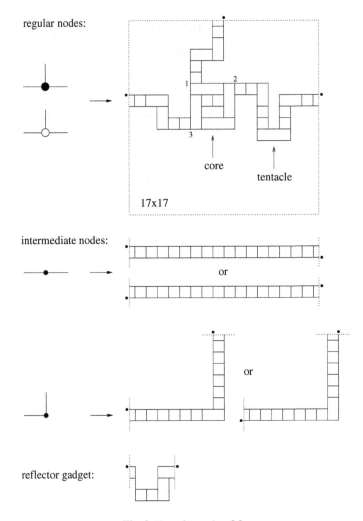

Fig. 3. Transformation M_3

G has a Hamilton circuit $\Longleftrightarrow$ G_3 has a Hamilton circuit.

"$\Rightarrow$": Starting with a Hamilton circuit p in G we construct a Hamilton circuit in G_3 by traversing strips and clusters as follows: beginning at corners of the cluster cores (Figure 4), strips and tentacles corresponding to edges in p are traversed by battlements paths (Figure 5). The remaining ones are covered by parallel paths (N.B.: the component markers indicate edges visited by battlements paths. Their positions determine the edges along the strips and tentacles that can be omitted to ensure that degrees do not exceed three). It remains to connect the nodes in the cluster cores. The core has been designed in such a way that a Hamilton path exists between each pair of three corners (and all nodes again have degree at most three). Thus, the two battlements paths ending in corners of each core can be connected by Hamilton paths. If parallel paths originate from the

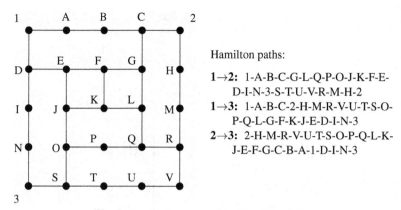

Hamilton paths:

1→2: 1-A-B-C-G-L-Q-P-O-J-K-F-E-
D-I-N-3-S-T-U-V-R-M-H-2

1→3: 1-A-B-C-2-H-M-R-V-U-T-S-O-
P-Q-L-G-F-K-J-E-D-I-N-3

2→3: 2-H-M-R-V-U-T-S-O-P-Q-L-K-
J-E-F-G-C-B-A-1-D-I-N-3

Fig. 4. The cluster core and its Hamilton paths

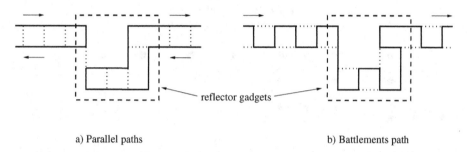

a) Parallel paths   b) Battlements path

Fig. 5. The two ways of covering strips

third corner of some cores, the corresponding corner edges have to be removed from the inter-core Hamilton paths. The result is a Hamilton circuit in G_3.

"⇐": Given a Hamilton circuit in G_3, we claim that tentacles and strips covered by parallel paths can be removed while maintaining a Hamilton circuit in the remainder of the graph. Once all these strips and tentacles have been removed from G_3, Hamilton paths in cores and battlements paths remain. These form a Hamilton circuit which corresponds to a Hamilton circuit in G because each cluster is connected to two neighboring clusters and clusters are the images of the original nodes of G. Figure 6 illustrates the parallel path scenario. To maintain a Hamilton circuit the parallel paths are replaced by the edges (A,B) and (C,D). At this point it is important to note that reflecting gadgets are necessary to prevent parallel paths (A..D) (B..C) which would invalidate this part of the proof.

□

Definition 2. *A collision path in a graph is an edge disjoint path $v_0 e_1 v_1 e_2 ... e_l v_l$ with at most one node repetition (i.e. $\exists\, i, j$ with $v_i = v_j$ and $i \neq j$) which ends right after the repetition, if there is one.*

Collision path examples are shown in Figure 7a).

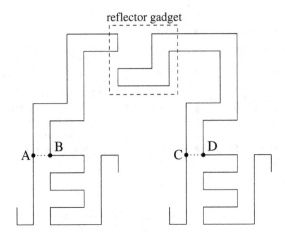

Fig. 6. Parallel path scenario

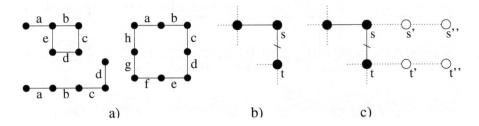

Fig. 7. Collision paths and graph adjustments

Corollary 1. *The sets of all cubic subgrid graphs G with the following properties are NP-complete:*

a) G has a Hamilton path with specified endpoints
b) G has a Hamilton path
c) G has a collision path of length $|V_G| - 1$ with specified starting point
d) G has a collision path of length $|V_G| - 1$

Proof. In all cases the NP membership is obvious. NP-hardness is shown by reducing HC3G.

Cases a) & b): Let G be a finite connected cubic subgrid graph without nodes of degree one. Then G has a "corner" node s of degree two (Figure 7b), i.e. s has no upper neighbor and there are no nodes in G to the right of s and its lower neighbor t. Such a node s can be found by first maximizing x coordinates of nodes in G and then maximizing the y coordinates on the resulting vertical line. Thus, G has a Hamilton circuit if and only if $G - (s, t)$ has a Hamilton path with endpoints s and t. Moreover, if two nodes s' and t' are added and connected to s resp. t (Figure 7c), it follows that G has a Hamilton circuit if and only if there is a Hamilton path in $G - (s, t) + (s, s') + (t, t')$.

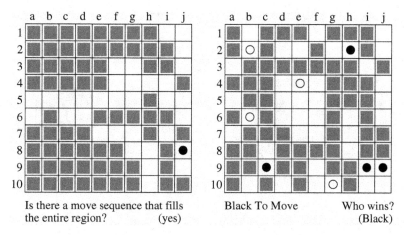

Fig. 8. An Amazons puzzle and a simple Amazons endgame

<u>Cases c) & d):</u> In these cases we extend G by four nodes s', s'', t', and t'' (Figure 7c) to form a new graph G'. There are no paths with a collision in G' of length $|V_{G'}|$ because collisions can only occur in the G part. However, if there is a collision in this part, the path ends there and its length is less than $|V_{G'}|$. Thus, the only collision paths of length $|V_{G'}| - 1$ are Hamilton paths from s'' to t''. This shows that G has a Hamilton circuit if and only if G' has a collision path of length $|V_{G'}| - 1$ (d). Since s'' is start or endpoint of all such paths, c) follows as well.   □

3   Simple Amazons Endgames

Definition 3. *A set of (vertically, horizontally, or diagonally) connected empty squares that is entirely surrounded by blocked squares or board edges together with amazons of one color placed inside the region is called an <u>Amazons puzzle</u>. An Amazons puzzle <u>solution</u> is a move sequence of maximum length.*

Definition 4. *Simple Amazons endgames are sequences of puzzles for amazons of both colors. Black is to move first. <u>Black wins</u> the simple Amazons endgame if the total solution length of Black's puzzles is greater than White's. Otherwise, <u>Black loses</u> (Figure 8 illustrates both definitions).*

Theorem 2. *The set $AP := \{(p, b) \mid$ Amazons puzzle p has solution length at least $b\}$ is NP-complete.*

Proof. We note that for a given position and solution length a move sequence can be guessed and verified in polynomial time. Hence, AP is an element of NP. We show AP's NP-hardness by mapping cubic subgrid graphs G into pairs (p, b) such that:

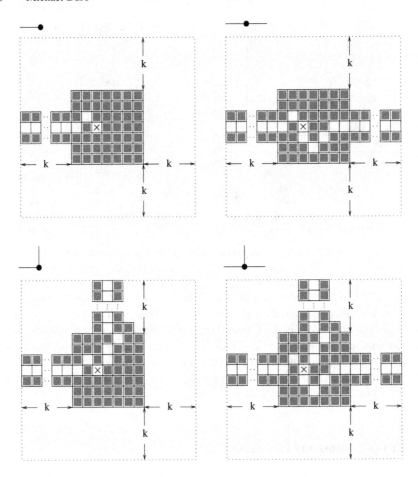

Fig. 9. Mapping parts of a cubic subgrid graph into Amazons board regions

G has a collision path of length $n - 1$ ($n = |V_G|$) starting in a specified node s if and only if the amazon can make at least b moves in position p. $\quad$ (*)

Nodes and their connections are mapped into board regions according to Figure 9. The amazon is placed on the marked center square in the image of the starting node. Figure 10 illustrates the transformation that defines p. The regions have been designed such that the amazon on her way from region to region must visit and block the marked central squares, and corridors can only be traversed once. Thus, the sequence of visited regions corresponds to a collision path in G starting with s.

Let m be the maximum number of moves the amazon can make in position p and l the maximum length of collision paths in G. We pick corridor length $12n$ (i.e. $k = 6n$) and claim

$$l \geq n - 1 \iff m \geq 12(n^2 - n). \tag{1}$$

The theorem follows by setting the move threshold b to $12(n^2 - n)$ in (*).

To prove (1) we consider upper and lower bounds on the number of moves in a maximum move sequence corresponding to a collision path of length l. Clearly, $m \geq$

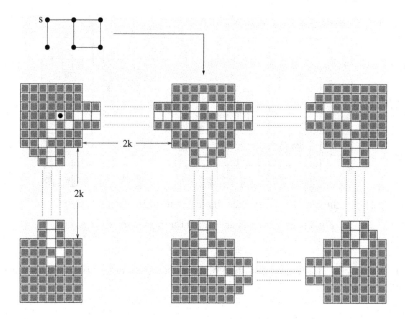

Fig. 10. Mapping example

$l(2k + 1)$ holds because the amazon can traverse at least the corridors square by square. On the other hand, $m \leq l(C + 2k + 1) + C$, where C is the maximum number of empty squares in the 7x7 region centers. Inserting $C = 11$ and $k = 6n$ leads to $m \geq l(12n + 1)$ and $m \leq l(12n + 12) + 11$. Therefore, we can conclude

$$l \geq n - 1 \Rightarrow m \geq (n-1)(12n+1) \quad = \quad 12n^2 - 11n - 1$$
$$l \leq n - 2 \Rightarrow m \leq (n-2)(12n+12) + 11 \quad = \quad 12n^2 - 12n - 13,$$

from which (1) follows.

$\square$

Corollary 2. *The set SAE := $\{p \mid$ Black wins simple Amazons endgame $p\}$ is NP-equivalent. Therefore, determining the winner of simple Amazons endgames in polynomial time is possible if and only if $P = NP$.*

Proof. (Following the terminology established in [5], a set S is called <u>NP-equivalent</u> if there are two NP-complete sets A and B with $A \propto_T S$ and $S \propto_T B$, where $\propto_T$ denotes oracle Turing reducibility in polynomial time) We show AP $\propto_T$ SAE $\propto_T$ AP.

<u>AP $\propto_T$ SAE</u>: from a given Amazons puzzle p and move limit b we construct a simple Amazons endgame by adding a strip of b empty squares that is surrounded by blocked squares. We place a white amazon on this strip and use black amazons in the puzzle region. It is easy to see that p has solution length $\geq b$, if and only if Black wins the endgame.

<u>SAE $\propto_T$ AP</u>: the solution length of each puzzle component of a given endgame can be found in polynomial time by a binary search that is guided by constant time queries

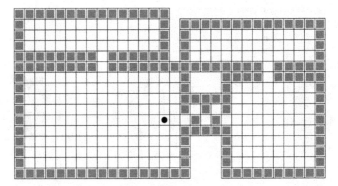

Is there a move sequence that fills the entire region? (yes)

Fig. 11. Easy for humans – hard for short-sighted programs

of the AP oracle. The winner can then be determined by comparing both players' total solution length.   □

4   Outlook

We have shown the hardness of Amazons puzzles by reducing an NP-complete graph problem. Since most computer scientists accept P $\neq$ NP as a working hypothesis, this result can be regarded as a "poor man's lower bound," meaning that (most likely) there is no polynomial time algorithm that can solve arbitrary Amazons puzzles. This limitation, however, applies only to the general problem. In particular, there exist types of large puzzles that humans find easy to solve, whereas current programs – being notoriously short-sighted – do not (Figure 11). This example clearly demonstrates the necessity of planning and reasoning in AI systems. Amazons, with its simple rule set and large branching factor, is therefore an ideal test-bed for future research on single agent and adversarial planning.

Acknowledgement

Many thanks to John Tromp who found an error in an earlier version of this paper.

References

1. M. Buro. How machines have learned to play Othello. *IEEE Intelligent Systems J.*, 14(6):12–14, 1999.
2. J. Culberson. Sokoban is PSPACE-complete. In *Proceedings in Informatics 4*, pages 65–76. arleton Scientific, Waterloo, Canada, 1999.
3. D. DeCoste. The significance of Kasparov versus Deep Blue and the future of computer chess. *ICCA J.*, 21(1):33–43, 1998.

4. G.W. Flake and E.B. Baum. RushHour is PSPACE-complete, or why you should generously tip parking lot attendants. *to appear in TCS*, 2000.
5. M.R. Garey and D.S. Johnson. *Computers and Intractability*. W.H. Freeman and Company New York, 1979.
6. A. Itai, C.H. Papadimitriou, and J.L. Szwarcfiter. Hamilton paths in grid graphs. *SIAM J. Comput.*, 11(4):676–686, 1982.
7. G. Kant. Drawing planar graphs using the canonical ordering. *Algorithmica*, 16(1):4–32, 1996.
8. R. Korf. Finding optimal solutions to Rubik's cube using pattern databases. *Fourteenth National Conference on Artificial Intelligence Ninth Innovative Applications of Artificial Intelligence Conference*, pages 700–705, 1997.
9. M. Müller. Computer Go: A research agenda. *ICCA Journal*, 22(2):104–112, 1999.
10. J. Plesnik. The NP-completeness of the Hamiltonian cycle problem in planar digraphs with degree bound two. *Information Processing Letters*, 8(4):199–201, 1979.
11. J. Schaeffer. *One Jump Ahead: Challenging Human Supremacy in Checkers*. Springer Verlag, 1997.

New Self-Play Results in Computer Chess*

Ernst A. Heinz

M.I.T. Laboratory for Computer Science (Room NE 43 – 228)
Massachussetts Institute of Technology
545 Technology Square, Cambridge, MA 02139, USA
heinz@mit.edu
http://supertech.lcs.mit.edu/~heinz

Abstract. This paper presents the results of a new self-play experiment in computer chess. It is the first such experiment ever to feature search depths beyond 9 plies and thousands of games for every single match. Overall, we executed 24,000 self-play games (3,000 per match) in one "calibration" match and seven "depth X+1 ⇔ X" handicap matches at fixed iteration depths ranging from 5–12 plies. For the experiment to be realistic and independently repeatable, we relied on a state-of-the-art commercial contestant: FRITZ 6, one of the strongest modern chess programs available. The main result of our new experiment is that it shows the existence of diminishing returns for additional search in computer chess self-play by FRITZ 6 with 95% statistical confidence. The average rating gain per search iteration shrinks by half from 169 ELO at 6 plies to 84 ELO at 12 plies. The diminishing returns manifest themselves by declining rates of won games and reversely increasing rates of drawn games for the deeper searching program versions. Their rates of lost games, however, remain quite steady for the whole depth range of 5–12 plies.

Keywords: diminishing returns, search vs. knowledge, self-play

1   Introduction

To the best of our knowledge, Gillogly and Newborn in 1978 independently reported the earliest attempts at modeling the relationship between the playing strength of chess programs on one hand and the available computing power or search depth on the other. Gillogly [4] introduced his "technology curve" that plotted the playing strength against what he called "machine power" on a logarithmic scale. Newborn [22,23] related the numbers of nodes as searched by different chess programs in three minutes (the average time per move in tournament games) to the playing strengths of the very same programs as derived from their actual performances in tournaments. Later on, Levy [18] and Levy and Newborn [17] refined Newborn's initial scheme by contrasting the highest rated tournament performances of the best chess programs with the years of their achievement. All these comparisons inevitably led to speculative extrapolations which Levy characterized as the "meta-science of prediction in computer chess" in his latest article [16] about the subject in 1997.

---

* An earlier report [7] on some intermediate results of our new self-play experiment appeared as M.I.T. LCS Technical Memo No. 608 (MIT-LCS-TM-608).

T.A. Marsland and I. Frank (Eds.): CG 2000, LNCS 2063, pp. 262–276, 2001.

Table 1. Timeline of published self-play experiments in computer chess. [TECHMATE and ZUGZWANG self-played with time handicaps.]

Year	Program	Experimenter	Depths (in Plies)	Δ ELO (per Ply)	No. Games All	Each
1982	BELLE	Thompson	3 – 8	+246	100	20
1983	BELLE	Condon, Thompson	4 – 9	+217	300	20
1988	TECHMATE	Szabo, Szabo	– –	– –	6,882	≥ 32
1990	HITECH LOTECH	Berliner et al.	4 – 9	+195 +232	1,056	16
1994	ZUGZWANG	Mysliwietz	– –	– –	450	50
1996	PHOENIX	Schaeffer	4 – 9	+228	120	20
1997	THE TURK	Junghanns et al.	3 – 9	+200	480	80

In the early 1980s, Thompson [3,30] pioneered the usage of self-play with his then reigning World Computer-Chess Champion machine BELLE. Self-play with handicaps in search depth, search speed, or search time between otherwise identical program versions represents a more rigorous approach of investigating the relationship of computing power and the strength of chess programs. A notable advantage of such matches is that the scoring rates quantify the differences in playing strength of the various participating versions of the same program. Despite unresolved questions regarding the magnitude of self-play rating differences [1], self-play seems to be the best of the available methods to resolve the old but still ongoing "search versus knowledge" debate [12,25,26]. Nearly everybody seems to agree with the intuitive notion that the positive effect of more search ought to taper off with increasing overall search effort. Yet, it is not obvious when and how such "diminishing returns for additional search" kick in.

In self-play matches, diminishing returns should lead to lower scoring rates of the deeper searching program versions with the progression towards higher depths. However, Thompson's experiments [3,30] led to the surprising result that the playing strength of BELLE increased almost linearly with search depth. For fixed-depth searches of 3–9 plies, the increase in playing strength amounted to roughly 200 ELO rating points per ply. Several other researchers later confirmed Thompson's findings by self-play experiments with their own chess programs HITECH, LOTECH, PHOENIX, and THE TURK [1,13]. In Figure 1 of their article [13], Junghanns et al. showed that the scoring rates of the program versions searching one ply deeper remained range-bound between 70%–80% in all cases. There are no clearly visible average downward trends at the end of these 9-ply data curves.

1.1 Previous Self-Play Experiments in Computer Chess

Table 1 presents an overview and timeline of self-play experiments in computer chess published up to now. Beside names, depths, and average ELO increases, the table also lists the overall numbers of games played in the experiments as a whole and for each single match. Unfortunately, all the experiments feature only very low numbers of games per match which do not allow for any confident quantification of rating differences between the opponents. Hence, we fully agree with Mysliwietz [20] who already criticized the statistical uncertainty of self-play experiments in computer chess back in 1994.

Based on this criticism, we re-assessed and carefully re-analyzed all experiments from Table 1 in our recent publications [6,9]. The outcome of our analyses showed that none of the previous self-play experiments provide any confident quantifications of the differences in playing strength. The experimental results are just not statistically significant, not even at a low confidence level of 90%. The experiments do not feature enough games per match to draw reliable conclusions. Based on rigorous analyses of hypothetical match results, we conjectured [6,9] that at least 1,000 games per match are necessary to assess diminishing returns in computer self-play with 95% statistical confidence. Further questions w.r.t. previous self-play experiments are the exact meaning of "fixed depth" in each case, the details of the experimental setups, and their repeatability.

The "fixed depth" question is not trivial because the modes of operation of the programs differ substantially depending on its real meaning. In the search-theoretical sense, "fixed depth" denotes true brute-force search with uniform path lengths from the root to all horizon nodes and no selectivity at all – neither by means of depth reductions or other kinds of forward pruning nor by any search extensions. In computer-chess practice, however, "fixed depth" usually equals "fixed iteration depth" which relates to the depth limit of iterative deepening [14,28] as performed by the top-level search control. Here, the programs operate with an iteration limit instead of a time bound but otherwise execute their sophisticated variable-depth search procedure as built in – with all kinds of depth reductions, forward pruning, and search extensions enabled.

The exact setups of the experiments are not only important for the purpose of repeatability. The engine settings and hash-table sizes, the opening books or positions used, the endgame databases, etc. may as well have non-negligible influences on the match results. Of course, the whole setup must be identical for all sibling versions of a program during handicap self-play. In particular, it is not admissible to increase the hash-table sizes of deeper searching versions to speed-up their times until completion.

As for repeatability, Table 1 shows that all self-play experiments published up to now featured proprietary chess program. Several of them also relied on special hardware (BELLE, HITECH, LOTECH, and ZUGZWANG). Hence, even assuming detailed knowledge of the exact setups, none of the experiments was independently repeatable by others in practice.

1.2   A New Self-Play Experiment in Computer Chess

We designed our new self-play experiment in such a way as to overcome the aforementioned drawbacks of its predecessors from Table 1. Our primary concerns were the rigorous analysis of the results (see Section 3) and their statistical significance. We played seven "depth X+1 $\Leftrightarrow$ X" handicap matches at fixed iteration depths ranging from 5–12 plies with 3,000 games per match (see Section 5). By extending the self-play depths beyond 9 plies for the first time ever, we sought to gain new information about potentially diminishing returns for additional search in computer chess at high depths.

Moreover, we intended our self-play experiment to be transparent and realistic at the time of execution and independently repeatable by others later on. To this end, we needed a state-of-the-art contestant featuring general world-wide availability, x86-PC compatibility, well-defined parameter control, and – last but not least – handicap self-play ability. The commercial chess program FRITZ 6 (written by Frans Morsch and Matthias Feist) met all our requirements because handicap self-play abilities were included in it upon our special request. FRITZ 6 is certainly one of the strongest modern chess programs available.

Further advantages of employing FRITZ 6 spring from its database capabilities, versatile chess-engine concept, and excellent opening book (composed by Alexander Kure). In particular, the wide and well-balanced opening book facilitates the automatic play of fair matches with thousands of games. Just to be sure, we checked the integrity and Black/White fairness of the opening book by means of a "calibration" match between two identical opponents (see Section 4.1). The various different engines available for the FRITZ 6 interface allow for the possibility to include other chess programs in our future self-play research.

2 Related Work

In the introduction, we already mentioned the attempts of Gillogly [4], Newborn [22,23], Levy [16,18], and Levy and Newborn [17] at modeling the relationship between playing strength and computing power. Newborn [21] introduced yet another technique to study this relationship in 1985. The rationale of Newborn's novel approach sprang from the assumption that new best moves as discovered by chess programs at higher search depths ought to represent better choices than the best moves preferred at shallower depths. To this end, Newborn tracked the behaviour of BELLE for searches to fixed depths of 11 plies on a set of 447 test positions from real games. Interestingly, his data closely correlated with Thompson's earlier self-play results of BELLE [3,30].

In 1997, Junghanns et al. [13] let PHOENIX and THE TURK search roughly 1,000 positions from self-play games to fixed depths of 9 plies while recording new best moves beside other information. Also during 1997, Hyatt and Newborn [11] conducted another behavioural experiment with Hyatt's chess program CRAFTY searching 347 new test positions to fixed depths of 14 plies. This experiment revealed the astonishing fact that the rate of new best moves as chosen by CRAFTY at high search depths of 9–14 plies remained quite steady around 15%–17% on average and hardly decreased anymore. Following up thereon, we confirmed Hyatt and Newborn's findings by repeating their "go deep" experiment with our own chess program DARKTHOUGHT in 1998 [6,10]. Recently, we pushed the limits of going deep to fixed depths of 16 plies [6,8] where the best-change rate of DARKTHOUGHT still remains steady at roughly 15%.

Self-play with handicaps in search depth, search speed, or search time between otherwise identical program versions is a valuable tool not only for computer chess but for computer strategy game-playing in general. Examples from other domains than chess include self-play experiments in computer checkers by Schaeffer et al. [12,13,24,25] with the reigning World Man-Machine Checkers Champion CHINOOK as well as self-play experiments in computer Othello by Lee and Mahajan [15] with their program BILL and by Brockington et al. [2,12,13] with his Othello program KEYANO. For all published self-play experiments in computer chess (see also Table 1) we provide further descriptions below.

1982: Thompson [30]. Thompson's initial experiment featured 100 self-play games with matches of 20 games each between versions of BELLE differing by exactly one ply in lookahead for fixed depths of 3–8 plies. The gain in playing strength averaged at 246 rating points per ply of search. The experiment showed no diminishing returns at any depth.

1983: Condon and Thompson [3]. In the second experiment, Condon and Thompson let BELLE self-play 300 games in round-robin style with matches of 20 games each between all program versions for fixed depths of 4–9 plies. The gain in playing strength averaged at 217 rating points per ply of search. The observed ratings slightly hinted at limited diminishing returns from a fixed depth of 6 plies onwards. Yet, the results of the experiment are not statistically significant.

1988: Szabo and Szabo [29]. The Szabos determined the technology curve of their chess program TECHMATE that self-played 6,882 games on two Atari ST computers. The number of games per match between longer and shorter searching versions of the program varied strongly from a minimum of 32 to a maximum of 1367. The gain in playing strength averaged at 156 rating points per doubling of available search time (computing power). The experimental data indicated slight diminishing returns at longer search times. However, the Szabos simply did not play enough games at long times to draw reliable conclusions.

1990: Berliner et al. [1]. The HITECH team made their chess machine self-play 1,056 games in a round-robin setting with matches of 16 games each between all program versions of HITECH and LOTECH (a variant of HITECH scaled down knowledge-wise) for fixed depths of 4–9 plies. The gain in playing strength averaged 195 rating points per ply of search for HITECH and 232 rating points per ply for LOTECH. The ratings showed possible signs of limited diminishing returns starting at a fixed depth of 6 plies. But there was no clear trend of diminishing returns at higher search depths and the experimental results are not statistically significant.

1994: Mysliwietz [20]. Mysliwietz let the parallel chess program ZUGZWANG self-play 450 games with 50 games per match between program versions that differed roughly by a factor of two in search speed due to varying numbers of allotted processors. The gain in playing strength averaged 109 rating points per doubling of search speed for 9 successive doubling steps. The observed ratings do not exhibit any diminishing returns at all.

1996: Schaeffer [13]. Junghanns et al. [12,13] briefly mentioned the results of a self-play experiment by Schaeffer with his chess program PHOENIX in 1996. The experiment comprised 120 self-play games with matches of 20 games each between program versions that differed by exactly one ply in lookahead for fixed depths of 3–9 plies. The gain in playing strength averaged at 228 rating points per ply of search. The result of the "9 ⇔ 8" match might be interpreted as an indication of diminishing returns. Yet, a single error-prone data point like this at the end of the curve really lacks significance.

1997: Junghanns et al. [13]. The self-play experiment with Björnsson and Junghanns' chess program THE TURK featured 480 games with matches of 80 games each between program versions differing by exactly one ply in lookahead for fixed depths of 3–9 plies. The gain in playing strength averaged around 200 rating points per ply of search. The scoring rates of the deeper searching versions of THE TURK actually increased steadily from fixed search depths of 6 plies onwards, thus even hinting at additional gains in returns for higher search depths rather than diminishing ones.

Junghanns et al. continued to look for diminishing returns by means of other metrics than self-play in [13]. They finally claimed to have found empirical evidence in this respect. According to their explanations, the low search quality of chess programs (i.e. their high error probability) and the abnormally large lengths of self-play games inadvertently hide diminishing returns in computer chess (which doubtlessly exist in their opinion). Although we greatly appreciate Junghanns et al.'s trial aimed at the better understanding of diminishing returns in computer chess, we are not convinced that their claims hold when subjected to rigorous methodological and statistical testing. Hence, the quest for indisputable and statistically significant demonstrations of diminishing returns for additional search in computer chess still remained to be concluded.

3 Statistical Analysis of Self-Play Experiments

In our recent publications [6,9] we introduced a general mathematical framework for the statistical confidence analysis of self-play experiments. Based on this framework, we scrutinized the self-play data published by other researchers for computer chess, computer checkers, and computer Othello (see Chapter 9 of [6]). Of course, we apply the same framework here to analyze our own new self-play results. For the sake of completeness, we briefly explain the underlying fundamentals and notations of the framework below.

We call $w = x/n$ the *scoring rate* which results from a score of $x \leq n$ points in a match or tournament of n games. The scoring rate $0 \leq w \leq 1$ estimates a player's real *winning probability* in games versus the respective opponents. Therefore, we may simply assume the scoring rate to be the sample mean of a binary-valued random variable that counts two draws as a loss plus a win. This enables the calculation of standard errors and %-level confident bounds for any match results by applying classical statistics [5,19] to the values of x and n.

Standard Errors of Scoring Rates. The standard error $s(w)$ of a scoring rate $w = x/n$ is given by $s(w) = \overline{w * (1 - w)/n}$.

Confident Bounds on Winning Probabilities. Let $z_{\%}$ denote the upper critical value of the standard $N(0, 1)$ normal distribution for any desired %-level of statistical confidence ($z_{90\%} = 1.645$, $z_{95\%} = 1.96$).

- $w \pm z_{\%} * s(w)$

places %-level confident lower and upper bounds on the real winning probability of a player with scoring rate $w = x/n$. [Remark: The bounds are accurate only if $x > 4$ and $n - x > 4$. Otherwise, the sample data does not provide enough information for the determination of statistically confident bounds. In such cases the approximate bound values as calculated by the given formula underestimate the real deviations possible.]

Confident Bounds on Differences of Winning Probabilities. From the above we derive %-level confident lower and upper bounds on the difference in real winning probability between two players with scoring rates $w_1 = x_1/n_1$ and $w_2 = x_2/n_2$ where $w_1 \geq w_2$.

- $l_\% = \max \left((w_1 - z_\% * s(w_1)) - (w_2 + z_\% * s(w_2)), \ -1 \right)$

- $u_\% = \min \left((w_1 + z_\% * s(w_1)) - (w_2 - z_\% * s(w_2)), \ +1 \right)$

For these bounds it holds that $-1 \leq l_\% \leq u_\% \leq 1$ and $u_\% \geq 0$. We denote the range $[l_\%, u_\%]$ by %-*level confident* Δw. The tables of this paper also refer thereto by "90%-C Δw" and "95%-C Δw" in their column heads.

The bounds allow for confident quantifications of differences in playing strength between two players as measured by their winning probabilities. Whenever $l_\% > 0$ we are %-level confident that the player with the higher scoring rate is indeed stronger than the other. If $l_\% \leq 0$, however, we cannot discriminate the two players' strengths with the desired confidence: the supposedly weaker player with the lower scoring rate might really be as strong as the other or even stronger.

Our self-play matches test the playing strengths of successive program versions on the scale of increasing search depths. We use the scoring rates w_1 of the match winners and $w_2 = 1 - w_1$ of the losers for our calculations of $l_\%$ and $u_\%$. After determining w and $s(w)$, we calculate the %-level confident ranges $[l_\%, u_\%]$ for all consecutive matches and call their intersection $[\Delta w]_\%$. If $[\Delta w]_\% = \emptyset$ (empty intersection) we are %-level confident that the differences in real winning probability and, thus, playing strength of successive program versions cannot be identical for all tested ones. Then, the overall results refute the notion of constant returns for additional search throughout the whole experiment with the desired %-level of confidence. Otherwise, the union $[L_\%, U_\%] = [l_\%, u_\%]$ of all confident bound ranges confirms constant or at least nearly constant returns for additional search of the tested program if $U_\% - L_\% < \epsilon$ for some small $\epsilon \geq 0$.

4   Experimental Setup

In our self-play matches the initial release version of FRITZ 6 competed against itself (engine date: November 10, 1999; size: 291,328 bytes). All opponents relied on the opening book "General.ctg" from the original FRITZ 6 CD-ROM with tournament mode and maximal variety of play activated. We disabled the book learning, tablebase access (no endgame databases installed), permanent brain, and early resign options. Moreover, we set the contempt values of the engines to zero. The detailed overall engine setup of FRITZ 6 for our self-play matches looked as shown in Table 2.

We executed the self-play matches on several different Windows-98/NT machines with 128 MB to 256 MB of RAM (300 MHz Pentium-II, 333 MHz Celeron, 450 MHz K6-2, 450 MHz & 500 MHz Pentium-III). We used the "Engine Match" function of FRITZ 6 to set up and play the matches at fixed iteration depths for both contestants. The engines with the higher fixed depths always appear as FRITZ 6A in the game scores. By activating the "Alternate Colours" option of the "Engine Match" dialogue,

Table 2. Detailed engine setup of FRITZ 6 for the self-play matches.

Book Choice	"General.ctg"
Book Options	use tournament book = on, use book = on
	minimum games = 2
	variety of play = maximal (++)
	influence of learn value = none (− −)
	learning strength = none (− −)
Use Tablebases	off
Engine Parameters	contempt value = 0, aggressiveness = 0
	selectivity = 2, tablebase depth = 0
Hashtable Size	32 MB

Table 3. Match details of FRITZ 6 self-play calibration results.

Depth	m	W : D : L / Total	Wins	Draws	Losses	Score
$8_W \Leftrightarrow 8_B$	68	924 : 1,288 : 788 / 3,000	30.80%	42.93%	26.27%	52.27%

we made sure that all opening positions chosen from the book at random served as sources for two games with the opponents playing reversed colours in each. We used the free CHESSBASE READER program to analyze the generated match databases.

4.1 Calibration Match

We checked the integrity and Black / White fairness of the FRITZ 6 opening book by means of a "calibration" match with 3,000 games between two completely identical opponents. The calibration match pitted FRITZ 6 at a fixed iteration depth of 8 plies against itself. During the calibration match, we disabled the "Alternate Colours" option of the "Engine Match" dialogue in order to avoid useless game repetitions. Thus, the 8-ply FRITZ 6 engine playing Black in the calibration match always appears as FRITZ 6A in the game scores.

Table 3 provides detailed information about the outcome of the calibration match: "m" gives the average number of moves per game and "W : D : L" presents the absolute overall numbers of wins, draws, and losses of the first player ("$8_{W/B}$" denotes 8-ply FRITZ 6 playing White / Black). The remaining columns of the table list the "W : D : L" data and the overall score of the first player as relative percentages of the total game count for the match. Table 4 subjects the results of the calibration match to our procedure of statistical analysis as introduced in Section 3. The close-to level score of the calibration match with a small tilt in favour of White (52.3% vs. 47.7%) and its good statistical confidence (less than 1% standard error) validate the suitability of both FRITZ 6 and its opening book for our self-play purposes.

Book Calibration. Assuming equal playing strength for FRITZ 6 regardless of colour at a fixed iteration depth of 8 plies, the calibration match verifies the integrity and Black / White fairness of the opening book.

Table 4. Statistical analysis of FRITZ 6 self-play calibration results.

Depth	Score	w	s(w)	90%-C Δw	95%-C Δw
$8_W \Leftrightarrow 8_B$	1,568.0 / 3,000	0.523	0.009	0.015, 0.075	0.010, 0.081

Table 5. Match details of FRITZ 6 self-play results.

Depth	m	W : D : L / Total	Wins	Draws	Losses	Score	ELO
6 ⇔ 5	63	1,686 : 915 : 399 / 3,000	56.20%	30.50%	13.30%	71.45%	+159
7 ⇔ 6	65	1,643 : 1,066 : 291 / 3,000	54.77%	35.53%	9.70%	72.53%	+169
8 ⇔ 7	67	1,457 : 1,212 : 331 / 3,000	48.57%	40.40%	11.03%	68.77%	+137
9 ⇔ 8	66	1,433 : 1,235 : 332 / 3,000	47.77%	41.17%	11.07%	68.35%	+134
10 ⇔ 9	68	1,252 : 1,451 : 297 / 3,000	41.73%	48.37%	9.90%	65.92%	+115
11 ⇔ 10	67	1,124 : 1,525 : 351 / 3,000	37.47%	50.83%	11.70%	62.88%	+92
12 ⇔ 11	68	1,059 : 1,592 : 349 / 3,000	35.30%	53.07%	11.63%	61.83%	+84

Engine Calibration. Assuming the integrity and Black / White fairness of the opening book, the calibration match verifies the equal playing strength of FRITZ 6 with Black and White at a fixed iteration depth of 8 plies.

[Remark: Strictly speaking, additional calibration matches are required in order to scale the calibration to other iteration depths. We did not deem the according effort worthwhile because it is very unlikely that the fairness and integrity of the opening book or the Black / White behaviour of FRITZ 6 at other depths differ substantially from those at 8 plies as observed in our single calibration match.]

5   Self-Play Results

We executed seven "depth X+1 ⇔ X" handicap matches with FRITZ 6 at fixed iteration depths ranging from 5–12 plies. The overall number of handicap self-play games amounts to 21,000 with 3,000 games per match. Table 5 provides detailed information about the outcome of the self-play matches (see Section 4.1 and Table 3 for an explanation of the format).

The stunning conclusion of our experiment is that it not only hints at but clearly shows the existence of diminishing returns for additional search in direct self-play by the chess program FRITZ 6. Beyond fixed iteration depths of 9 plies, the scoring rates of the deeper searching program versions decline substantially from 68.35% for "9 ⇔ 8" to a mere 61.8% for "12 ⇔ 11". The average rating increase for an additional ply of search (measured in iteration depth) shrinks dramatically by half from 169 ELO for "7 ⇔ 6" to just 84 ELO for "12 ⇔ 11".

Further evidence of diminishing returns is visible from the "W : D : L" data. The changes in the rates of games won and drawn by the deeper searching program versions are of particular significance in this context. While the rates of lost games stay fairly constant around 11%, the rates of won games decrease steadily from 56.2% for "6 ⇔ 5" to 35.3% for "12 ⇔ 11". Reversely, the rates of drawn games increase from 30.5%

Table 6. Statistical analysis of FRITZ 6 self-play results.

Depth	Score	w	s(w)	90%-C Δw	95%-C Δw
6 ⇔ 5	2,143.5 / 3,000	0.715	0.008	0.402, 0.456	0.397, 0.461
7 ⇔ 6	2,176.0 / 3,000	0.725	0.008	0.424, 0.477	0.419, 0.483
8 ⇔ 7	2,063.0 / 3,000	0.688	0.008	0.347, 0.403	0.342, 0.409
9 ⇔ 8	2,050.5 / 3,000	0.683	0.008	0.339, 0.395	0.334, 0.400
10 ⇔ 9	1,977.5 / 3,000	0.659	0.009	0.290, 0.347	0.284, 0.352
11 ⇔ 10	1,886.5 / 3,000	0.629	0.009	0.229, 0.287	0.223, 0.292
12 ⇔ 11	1,855.0 / 3,000	0.618	0.009	0.207, 0.266	0.202, 0.271
[Δw]	–	–	–	$\emptyset$	$\emptyset$

for "6 ⇔ 5" to 53.0% for "12 ⇔ 11" (see Figure 3 later on). Although the deeper searching program versions apparently do not lose more games, they show clear signs of diminishing abilities to win with progressing search depth. Interestingly enough, the average length of the self-play games hardly changes throughout the whole depth range.

5.1 Statistical Analysis of the Results

Table 6 presents a rigorous statistical analysis of the match results based on the framework introduced in Section 3. The experimental data allows us to conclude with 95% statistical confidence that the differences in playing strength of FRITZ 6 in handicap self-play at fixed iteration depths of 11–12 plies are indeed smaller than those at 6–9 plies: $[\Delta w]_{90\%} = [\Delta w]_{95\%} = \emptyset$ and $0.292 = \max(u_{95\%}) < 0.334 = \min(l_{95\%})$ holds for these two sets of depths.

Yet, the non-empty intersection $[0.397, 0.400]$ for the $\Delta w_{95\%}$ ranges up to "9 ⇔ 8" shows that the differences in playing strength could still be constant from 6–9 plies. The same holds for the successive pairs of iteration depths from 10–11 plies ($[\Delta w_{95\%}] = [0.284, 0.292]$) and 11–12 plies ($[\Delta w_{95\%}] = [0.223, 0.271]$). However, we deem such a scenario of constants to be highly unlikely because of the steadily decreasing scoring rates and Δw bounds from "7 ⇔ 6" onwards.

Unfortunately, the available data does not directly allow for the confident quantification of diminishing returns at successive iteration depths. Still, it does provide interesting bounds on the differences in rating gains per iteration at high and low depths respectively. For instance, from $u_{95\%} = 0.271$ for "12 ⇔ 11" and $l_{95\%} = 0.424$ for "7 ⇔ 6" we may conclude with 95% statistical confidence that the difference in playing strength between 12-ply and 11-ply FRITZ 6 is at least by $0.424 - 0.271 = 0.153$ (i.e., 15.3 percentage points) smaller than the one between 7-ply and 6-ply FRITZ 6. This translates to a 95%-confident drop of the scoring rate by at least $0.153 / 2 = 0.0765$ (i.e., 7.65 percentage points) when going from "7 ⇔ 6" to "12 ⇔ 11" with FRITZ 6.

6   Discussion and Conclusion

Our past conjectures as published in [6,9] (see Section 1.1) were right on track. We really needed thousands of games per match in order to demonstrate the existence of

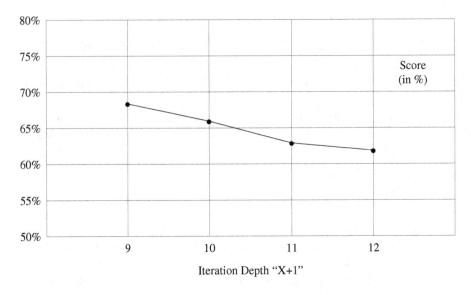

Fig. 1. Self-play of FRITZ 6 at high depths from "9 ⇔ 8" to "12 ⇔ 11"

diminishing returns for additional search in computer chess self-play with 95% statistical confidence. But more than 20,000 self-play games of FRITZ 6 still did not suffice to quantify the diminishing effect at successive iteration depths with good confidence. Altogether, the 24,000 games required an equivalent of roughly six month of CPU time on a fast 450 MHz K6-2 (the "12 ⇔ 11" match alone required almost half of the whole time). By distributing the games over several different machines, we were able to conduct the full experiment within a wall-clock time frame of three months.

The effects of diminishing returns for additional search seem to kick in strong with FRITZ 6 at iteration depths beyond 9 plies (see Figure 1 and Table 5). These depths lie just outside the range covered by previous self-play experiments (see Section 2 and Table 1). If we ignore the matches beyond "9 ⇔ 8", our results as depicted in Figure 2 actually resemble the past scores obtained by other researchers. According to Table 6, $[\Delta w_{95\%}] = [0.397, 0.400]$ for our matches up to "9 ⇔ 8". As in the case of previous self-play experiments, our data does not confirm the existence of diminishing returns for the depth range of 5–9 plies alone with 95% statistical confidence. Based on the large number of 3,000 games per match in our experiment, we may therefore conclude with good confidence that diminishing returns remain quite subdued for FRITZ 6 in self-play at iteration depths of 5–9 plies. The results of all previous self-play experiments suggest this to be a general phenomenon of computer-chess programs performing iteratively deepened alpha-beta searches.

However, Table 5 offers some interesting evidence that diminishing returns do affect FRITZ 6 also at low search depths. As described in Section 5 and plotted in Figure 3, the rates of games won by the deeper searching program versions decline steadily for the whole range of iteration depths while the rates of draws reversely increase in accordance even at low depths. Hence, diminishing returns with respect to win / draw rates (instead

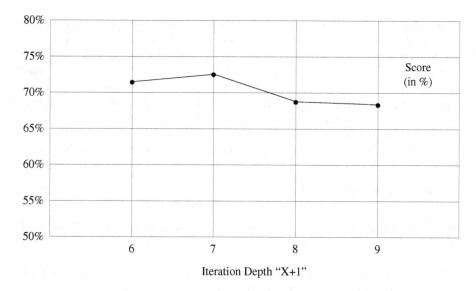

Fig. 2. Self-play of FRITZ 6 at low depths from "6 ⇔ 5" to "9 ⇔ 8"

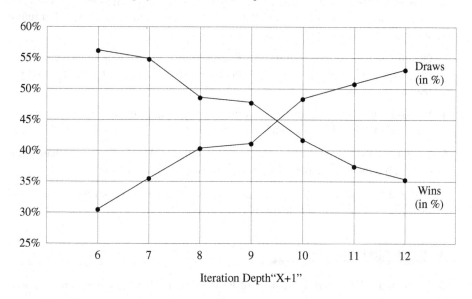

Fig. 3. Self-play win / draw rates of FRITZ 6 from "6 ⇔ 5" to "12 ⇔ 11"

of the scoring rates) are clearly visible for all iteration depths. This suggests that the detailed win / draw rates might be much better indicators for diminishing returns than the overall scoring rates.

Last but not least, we like to mention that FRITZ 6 does only very limited pre-processing at the root node (according to personal communication with its author,

Frans Morsch). Hence, the effects of diminishing returns observed in our experiment are not simply caused by extensive pre-processing. Schubert's recent report [27] in a German computer-chess periodical provides further evidence for this. After our publication of some intermediate results [7], Schubert conducted his own self-play experiments with HIARCS 7.32 and an enhanced version of FRITZ 6 which pre-processes even less than the original release version used by us. HIARCS 7.32 is known to be a slow and knowledge-intensive chess program that hardly relies on any pre-processing but rather expands much effort to analyze each node during the search in great detail. Diminishing returns were visible for both FRITZ 6 and HIARCS 7.32 in Schubert's experiment.

7 Future Work

Based on the experiments and results discussed in this paper, the course of future self-play research in computer chess seems quite obvious.

1. Include the detailed win / draw rates of all matches in the analyses and further investigate their suitability as indicators for diminishing returns.

2. Develop mathematical models of the scoring rates and the win / draw rates (e.g. by means of least-square error interpolations) in order to extrapolate and predict future self-play results at even higher search depths.

3. Quantify the diminishing returns and differences in playing strength at fixed iteration depths up to 12 plies with good statistical confidence $\Rightarrow$ play even more games per match at these depths.

4. Perform self-play experiments with other chess programs in order to ensure that we are not only measuring weird artifacts of FRITZ 6.

5. Execute statistically significant self-play matches with speed and time handicaps that complement the depth-handicap results in a meaningful way.

6. Push the fixed iteration depths of self-play in computer chess beyond 12 plies and then do the same as before and suggested above.

Acknowledgements

ChessBase GmbH donated free copies of FRITZ 6 to us after including handicap self-play abilities in the standard retail version, a feature we had already suggested long ago. Matthias Wüllenweber of ChessBase GmbH proved to be an especially avid supporter of our new self-play experiment, providing various engine binaries for test purposes among other things.

The availability and exclusive usage of several fast x86-PC compatible machines for a period of some months starting in December 1999 was equally important for the overall success of the experiment. These PCs were kindly provided by the Supertechnologies Group of the Laboratory for Computer Science at the Massachussetts Institute of Technology (M.I.T.), headed by Prof. Leiserson.

References

1. Berliner, H. J. and Goetsch, G. and Campbell, M. S. and Ebeling, C. (1990). Measuring the performance potential of chess programs. *Artificial Intelligence*, Vol. 43, No. 1, pp. 7–21.
2. Brockington, M. G. (1997). KEYANO *unplugged – The construction of an Othello program.* Technical Report TR 97–05, Department of Computing Science, University of Alberta.
3. Condon, J. H. and Thompson, K. (1983). BELLE. *Chess Skill in Man and Machine*, P. W. Frey (ed.), pp. 82–118, Springer, 2nd ed. 1983, ISBN 0-387-90790-4 / 3-540-90790-4.
4. Gillogly. J. J. (1978). *Performance Analysis of the Technology Chess Program.* Ph.D. Thesis, Carnegie-Mellon University [printed as Technical Report CMU-CS-78-189, Computer Science Department, Carnegie-Mellon University].
5. Heinhold, J. and Gaede, K.-W. (1964). *Ingenieur-Statistik.* Oldenbourg, 3rd ed. 1972, ISBN 3-486-31743-1 (in German).
6. Heinz, E. A. (2000). *Scalable Search in Computer Chess.* Vieweg / Morgan Kaufmann, ISBN 3-528-05732-7.
7. Heinz, E. A. (2000). A New Self-Play Experiment in Computer Chess. Technical Memo No. 608 (MIT-LCS-TM-608), Laboratory for Computer Science, Massachussetts Institute of Technology.
8. Heinz, E. A. (2000). Modeling the "go deep" behaviour of CRAFTY and DARKTHOUGHT. *Advances in Computer Games 9*, Proceedings, H. J. van den Herik and B. Monien (eds.), to be published.
9. Heinz, E. A. (2000). Self-play experiments in computer chess revisited. *Advances in Computer Games 9*, Proceedings, H. J. van den Herik and B. Monien (eds.), to be published.
10. Heinz, E. A. (1998). DARKTHOUGHT goes deep. *ICCA Journal*, Vol. 21, No. 4, pp. 228–244.
11. Hyatt, R. M. and Newborn, M. M. (1997). CRAFTY goes deep. *ICCA Journal*, Vol. 20, No. 2, pp. 79–86.
12. Junghanns, A. and Schaeffer, J. (1997). Search versus knowledge in game-playing programs revisited. *15th International Joint Conference on Artificial Intelligence*, Proceedings Vol. I, pp. 692–697, Morgan Kaufmann, ISBN 1-558-60480-4.
13. Junghanns, A. and Schaeffer, J. and Brockington, M. and Björnsson, Y. and Marsland, T. A. (1997). Diminishing returns for additional search in chess. *Advances in Computer Chess 8*, H. J. van den Herik and J. W. H. M. Uiterwijk (eds.), pp. 53–67, University of Maastricht, ISBN 9-062-16234-7.
14. Korf, R. E. (1985). Iterative deepening: An optimal admissible tree search. *Artificial Intelligence*, Vol. 27, No. 1, pp. 97–109.
15. Lee, K.-F. and Mahajan, S. (1990). The development of a world-class Othello program. *Artificial Intelligence*, Vol. 43, No. 1, pp. 21–36.
16. Levy, D. N. L. (1997). Crystal balls: The meta-science of prediction in computer chess. *ICCA Journal*, Vol. 20, No. 2, pp. 71–78.
17. Levy, D. N. L. and Newborn, M. M. (1991). *How Computers Play Chess.* Computer Science Press, ISBN 0-716-78121-2 / 0-716-78239-1.
18. Levy, D. N. L. (1986). When will brute force programs beat Kasparov? *ICCA Journal*, Vol. 9, No. 2, pp. 81–86.
19. Moore, D. S. and McCabe, G. P. (1993). *Introduction to the Practice of Statistics.* W. H. Freyman, 2nd. ed., ISBN 0-716-72250-X.
20. Mysliwietz, P. (1994). *Konstruktion und Optimierung von Bewertungsfunktionen beim Schach.* Dissertation (Ph.D. Thesis), University of Paderborn.
21. Newborn, M. M. (1985). A hypothesis concerning the strength of chess programs. *ICCA Journal*, Vol. 8, No. 4, pp. 209–215.

22. Newborn, M. M. (1979). Recent progress in computer chess. *Advances in Computers*, Vol. 18, pp. 59–117 [reprinted in *Computer Games I*, D. N. L. Levy (ed.), pp. 226–324, Springer, ISBN 0-387-96496-4 / 3-540-96496-4].

23. Newborn, M. M. (1978). Computer chess: Recent progress and future expectations. *3rd Jerusalem Conference on Information Technology*, Proceedings, J. Moneta (ed.), North-Holland, ISBN 0-444-85192-5.

24. Schaeffer, J. (1997). *One Jump Ahead: Challenging Human Supremacy in Checkers*. Springer, ISBN 0-387-94930-5.

25. Schaeffer, J. and Lu, P. and Szafron, D. and Lake, R. (1993). A re-examination of brute-force search, *AAAI Fall Symposium*, Proceedings (AAAI Report FS-93-02: *Intelligent Games – Planning and Learning*), S. Epstein and R. Levinson (eds.), pp. 51–58, AAAI Press, ISBN 0-929-28051-2.

26. Schaeffer, J. (1986). *Experiments in Search and Knowledge*. Ph.D. Thesis, University of Waterloo [reprinted as Technical Report TR 86–12, Department of Computing Science, University of Alberta].

27. Schubert, F. (2000). Das Ende der Fahnenstange? Über den fallenden Grenzwertnutzen im Computerschach. *Computer-Schach & Spiele*, Vol. 18, No. 4, pp. 50–54.

28. Slate, D. J. and Atkin, L. R. (1977). CHESS 4.5 – The Northwestern University chess program. *Chess Skill in Man and Machine*, P. W. Frey (ed.), pp. 82–118, Springer, 2nd ed. 1983, ISBN 0-387-90790-4/3-540-90790-4.

29. Szabo, A. and Szabo, B. (1988). The technology curve revisited. *ICCA Journal*, Vol. 11, No. 1, pp. 14–20.

30. Thompson, K. (1982). Computer chess strength. *Advances in Computer Chess 3*, M. R. B. Clarke (ed.), pp. 55–56, Pergamon, ISBN 0-080-26898-6.

SUPER-SOMA – Solving Tactical Exchanges in Shogi without Tree Searching

Jeff Rollason

47 Rickmansworth Road, Pinner, Middx, HA5 3TJ, England
jeffrollason@cs.com

Abstract. A key feature of programs that play games such as Chess and Shogi is the ability to evaluate the outcome of threatened tactical moves. In Chess this is usually solved using a combination of tactical and capture search. This works well as exchanges rapidly simplify and a solution can usually be quickly found. In Shogi (Japanese Chess) the problem is not so simple as captured pieces are immediately available for tactical drops and so tactical threats do not quickly simplify. Since the number of tactical threats in Shogi also tends to be much larger than in Chess, then this makes solving threats using tactical and capture search much more difficult. In the Shogi-playing program SHOTEST I have taken a different approach to this and created a tactical exchange evaluator which can statically do the work of a tactical search. This approach has its ancestry in the well-known and simple SOMA algorithm used to determine single square exchanges. However the algorithm SUPER-SOMA described in this paper can also deal with multi-square captures, pins, ties, discovered attacks, promotions, defensive play, mate threats, mate ties and even positional moves.

Keywords: shogi, shotest, soma, super-soma, evaluation, search

1 Introduction

The game of Shogi has much in common with Western Chess (see section 4). However it poses problems for the AI-programmer that make it difficult to solve. In Chess there are two competing methods for controlling tree search. These are broadly:

- Brute force, which generally looks at most moves and uses tree-search techniques to reduce the number of nodes evaluated. Each node is a simple and quick evaluation.
- Selective search, which uses Chess knowledge to direct search down selected paths, performing more complex evaluations of a much smaller number of nodes.

In the early days of Chess the Brute force method was generally more successful as selected techniques proved unreliable for controlling search. In more modern implementations there is now higher selectivity, but the search still examines a high proportion of all moves at the shallower plies. The number of moves examined per ply is still largely controlled by search cutoffs. This tendency to consider most legal moves is only possible because the number of legal moves is usually only between 30 and

T.A. Marsland and I. Frank (Eds.): CG 2000, LNCS 2063, pp. 277–296, 2001.

40. In Shogi this figure is nearer 106. This figure was derived from a test I performed on 100 games between two different computer programs that had progressed to 300 or more moves. This gave a mean of 106, a median of 90 and maximum of 340 moves (the theoretical maximum is 593). I only tested moves 1 to 300, and not beyond. The distribution of these showed two distinct peaks at around 31-40 moves and 71-80 moves. Since in order to solve Shogi it is necessary to be able to work at all stages of the game, it is reasonable to find a value for the number of legal moves which will encompass the larger proportion of positions. If this boundary is set at the bottom 90%, then a figure of 150 legal moves fulfils this. Using this we can calculate the impact this will have on conventional alpha-beta search.

Calculating the minimum number of nodes for a depth-first, fixed depth alpha-beta search (Knuth and Moore, 1975):

$$N = 2 \times w^{(d/2)} \qquad \text{width "w" and depth "d"} \qquad (1)$$

# Moves	Depth	# Nodes
30	8	1,620,000
150	8	1,012,500,000
300	8	16,200,000,000
30	12	1,458,000,000

From the table we can see that at depth 8 increasing the number of legal moves from 30 to 150 is the equivalent of increasing the depth to 12 for the same width. Shogi still needs deep searches as Chess, but these are inevitably harder.

For this reason it is almost certain that to implement a Shogi-program it is necessary to consider selective techniques if the intended Shogi program expects to evaluate deep search trees. With this in mind I determined to create a Shogi program that used highly selective search methods. To reduce 300 candidate moves to a more manageable number would require a sophisticated evaluation with a strong idea of which moves were viable to consider.

In this paper I have, conforming to my Chess background, assumed white is to play next, which is contrary to normal practice in Shogi articles where Black normally plays first.

2   Design Plan for SHOTEST

The primary design plan is to create Shogi program with an evaluation function to do two key things:

1. Reduce the need for search wherever possible
2. Use the evaluation to direct the search

As a part of this I planned to create an evaluation function that would achieve lookahead of captures and tactical moves, but without performing a tree search. This would do part of the job of a capture and limited tactical search. The expectation was that such an evaluator would be less expensive than search, but less accurate. This limitation need not compromise the ability of the program as this evaluator could be used within a limited tree search. Where the evaluator could see that the position was complex it could direct the search to examine the position for further plies. Since the evaluator should have a good understanding of the threats in the position it should be able to make a good choice of moves to search, and have a good idea when a position has stabilised.

If it is viable to substitute static evaluation for conventional capture search, then such a system would replace the evaluation of a large number of simple nodes by fewer more complex evaluations. This should have extra benefits. For example in Shogi positional components often exceed material values. Having a more complex node analysis is likely to make it easier to reliably assign such high values. In practice Shotest examines 50 times fewer nodes than (for example) the program YSS in the same time (Yamashita, H - 1997). If the search can afford to examine 1/50th the number of nodes, then the positional evaluation can afford to take 50 times longer than it would in a simpler search.

As a final component, if the tactical evaluator was successful, then it might also be possible to create a predictive engine that would combine both positional and tactical moves.

3   Design Plan for SUPER-SOMA

The core of the proposed work is to attempt to predict exchanges across the whole board. There is already an algorithm called SOMA that does this (Michie, D - 1966), but only considers each square in isolation. This does not allow whole board situations to be assessed. SUPER-SOMA needs to find a mechanism to cross-reference these potential exchanges in such a way that sequences of moves from different parts of the board can be predicted. This will require each SOMA exchange to be linked in such a way that choices of captures can be prioritised. The following scheme does this.

A   Generate the table "XREF" to contain a list of tactical moves

    A1. Do a basic scan to determine all attacks on all squares.

    A2. Identify all pins, ties and discovered attacks. These would include ties to defending material, prevention of promotion and threats of mate.

    A3. Create the table XREF with all exchanges on the board. This would use a variant of SOMA that would understand where pieces are pinned tied or activate discovered attacks.

    A4. Add further threats to XREF including forks, attacks on immobile pieces and promotions.

B   Apply SUPER-SOMA to XREF

    B1. Apply an initial weight to each XREF entry depending on its expected net value as an isolated tactical move.

    B2. Cross reference each XREF entry, applying an enhanced weighting for each table entry based on its influence on other XREF entries. e.g. The move BxP might have the initial weight of one pawn, but if the bishop is also vulnerable to capture, then the BxP entry would be weighted as value of pawn+bishop.

B3. Choose the best move from XREF for the side to play next. This might be a capture, promotion, fork or even a move to neutralise an opponent's threat. If there are no moves with a positive weighting then generate a "pass" move.

B4. Update the XREF table after making the move, and change the side to play (if appropriate)

B5. Repeat from step (B1) until both sides have played a pass move.

This complete process is quite complex. To make it comprehensible it is necessary to demonstrate some of the components in simple examples. A core component in this is the simple SOMA algorithm, which is described in section 5.

4   Basic Rules and Notation of Shogi

Before proceeding with the body of this paper it is necessary to outline the rules of Shogi so that the examples can be understood. As indicated before, Shogi is in many respects similar to Chess with the same game objective. The key differences are: (1) the drop rule, which allows captured pieces to be dropped on the board as a move, (2) the similar but differing piece sets, (3) The 9x9 board and (4) the three rank deep promotion zone.

The examples that follow show Shogi positions using Japanese Kanji.

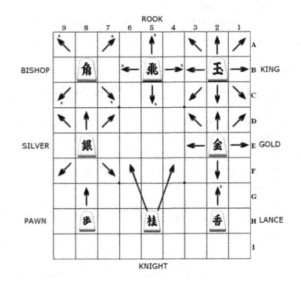

Fig. 1. Shogi pieces and their moves

This is hard on Western eyes but is the normal representation. I have chosen to use this in preference to westernised notation as the latter is not widely used, even by western Shogi players. I have made this slightly easier by showing actual pieces as they would be shown on a Shogi board rather than the plain Kanji notation used in Japanese Shogi texts. The latter assumes that the reader can distinguish between inverted and non-inverted Kanji characters, which is hard for new readers. The pieces and their moves are shown in Figure 1. Move arrows marked by "*" indicate multiple square moves are possible. I have not shown the pieces or moves of promoted pieces, as the examples below do not use any. The promoted versions of pawns, lances, knights and silvers all move in the same way as golds. Promoted bishops and rooks move as kings, in addition to the moves available to their unpromoted counterparts. I have also adopted text notation in my discussion based on that used by Western Shogi players. It uses a coordinate system that reverses Chess by starting with a number before a letter, e.g. squares 4e, 5a etc.. rather than e4 and a5. Pieces are assigned letters, e.g. "P" pawn and "N" knight etc. Promoted pieces are shown as the original piece letter followed by a "+". Unlike Chess there are no real black and white pieces, but the idea of black and white sides is adopted to denote the side to move first and second respectively. In Shogi the same pieces are used for both sides and are distinguished on the board by the direction that they face. The pieces are pointed and lie flat on the board (rather than erect, as in Chess): A player's pieces face away and their opponent's towards the player. Promotion is achieved by turning the pieces over while keeping the direction the same. Promotion can occur when a pawn, lance, knight, silver, bishop or rook either move into, from or within the opponent's back three ranks. On the boards below, black plays from the bottom with the pieces pointing up the board.

In the text notation I have also applied a small modification that shows white pieces in uppercase and black pieces in lowercase. This makes some of the discussion slightly easier to follow. Play proceeds in the same way as Chess with the object being to deliver checkmate. The drop rule allows a player to play (drop) a previously captured pieces on any part of the board as a move. Such a piece will be dropped in its unpromoted state. Knights cannot be dropped on the opponent's back 2 ranks, or lances on the opponent back rank. Pawns can also not be dropped in a file that an existing unpromoted pawn already occupies, i.e. pawns can never be doubled. These basic rules result in a game that is lively compared to Chess, with much less dependence on material values.

5   Basic Operation of SOMA

A critical component in SUPER-SOMA is the fundamental SOMA algorithm. It is necessary to first understand how this works. The SOMA evaluation of exchanges on a single square require swapping off the pieces on that square, with each side having the option to stop the exchange at their most beneficial moment. For example in the position in Figure 2 the black pawn at 5d is attacked by the knight at 6b, rook at 5c and bishop at 4c, and defended by the knights at 6f and 4f. If the simplified values of P=1, N=2, S=3, G=3, B=5 and R=6 are used then the full exchange in this position would be:

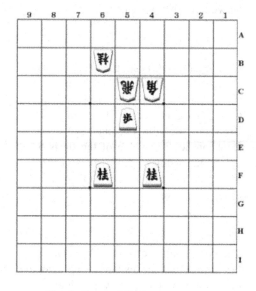

Fig. 2. Simple SOMA exchanges

```
Move   Net-exchange-value
Nxp    +1
nxN    -1
Bxn    +1
nxB    -4
Rxn    -2
```

From the exchange one can see that white is losing material. It has won 1 pawn + 2 knights, but lost a bishop + knight, leaving white 2 points down. White could stop earlier by rejecting the exchange Bxn. this leaves white 1 point down, which is worse. White can also stop earlier by not performing the first Nxp, leaving 0 exchanges points. Therefore white is better off not making any exchange at all.

The underlying idea behind this is that each side can stop the exchange early by not making a capture. This is only worthwhile when stopping early improves on the current exchange total. Determining where the endpoint is requires iterating the exchange until neither side can find an improvement.

In Shotest this simple mechanism is made more complex because pieces change their value as they move, therefore an exchange may end with a Gold capture which leaves the Gold on an inferior square and so this results in a loss of material. An exchange may therefore gain only because it has disrupted the positions of the pieces.

6   Enhancements to Simple SOMA

Shotest can also account for pins and ties in the SOMA evaluation, for example in Figure 3. Considering square 5g we have the exchange:

```
Nxp +1
gxN -1
```

Treating this as a simple SOMA exchange it is easy to see that no profitable capture is possible. However the Gold on 4g is tied to defending the Bishop on 3g, which is attacked by the Rook on 3e. The value of the tie is 5 points, the value of the Bishop. If this is incorporated into the SOMA exchange we have:

```
Nxp +1
gxN -1 +5 = +4
```

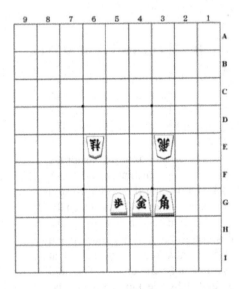

Fig. 3. Pins and Ties with SOMA

The consequence of this is that black's final move gxN leaves black with a net loss of 4 points, which is worse than no recapture. Applying the SOMA principle the exchange ends with Nxp with a net gain of 1 point for white. The tie to protecting the bishop has prevented the re-capture.

There are many subvariations possible here requiring special processing. In particular tied pieces and targets may be involved in the exchange square being considered. In these cases the tie or pin may be broken during the exchange.

Another consideration is that pieces that are pinned or tied may effect the ordering of the exchange. If the first capturing piece is a tied knight and the second piece is a free rook, then the capture sequence may be re-ordered. Finally the target of a capture may be tied, in which case the XREF entry may generate a second triggered entry which only becomes active when the parent capture occurs. Discovered attacks are also processed in a similar fashion.

7   Board Analysis Using SUPER-SOMA with the XREF Table

The examples above consider only captures around single squares. The SUPER-SOMA algorithm allows separate exchanges on the board to be linked and prioritised. The two examples below start with a simple case to demonstrate the principle. In these examples I have turned off the variable value of piece material as would make the examples harder to understand. I have instead assumed the following piece values:

P	L	N	S	G	B	R	K
52	120	120	200	220	360	480	3999

P+	L+	N+	B+	R+
240	230	230	460	550

All examples assume white is to play next

8   Example 1 of Whole Board Analysis

In the example in Figure 4 SUPER-SOMA finds 6 moves for the XREF table, as follows:

Move	Value	Type of move
l 6fxR6d	960	capture
s 4dxB4c	740	capture and promote
R 6dxs4d	-560	capture
R 6dxl6f	-720	capture
B 4cxp8g	152	capture and promote
s 4d- 5c	20	promote

From this we can see that the greatest value move is L 6fxR6d with an exchange value of 960 points (value of black losing rook plus value of white gaining rook in hand). This is a move for black though and white is to play next. White's most valuable move is the capture and promotion move B 4cxp8g. White's other moves are R 6dxl6f, which loses a Rook for a Lance and R 6dxs4d losing a Rook for a Silver, so both have negative values.

If you apply SUPER-SOMA to this position it predicts the following:

```
R 6dxs4d     400
n 3fxR4d     960
B 4cxp8g     152
  pass
  pass
Net value -408
```

SUPER-SOMA has predicted that the best move is the sacrifice of the Rook for the Silver. In Shogi terms this makes sense as simply moving the Rook away from the threat by the lance will result in black then capturing the Bishop with the Silver, whereas sacrificing the Rook for the Silver also prevents the capture of the Bishop, so this is the correct result. We can look to see how SUPER-SOMA does this.

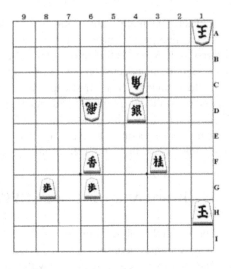

Fig. 4. Example 1 - Whole board analysis and SOMA

8.1   First Iteration of SUPER-SOMA - Example 1

The following is the table XREF used by Shotest as it would appear after the first step "B3" above. It needs some explanation:

```
              Key for table XREF:
The header BDRI+SRDX below shows "-" for "off" and lowercase
letter for "on"

B = Threat can be blocked
D = Threat can be defended by some other piece
R = Threatened piece can run from threat
I = Threatened piece has no safe moves (Immobile)
+ = Move gives check
S = Threat can be neutralised by some means (BDR above)
R = Target is tied to defending another piece
D = Threat does not immediately capture, e.g. Fork
X = Entry is currently active (may be triggered later)

Val   = Net value of exchange
1st   = Value if exchange after first capture
Weight = Weighted priority after cross-reference "B2" above
         If followed by "n" then this is a neutralising step
         rather than a capture
```

```
                   Table XREF:
BDRI+SRDX            Val  1st Weight
--r--s--x l 6fxR6d   960  960    960   n
--r--s--x s 4dxB4c   740  740    944   n
--r--s--x R 6dxs4d  -560  400   1160
--r--s--x R 6dxl6f  -720  240    240
--r--s--x B 4cxp8g   152  152    892
-d---s--x s 4d- 5c+   20   20     20   n
```

From the table above we can see how the top weighted move is R 6dxs4d with 1160 points, despite its low initial value. If we examine the cross-referencing that occurred in step "B2" we can understand how this value was derived.

```
Step B2:  Cross-referencing of table

B 4cxp8g gives  152 to s 4dxB4c  capturing piece is our target
l 6fxR6d gives  960 to R 6dxs4d  our piece is target of other capture
s 4dxB4c gives  740 to R 6dxs4d  capturing piece is our target
s 4d -5c gives   20 to R 6dxs4d  capturing piece is our target
l 6fxR6d gives  960 to R 6dxl6f  our piece is target of other capture
                                 capturing piece is our target
s 4dxB4c gives  740 to B 4cxp8g  our piece is target of other capture
```

The first choice is R 6dxs4d as previously indicated, but the second choice is to simply neutralise the move l 6fxR6d by moving the Rook away. The first "R" column is marked by "r" indicating that the Rook can safely run and so this neutralising move is possible. In some cases a threat cannot be fully neutralised, e.g. if the piece has no safe squares then it may simply need to be defended. This might prevent the capture, or may just allow re-capture of the attacking piece, for example LxR above could be defended by a pawn drop behind the Rook. This would not stop the loss of the Rook, but would allow capture of the Lance in compensation. This partial defence would be reflected in the cross-reference weighting given. The third choice above is the initially positive capture B 4cxp8g. SUPER-SOMA will therefore predict the move R 6dxs4d above and proceed to step "B4" to update the table.

8.2   Second Iteration of SUPER-SOMA - Example 1

After step B4 above and iterating B1, B2 and B3 again, the XREF will now contain the following:

BDRI+SRDX	Val	1st	Weight	
--r--s--x n 3fxR4d	960	960	960	
--r--s--x B 4cxp8g	152	152	152	n

Now four of the table entries have gone as they are now void, for example the capture s 4dxB4c is no longer possible as the Silver has been captured. Black is now the side to play. There are only two moves and these are not linked. Black can either neutralise the capture B 4cxp8g by moving the pawn or capture the Rook on 4d. The latter has a much higher weight and so is chosen.

8.3   Third Iteration of SUPER-SOMA - Example 1

After n 3fxR4d the table contains just one move. This has a positive value and is selected. After this move both black and white will play pass moves and the sequence ends.

BDRI+SRDX	Val	1st	Weight
--r--s--x B 4cxp8g	152	152	152

8.4   Discussion of Example 1

The example above is relatively simple, but demonstrates the basic mechanism of SUPER-SOMA. This position is very easy to resolve using a simple capture search.

9   Example 2 - Whole Board Analysis

The following example looks at a position with many more tactical features. In Figure 5 we have a much more complex example, which includes defensive and forking moves and ties to mate and promotion. This rather artificial example, which does not pretend to resemble a realistic game position, has been contrived to demonstrate a wide range of SUPER-SOMA features within one compact example. This makes it possible to show how SUPER-SOMA links these features together. In this particularly volatile example there is a 14-ply mate threat, which SUPER-SOMA does not stati-cally detect, but it would not be reasonable to expect SUPER-SOMA to be able to do this. A key feature in this position is the Gold on 5h which is tied to defending the threat of a Gold drop on 4h giving mate. Therefore the Gold cannot capture the Bishop on 6h or defend the Silver on 6g. This uses a static assessment of the mate threat based on the simple use of bitmaps. To a limited extent some multiple move mates can also be detected, and also unproven or incomplete mate threats, e.g. a square may be threatened by a Gold drop which leaves the king with two free squares. The evaluator may tie the defender to this square with a value less than mate (e.g. a value of a Silver).

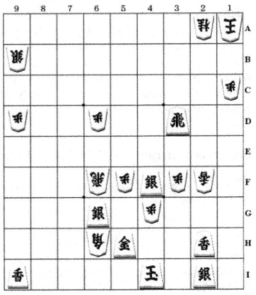

Fig. 5. Example 2 – Whole board analysis White has 1G and 1B in hand

This assessment of mate threats could use a tree search, but this an expensive com-ponent to build into a node evaluator. There are no less than 9 pins and ties in the position. These are:

```
19i --> P9d --- S9b     Pin (Pawn cannot move outside pin plane)
r3d --> P6d --- R6f     Tie
s6g --> P5f --- R6f     Tie
R6f --> s6g --- g5h     Tie
G*4h--> -4h --- g5h     Tie to mate threat
P5f --> -5g --- s4f     Tie to prevent promotion
P3f --> -3g --- s4f     Tie to prevent promotion
```

```
L2f --> L2h --- s2i Pin
L2f --> L2h --- s2i Tie
```

SUPER-SOMA predicts the following move sequence:

```
R 6fxs6g+    470
r 3dxP3f     104   <-- this is the key move
L 2fxl2h     240
s 2ixL2h     240
(1 9ixP9d)     0   neutralise threat pass
B 6h- 7i+    100
  pass
  pass                  Net value 456
```

9.1   First Iteration of SUPER-SOMA - Example 2

We look at the moves in XREF ordered by initial value. Since there are many moves I have separated the white moves out first.

BDRI+SRDX		Val	1st	Weight	
----+s--x	B*1fxr3d	960	960	960	White
-d---s--x	B 6hxs4f	500	500	876	
--r--s--x	R 6fxs6g	470	470	1430	
--r--s---	P 3f- 3g	376	376	188	(inactive)
--------x	B 6h- 7i+	100	100	100	
--r--s--x	L 2fxl2h	0	240	240	
--------x	R 6f- 6e	-104	0	0	

The top move here is the fork B*1fxr3d (B*1f, R1fxr3d) forking Rook and King. This move has no cross-reference with other moves so its final weight and initial value are the same.

The move B 6hxs4f is weighted from 500 to 876 because it is removing the Silver at 4f which is tied to preventing the promotion of the Pawn at 3f to becoming a Tokin at 3g. If this move gets played then the 4th entry P 3f- 3g above would become activated in the table.

The third move R 6fxs6g is the actual move chosen. It is weighted from 470 to 1430 because it stops the capture s 6gxR6f.

The remaining moves are not interesting. The only cross-referenced entry is L 2fxl2h which is weighted because it prevents l 2hxL2f.

We can now look at the black moves in the table:

BDRI+SRDX		Val	1st	Weight		
-d---s--x	s 6gxR6f	960	960	1430	n	Black
bdr--s--x	l 2hxL2f	240	240	240	n	
bdr--s--x	l 9ixP9d	104	104	104	n	
bd---s--x	r 3dxP3f	104	104	1564	n	
bd---s--x	r 3d- 3a	70	70	1030	n	
--r--s--x	s 6gxP5f	-192	104	278	n	
--------x	g 5hxP4g	-312	104	-312		
--r--s--x	r 3dxP6d	-752	104	208	n	

The first move s 6gxR6f is the opposite of R 6fxs6g above and receives the same weighting. Note that to neutralise this move by simply moving the Rook away would be a viable choice for white, although the rook is both tied to defending the pawns on 6d and 5f. In this instance white has other better moves.

The second and third moves have no cross-reference linkage. The fourth r 3dxP3f is the top weighted move and also the most interesting as it is weighted for stopping the fork B*1fxr3d and also defending the Silver on 4f, preventing B 6hxs4f. This is the first example of a move being credited for performing a defensive move. This is complex behaviour way beyond simple SOMA. The fifth move r 3d- 3a promotes the Rook and prevents the fork. The remaining moves are not interesting SUPER-SOMA will select the top white move R 6fxs6g and update the table.

9.2   Iterations of SUPER-SOMA - Example 2

The second iteration retains the same top black move as found after iteration 1, the move r 3dxP3f defending the Silver and avoiding the fork.

Iteration 3 selects the move L 2fxp2h which generates a threat against the Silver on 2i. The XREF table entry for this white move is converted into the black move s 2ixL2h for the next iteration.

Iteration 4 selects the re-capture above. Iteration 5 find white with no good attacking moves and so the black move l 9ixP9d is neutralised by moving the Silver to 9c. Iteration 6 and black has no moves, so passes. Iteration 7 and white simply promotes the Bishop by B 6h- 7i, and the sequence is ended after both sides pass. You may like to input this position into your own Shogi program. If this position is analysed by Shotest with tree search it predicts the forced win, after examining 4857 nodes.

```
 1.  [   1]  B  *  1f
 2.  [   1]  S     2i-3h
 3.  [   1]  L     2fx2h+
 4.  [   1]  G     5hx4g
 5.  [   4]  R     6fx6g+
 6.  [   1]  P  *  1b
 7.  [   1]  K     1ax1b
 8.  [   1]  R     3d-3b+
 9.  [   1]  K     1b-1a
10.  [  10]  R+3bx2a
11.  [   1]  K     1ax2a
12.  [   6]  N  *  3c
13.  [   2]  K     2a-3b
14.  [   1]  K     4i-4h
```

The numbers in brackets are the position in the search that that move was considered. For example 10. [10] R+3bx2a above indicates that 9 other moves were considered before R+3bx2a.

10   Extensions to SUPER-SOMA

The two examples above demonstrate the most important features of the SUPER-SOMA algorithm and show how the general mechanism is used to cross reference and predict tactical sequences. However there are many other features that are not shown in these examples, because they would have made them unnecessarily complex.

Some of these extra features are demonstrated below within simpler examples. In all cases there are no pieces in hand.

10.1 Checks

A check is a special kind of tactical move that requires a forced response. This complicates the assessment of the move to be chosen. In the position in Figure 6 there are 4 moves in the table:

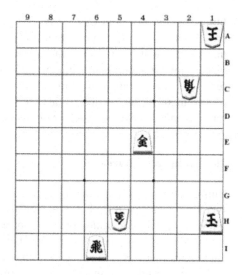

```
BDRI+SRDX            Val

--r--s--x  G 5hxr6i   960
-d--+s--x  B 2cxg4e   440
--------x  g 4ex-3e  -152
----+---x  r 6ix-6a+   70
```

The only cross reference is G 5hxr6i prevents the promotion move r 6i-6a+. This leaves G 5hxr6i as the top move. However SUPER-SOMA chooses B 2cxg4e before G 5hxr6i because the

Fig. 6. Dealing with check

bishop capture gives check which, after the king moves, leaves white with the next move. The cross-reference recognises this and forces the move selection by simply adding a large weight to B 2cxg4e.

10.2 Secondary Captures

A move which captures a piece may then also generate a further capture threat which can be entered into the table. This is illustrated in Figure 7.

This generates the following table entries.

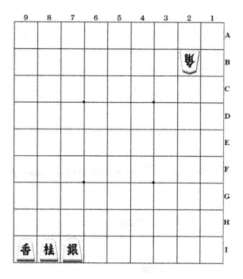

```
BDRI+SRDX            Val

--------x  B 2bxl9i+  330
--r--s---  I 9ixn8i   240
     (inactive entry)
--r--s--x  1 9ix-9a   110
```

In this case the table starts with two active and one inactive entry. When the move B 2bxl9i+ with promotion is selected the follow-on potential capture entry I 9ixn8i is activated (triggered). During cross-reference the

Fig. 7. Generation of further captures

initial move is credited with the value of the triggered entry. Once activated the new move is thereafter fully cross referenced as any other table entry. This mechanism is limited as the potential extra move B+8ixs7i is not recognised, as this mechanism only examines to a depth of one level. To try and extend this further does not really make sense as it becomes increasingly complex to detect deeper captures because of pieces moving, gaining or losing protection and blocking / unblocking of attacks. The results of a deeper application of this idea would be increasingly susceptible to errors.

The more recent version of this mechanism is also used to create secondary captures when a discovered attack or pin/tie is activated.

10.3 Multiple Attacks

This is perhaps obvious, but a capture on a square may be by one of several pieces, therefore each possible first attacking piece is considered as a separate entry. This allows a heavily committed piece to not be selected to capture, even though it is the smallest piece. This becomes critical when the first obvious choice to capture can also capture another piece, in which case an uncommitted piece captures first.

In the example in Figure 8 the silver might capture first, but it is committed to attacking the gold, therefore the bishop captures first. A variation on this occurs where a piece is tied to defending another piece, but in this instance the favorite choice to make the capture is selected first anyway and would generally work even with a single entry in the table per capture square.

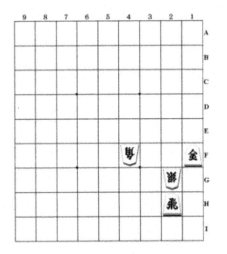

Fig. 8. Multiple attacks   **Fig. 9.** Defending moves

10.4 Defending Moves

Just as any move can create a secondary attack, a move can also have defensive properties. It is useful to examine tactical moves for a secondary defensive role after a complete XREF table has been created. For example in the board in Figure 9 the bishop on 4f might capture the gold on 2h. If instead it captures the knight on 6h, it also defends the silver on 9e, so this will be the preferred move.

10.5 Running to Safer Squares

A piece may be attacked but have no safe square to run to. In this instance it may be beneficial to move to a square where a recapture is possible.

In the position in Figure 10 the rook on 3c is attacked by the silver on 2b and is not defended. The rook can legally move to 3b, where it will still be captured without recapture. However it can also move to 3a, which is defended by the bishop. Here 3a is the selected move since it results in some material compensation after the rook is lost.

This move would be hard to generate in a conventional capture.

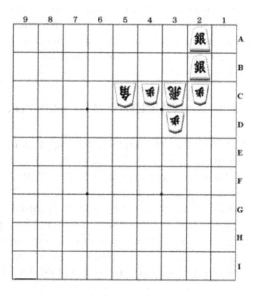

Fig. 10. Running to a safer square

11 Correcting Errors in SUPER-SOMA

A clear flaw in the mechanism described so far is its dependence on getting the correct choice of move at each iteration. A tree-search can experiment with moves selected without any strong need for picking the correct move as its first choice to search. It is clear that it would not be hard to create example positions where SUPER-SOMA picks the wrong line.

The first consideration here might be to find out how often the selected first move is actually the best choice. The correct way of doing this would be to run a series of tests that compared the choice made by the algorithm with the results of a full search. Note that this would not be a simple capture search, but would instead have to include non-capturing tactical moves. It would therefore be a full search. This experiment has not yet been done, but an intermediate test has been performed.

The current version of this algorithm runs twice for each node. The second run simply rejects the first choice made by the first run and then predicts a second sequence based on a different first move. The program then compares the results of these two tests and in minimax style chooses the move with the highest score.

In tests run over a complete game the first choice move was retained in about 95% of the positions analysed. This only considered positions where there were at least two moves that seemed worth playing, i.e. it would not compare a situation where the first move was obviously good, but the second choice was a meaningless sacrifice. The result indicates that the algorithm seems to be happy with the first choice most of the time. Of course this result depends on an incestuous test as the testing mechanism has the same intrinsic flaws as the mechanism being tested. However this test is still valid as the tree-search is much more exhaustive and will almost always find the correct move, even if errors occur in the tree.

12   Using SUPER-SOMA with Tree Search

Limitations on the length of this paper do not allow a thorough explanation of how this operates, so the following gives a general outline of how SUPER-SOMA is used.

The tree search used with SUPER-SOMA uses a plausibility analysis that orders moves to examine at the next ply. SUPER-SOMA adds an ordered list of likely moves to this, which may include defensive as well as capturing and other tactical moves. These suggested moves get very high priority in move choice. In addition to this the search frequently chooses to terminate early. SUPER-SOMA provides a estimation of the likely outcome of the exchanges in a position and this is used to assess whether the position is likely to fall outside of the alpha-beta window. The complexity of the position is also assessed by a linear value which sums the complexity of the line predicted by SUPER-SOMA. This is also used to determine whether it is safe to attempt to terminate the search.

SUPER-SOMA is particularly effective in generating defensive and blocking moves. A capture search cannot do this and a full width search may be overwhelmed by huge moves lists that are just too expensive to search. A general purpose search may be able to spot all defensive moves, thus narrowing the full list, but to routinely search all of these is still too expensive. SUPER-SOMA will decide whether active tactical or defensive moves are the best choice and thereby reduce the chance that inappropriate lines are searched.

13   Comparing SUPER-SOMA with Other Techniques

From the examples above it can be seen that SUPER-SOMA is sophisticated. It can obviously detect complex features, but is limited in its capacity to correct errors made in choosing moves. This means that each choice of move is very critical as there are limited options to try out alternatives.

It is not easy to compare it with other techniques because its function does not exactly match any alternative method. Also it is not used as a complete alternative to search-based methods, as indicated in the previous section, but is intended to be used in co-operation with a searching mechanism.

On a simple level the basic SUPER-SOMA can be compared to capture search and also to general tactical search. I will consider these in turn.

13.1 Capture Search

In positions where simple captures dominate then SUPER-SOMA looks inadequate as the move selection is not very deep. SUPER-SOMA cannot deal with captures that trigger a long chain of captures. The later version of the algorithm is capable of generating secondary captures, but only for one generation. For example, a promoted bishop capturing on 2b can then detect the possibility of a further capture on 1a, but not the following capture than might occur on 2a.

In compensation for this it can predict other types of moves such as forks, mates, promotions and defensive moves; predicting plausible move sequences. These are common situations which capture search cannot easily deal with.

13.2 Tactical Search

A normal tactical search will out-perform SUPER-SOMA alone, since it is limited to considering a single line of moves and will make many mistakes. However it can make a reasonable assessment very quickly. If used inside a selective search then it may well be much more effective than tactical search because at any node in the tree it can still make reasonable predictions for the outcome of tactical sequences.

14   Development of SUPER-SOMA

The design of this mechanism has been driven by the nature of the position types that it expects to have to analyse in normal Shogi games. For example I have not tried to create an algorithm that can deal reliably with situations with very long chains of linked captures, which would test my algorithm to the limits. I have instead considered the types of position that commonly occur and tried to deal with these. The guiding consideration is therefore a question of probability: A simple feature might be analysed in great depth simply because it is very commonly met. This not a purist approach to development. It has allowed compromises to be made to simplify the mechanism at the expense of true generality. A more general approach would be more satisfying but would probably make the algorithm too complex to be practical and therefore would not result in the creation of a viable playing algorithm.

15   SUPER-SOMA in Use

SUPER-SOMA has been in use in the program Shotest since version 2.0. This program is unbalanced as it still needs much more Shogi knowledge. It also lacks a tsume (mate) search, depending instead on general purpose search to detect mates. Despite these deficiencies, Shotest came 3rd twice in the 1998 and 1999 CSA World Computer Championships and 7th in 2000, competing with between 35 and 45 programs. The experience of these events indicates that Shotest currently performs badly in the opening and endgame, where Shogi knowledge is important, but that it performs well in the middlegame where positions can become very complex. Even though Shotest examines some 50 times fewer nodes than the other top programs, analyzing 3000 nodes/sec on a PII-400, it seems to play well.

```
                                                                      Frequency
11  *                                                                         6
10  **                                                                       14
 9  ***                                                                      26
 8  ******                                                                   54
 7  ************                                                            101
 6  *********************                                                   168
 5  ****************************                                            214
 4  ***************************************************                     329
 3  ****************************************************************        406
 2  **********************************************************************  457
 1  *****************************************************                    357
 0  *****************************                                            233
```

Fig. 11. Number of captures, promotes and forks from move 50 in 50 human games

A likely large part of Shotest's success is owed to the ability of SUPER-SOMA to handle complex positions without needing to grow enormous game trees. This is important in Shogi as late middlegame and endgame positions commonly leave large numbers of pieces hanging. This is shown in Figure 11 after examining 2365 positions from 50 randomly selected human games, ranging from amateur to top profession level.
This calculates the number of squares where a profitable captures, promotions and forks can be made, starting from move 50 to avoid the quiet opening phase.

From the frequency data in Figure 11 it can be calculated that 90% of positions have profitable captures. Over 75% of positions have 2 or more captures and 38% have 4 or more. This contrasts sharply with Chess where a limited hand examination of just 20 games gave a figure of 24% for positions with profitable captures with only 5% with 2 or more profitable captures.

Figure 12 presents a similar test over 50 Shotest games against other programs, which interestingly shows that positions with no captures are much more common than in the human games of Figure 11. These tests show that 79% of the positions have profitable captures, 59% have 2 or more and 28% have 4 or more. This may reflect Shotest's inclination to resolve tactical threats rather than leave them open.
The broad conclusion from this is that tactically active positions are the norm in Shogi. If the evaluation can deal with these then this could greatly reduce the burden on tree search.

Whether or not SUPER-SOMA works well depends on how reliably it can predict moves. In the combined sample of the 100 games above it predicted the next move in the game score 65% of the time. This only considered moves where either SUPER-

SOMA or the game record predicted an active move (i.e. a defensive or attacking move). Of the failures, 37% occurred when a pawn capture was predicted, but was left hanging in the actual game record.

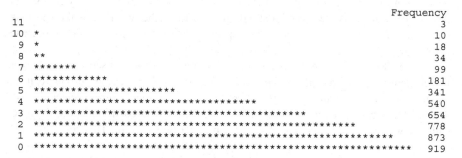

```
                                                                  Frequency
11                                                                        3
10  *                                                                    10
 9  *                                                                    18
 8  **                                                                   34
 7  ******                                                               99
 6  ***********                                                         181
 5  *********************                                               341
 4  ***********************************                                 540
 3  ***********************************************                     654
 2  ***************************************************************     778
 1  *********************************************************************  873
 0  ************************************************************************  919
```

Fig. 12. Number of captures, promote and forks from move 50 in 50 computer games

In use, SUPER-SOMA averages a lookahead of 4.2 plies at the end of the primary continuation over a large number of games. This is quite deep considering that in the early part of the game in many positions there are no predicted captures. It commonly predicts sequences of 10 moves or more. When playing to a time limit of 3 seconds per move (on a PII-400) this average depth extends the average 6 ply primary depth to 10.2 plies.

16   Conclusions

As far as I am aware, SUPER-SOMA represents the first attempt to generate an evaluation function that can completely analyse a position in chess-like games, which is accurate enough to allow plausible play without speculative tree-search. Many programs exist that can evaluation positional features, but such programs cannot cope with exchange threats without search.

I originally tested this idea 20 years ago with the Chess-playing program Merlin. The mechanism then was crude and it seemed that the narrow trees generated by Chess favoured other methods, so the idea fell into disuse. The new implementation for Computer Shogi is much more elaborate and has not been exhaustively compared with other methods, but it at least seems to perform well. The existing interface with the tree-search needs improvement. The implementation is still young and so the full potential of this mechanism has not yet been fully realised. When SUPER-SOMA is more developed I will be able to perform more rigorous comparative tests.

In conclusion I would assess that SUPER-SOMA with directed search looks as if it can make a viable alternative to conventional tactical search. Its sophistication makes it very good for reliably detecting quiescence in a position. It also makes it much easier to generate good moves deep in the tree. Conventional brute force methods for doing this depend on exhaustively detecting good moves in one branch so that they can be tried in neighbouring branches. SUPER-SOMA can reach a completely new position deep in the tree and immediately have good candidate moves to search.

The greatest weakness is its limited single lines of analysis. Currently it allows two alternative moves to be considered for the first move. This is a somewhat limited concession to correcting errors, but this mechanism is usually backed by real search. As SUPER-SOMA is developed it should be able to predict where it is likely to make errors and indicate to the search that the position is unstable and should be searched further.

Since SUPER-SOMA predicts a sequence of moves, it can be used with some positional elements. At present each piece has a value assigned for each square it can occupy, both for its material value and also its value in the currently selected castle. These values can be traced, so that an exchange will also change positional values. This allows it to predict sequences that disrupt a castle structure. This could probably be extended in many ways, as yet unexplored. At present the castle mechanism is badly tuned, so this extension currently has limited value. Shotest also has other features that strongly contribute to its success, but these are outside the scope of this paper.

As more work is done it will be easier to assess whether this or more conventional methods are the way to go for Computer Shogi.

References

1. Knuth, D.E. and Moore, R.W. (1975), An Analysis of Alpha-Beta Pruning, *Artificial Intelligence*, Vol. 6, pp. 293-236. ISSN 0004-3702
2. Yamashita, H and Matsubara, H (1998).Computer Shogi – ISBN4-320-02892-9\2300
3. Michie D (1966): Game Playing and Game Learning Automata. In Advances in Programming and Non-Numerical Computation, Ed. Fox L, pp 183-200. Oxford, Pergamon.

A Shogi Processor
with a Field Programmable Gate Array

Youhei Hori[1], Minenobu Seki[1], Reijer Grimbergen[2],
Tsutomu Maruyama[1], and Tsutomu Hoshino[1]

[1] Institute of Engineering Mechanics and Systems, University of Tsukuba,
1-1-1, Ten-ou-dai, Tsukuba, Ibaraki, 305-8573 Japan
hori@darwin.esys.tsukuba.ac.jp
[2] Electrotechnical Laboratory,
1-1-4, Umezono, Tsukuba, Ibaraki, 305-8568 Japan
grimbergen@fu.is.saga-u.ac.jp

Abstract. In this paper we describe the architecture of a shogi processor based on reconfigurable hardware. For our implementation, we have used Field Programmable Gate Arrays (FPGAs), which can be reconfigured dynamically by downloading configuration data from host computers. Because of this reconfiguration flexibility, it is possible to implement and evaluate new algorithms quickly and to make small subsystems (of very low cost) that can be used on demand. For shogi these two features are especially important, as there are no stable subsystems that can be ported to special purpose hardware. Also, in shogi different modules are needed for different stages of the game. To test the feasibility of using FPGAs for shogi, we have implemented two modules that are general for all strong shogi programs on one off-the-shelf PCI board with one FPGA. The piece cover module on an FPGA is 62 times faster than the software module, while the module for finding mate on an FPGA is 9 times faster than the software module.

Keywords: Reconfigurable hardware, FPGA, Shogi.

1 Introduction

In chess, DEEP BLUE demonstrated that a large system with special purpose hardware can beat the human world champion [9]. DEEP BLUE's performance is often credited to its hardware power, which is indeed impressive. However, the number of positions that needs to be searched for strong chess play is too large to rely on hardware speed alone [5]. Computation power is important, but the key to the success of DEEP BLUE is the combination of hardware speed and sophisticated algorithms for tree search, position evaluation, move generation and so on. It was only possible to make special hardware for chess because there had been a long history of research into chess algorithms and there was consensus about which algorithms were most effective.

This consensus does not necessarily exist in other games. An example is *shogi*, a Japanese game that is similar to chess. Even though the objective of shogi is the same as for chess (capture of the king of the opponent), there are some important differences between the games. There are more pieces in shogi than in chess (40), there are different

T.A. Marsland and I. Frank (Eds.): CG 2000, LNCS 2063, pp. 297–314, 2001.

pieces (no queen, but golds and silvers) and the board is slightly bigger than for chess (9x9 instead of 8x8). The most important difference between chess and shogi is the possibility of re-using captured pieces in shogi. Pieces that have been captured from the opponent are put next to the board at the side of the player that captured them (these pieces are called *pieces in hand*). If it is a player's turn to move, either a move with a piece on the board can be played, or one of the pieces that was previously captured can be put back on an empty square of the board (this type of move is called a *drop*). As a result of this rule, the average number of legal moves in shogi is much larger than in chess. Therefore, using the same algorithms as in chess does not lead to programs that are strong enough to compete with human experts.

Modifications of chess algorithms and even completely new search algorithms have been proposed for shogi with mixed results. Descriptions of two of the best programs can be found in [6][14]. So far, there has been no attempt at making specialized hardware for shogi. One of the reasons for this is that there is no sponsor like IBM to fund such a costly project. However, the second problem is that there is no consensus about the algorithms that work best for shogi. Hardware based on fixed chips like those used in DEEP BLUE can not be used for shogi, because the chips will already be outdated by the time they have been manufactured. A third problem of the DEEP BLUE approach is that it is very hard for others to improve upon the state-of-the-art. The DEEP BLUE chip is not available to others for further improvements.

We think that shogi will benefit from the use of high speed hardware. In this paper, we propose a solution to the problems of outdated and inaccessible hardware. We describe how a shogi program can be implemented on reconfigurable hardware devices called *Field-Programmable Gate Arrays* (FPGAs). FPGAs are relatively cheap, flexible and have a short turn-around-time as it takes only a couple of hours to reconfigure an FPGA.

This paper is organized as follows. Section 2 introduces the features of systems with reconfigurable hardware. Section 3 describes the shogi program which we are currently implementing on FPGAs. Section 4 gives the architecture of the shogi processor. Section 5 describes the implementation issues of the shogi processor. In Section 6 and Section 7, details and results for two modules of the shogi processor are given. Finally, conclusions are given in Section 8.

2   High Speed Computation with Reconfigurable Hardware

The basic structure of a shogi program is the same as that of a chess program, and most time for finding the next move is spent on search. Therefore, improvements of hardware speed have the same impact on shogi programs as on chess programs and we expect considerable improvements of the playing strength of shogi programs by using special purpose hardware.

However, typical special purpose hardware like ASIC takes several months to be manufactured, after which a month or so is needed for modification. Such a long system development time is undesirable in shogi, as most of the important modules are still being revised often. Therefore, a shogi chip would be out of date at the time it can be used. A possible solution is to produce multiple chips, replacing only the modules that have been revised. A problem with this approach is that the circuit for a shogi program requires wide memory bandwidth and a high data transfer rate. When a shogi program is implemented on several chips, data has to be transferred between these chips. However, the number of I/O pins on the chips is smaller than the required data width and the operation speed of I/O pins is much slower than that of the internal processor. For high speed computation, it is best to have all modules of a shogi program on a single chip.

We think that hardware systems that are based on reconfigurable devices are a more promising approach to speed up shogi programs. In our research, we have focused on *Field Programmable Gate Arrays (FPGAs)*, which are very popular reconfigurable devices. A detailed hardware description of an FPGA is outside the scope of this paper, but more information about FPGAs can be found in the proceedings of several international conferences on FPGA-based systems that have been held [16][17][18]. Here we limit ourselves to the way an FPGA can be used for implementing a shogi program.

An FPGA chip consists of logic cells, internal memory and the connections between these cells and memory [15][1]. The function of the logic cells and the connections between cells and memory can be changed by downloading configuration data to the chip. The configuration data is generated by compiling a program written in a hardware description language. Compiling programs from the hardware description language takes several hours, while the actual reconfiguration of the FPGAs takes less than 100 msec in general.

The reconfigurability of FPGAs makes it possible to modify circuits for new algorithms relatively quickly. Minor modification of the circuits on the FPGAs can be finished and tested on the hardware in several hours. The size of FPGAs as well as the speed has improved drastically in the last couple of years and we expect that it will soon be possible to have a complete shogi program running on a single FPGA chip.

The flexibility of the FPGAs also makes it possible to realize a small but high performance shogi processor by removing unnecessary modules and reconfiguring required modules. This is important for shogi, as the behavior of shogi programs depends on the stage of the game. In the opening, the system has to perform efficient pattern matching between the current position and positions in opening libraries. In the middle game, it is necessary to search the game tree efficiently to find the best move. In the endgame, attack and defense of the king is vital, and special circuits to find mate are required. Also, different parameters and different evaluation functions may be required depending on the strategies that have been employed in the game.

Of course, FPGAs will not be as fast as the specialized hardware used by DEEP BLUE, but FPGAs are relatively cheap and flexible. Once there is consensus about the most effective software and hardware for shogi, it is still possible to produce a special purpose chip for further speed improvements.

3   The SPEAR Shogi Program

As said, the basic structure of a shogi program is similar to a chess program. The next move to play is determined by exploring a mini-max tree [13] with iterative alpha-beta search. Other common techniques taken from chess programs are transposition tables, quiescence search [2], principal variation search [10] and the history heuristic [11]. The main difference between programs for chess and shogi is that in shogi programs considerable effort is spent on dealing with the high number of legal moves, which on average is about 80 [7]. This is more than twice the average number of legal moves for chess, which is about 35 [8].

Therefore, to search deeply it is important to discard moves early, often without any search. Most strong shogi programs make decisions about which move to play by the following steps: (1) update the position, (2) generate all legal moves, (3) evaluate the position after each move, (4) discard bad moves, thereby keeping the plausible ones, (5) search plausible moves. This procedure has the disadvantage that moves that are obviously bad still need to be evaluated. Evaluation is expensive, so this extra evaluation is not efficient.

SPEAR is a shogi program that aims at generating plausible moves more efficiently by using the evaluation after the previous move to generate moves [3][4]. During evaluation, information about *game conditions* like piece shape, castle shape, king danger and so on are collected. Plausible moves are then generated based on the current position and the game conditions.

In this section, we describe the data structures that are being used in SPEAR and how SPEAR finds the best next move.

3.1   Data Structures

The data structures in SPEAR are divided into data structures for positional data and data structures for the game conditions. Here are the data structures for positional data:

- *Board data*: information about which piece is on which square.
- *Piece cover table*: information about which piece can move to which square.
- *Piece info table*: information about which side the piece belongs to, where it is located and if the piece is promoted or not.

Here are the data structures for the *game conditions*:

- *Weak piece table*: information about which piece is not active, undefended, threatened by opponent pieces and so on.
- *Strong piece table*: information about which piece is active, able to promote, attacking opponent pieces and so on.
- *Piece square table*: the relative value of the pieces on each square.
- *King danger*: the evaluated value of king danger.
- *Soma value*: the expected profit of capturing a piece on a square.

3.2   Procedure for Deciding Upon the Next Move

In the program, the following steps are repeatedly executed to decide which move to play next: (1) update the position, (2) evaluate the position and update game conditions, (3) generate plausible moves based on the game conditions, (4) search plausible moves.

Figure 1 shows the block diagram of this program. Each module directly corresponds to one of the stages above. We will now give a more detailed explanation for each module.

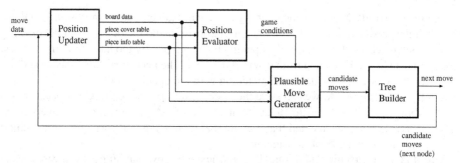

Fig. 1. A block diagram of the SPEAR program

Position Updater Based on the move data, this module updates *board data*, *piece cover table* and *piece info table*. This data is then used for position evaluation and plausible move generation.

Position Evaluator This module evaluates the position and updates the game conditions. The evaluation module has many sub-modules such as *piece shape evaluator*, *castle evaluator*, *king position evaluator* and so on. Each sub-module evaluates the position and adds the evaluation to the game conditions. For example, *piece shape evaluator* evaluates the weakness or strength of a piece formation, and this evaluation is added to *weak piece table*, *strong piece table* and *piece square table*.

Plausible Move Generator This module generates plausible moves based on the game conditions. Using the game conditions (as well as board data, piece cover table and piece info table), only promising looking moves are generated. A value is given to each plausible move, and the best N moves are selected and added to the game tree (in the latest tournament version of SPEAR N was set to 20).

Tree Builder This module adds the remaining plausible moves to the game tree. Game tree search in shogi is similar to the search in chess programs.

4   Architecture of the Shogi Processor

As pointed out earlier, for efficient hardware implementation a single chip is best. Since a complete shogi program does not fit on a single FPGA chip yet, we must reduce the size of the circuit by reconfiguring the chip for different stages of the game and use different configurations for different strategies. In this section we discuss the issues in designing a shogi processor. First, we will explain the circuits for each game stage. Second, we will give the architecture for the middle game and endgame circuit for which we have made a partial implementation.

4.1   Circuits for Different Game Stages

An FPGA chip can be reconfigured in about 100ms, so deleting obsolete modules and reconfiguring other modules for each stage of the game is a viable option for reducing the size of the program. The following observations for the circuits in different game stages are not specific to SPEAR, but apply to most shogi programs.

Opening Circuit Moves in the opening are usually generated by an *opening book*, so high speed matching between the current position and the positions in the opening book is required. However, the opening book is typically too large to fit in memory, so data has to be loaded from external memory. The necessary data transfer and I/O operations will slow down the matching with the opening book.

By using an FPGA, we can store the opening library on the processor because this library (which will occupy a large part of the FPGA memory), can be overwritten if necessary. It is expected that this internal processing will speed up the matching with the opening book considerably.

Middle Game Circuit This circuit finds the next move by searching the game tree. Most time during this stage is spent on position evaluation and candidate move generation. With an FPGA, we can expect high speed computation by processing different sub-modules in parallel and in pipeline. For example, the evaluation of a position has many different features like piece activity, piece formation shape, capturing profit and so on. Some of these features are independent and can be processed in parallel. Moreover, we will later explain how different features can be processed in pipeline, leading to further speedups.

Move generation can also be performed in parallel and in pipeline. The plausible moves have many different categories like attack moves, defense moves, captures, promotions and so on. These categories can be generated in parallel and each move can be generated by pipeline processing.

Endgame Circuit In the endgame, there is a dramatic change of the weights of the evaluation function features. Mating the king becomes the most important feature and material gain is only a secondary consideration. One of the modules that most strong shogi programs use is a *tsume shogi solver* to find mate (more details will be given in Section 7). Because of the reconfigurability of an FPGA, this module can be prepared on demand.

Library For/Against Strategies Using a library to deal with different strategies is a good way to guide the search to promising parts of the game tree. For example, different castles have different strengths and weaknesses, so an attack should be aimed at the weak point of the castle. There are many different strategies and castle formations in shogi, so such a library would be too large to keep in memory on the chip. With an FPGA, we can load and configure only the required libraries.

4.2   Data Structures

For a hardware implementation, we need the following data structures. Even though these data structures have been taken from SPEAR, similar data structures are used in most shogi programs.

- The *move data* contains information about a move. It shows which player played the move, if the moving piece promoted, which square the piece came from and which square the piece moved to.
- For each piece, there are 10 bits of *Piece data*. The bit assignment, shown in Figure 2, has the following meaning.
  - The 3 bits of *piece_type* are used to store the kind of piece (King, Rook, Bishop and so on). In shogi there are 8 different pieces, so 3 bits are needed.

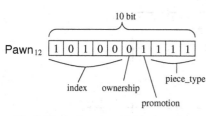

Fig. 2. Bit assignment of the piece data

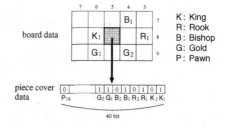

Fig. 3. The piece cover data of the square$(5, 8)$

- The *promotion* bit is set if the piece is promoted.
- The *ownership* bit is set if the piece is owned by black.
- The 5 bits of *index* distinguish pieces of the same type. For example, there are 18 pawns in shogi, which can be uniquely identified by the index. The *piece_type* and *index* are used to find the piece in the piece cover table and the piece info table.

The piece data in Figure 2 means that the piece is Pawn$_{12}$, promoted, and owned by white.

- The *board data* is a set of piece data and contains the location of pieces. For the complete description of the board, 810 bits (=81 squares×10 bits of piece data) are needed.
- The *piece cover table* stores which piece can move to which square. One square has 40 bits of piece cover data where each bit corresponds to one of 40 pieces (40 is the total number of pieces in shogi). If a piece can move to a square, the corresponding bit is *1*. If not, the corresponding bit is *0*. In Figure 3, the piece cover data shows that King$_1$, Rook$_1$, Bishop$_1$, Gold$_1$ and Gold$_2$ can move to square $(5, 8)$.

4.3 Architecture of the Middle Game Circuit

Before implementing a shogi program in hardware it is important to think about the architecture of the processor. Not all applications are speeded up by the use of an FPGA. If an application has no possibilities for parallel and pipeline processing, the performance will be disappointing, sometimes even slower than the corresponding software application.

With this in mind, we designed a middle game circuit for shogi that is highly parallelized and designed to perform pipeline processing. The architecture directly corresponds to the structure of the software program given in Section 3. Therefore, it has four major modules that we will now describe in more detail.

Position Updater In *Position Updater*, board data, piece cover table and piece info table are updated for each move. Figure 4 shows the structure of the circuit for *Position Updater*. The modules *board data updater* and *piece info table updater* can be processed in parallel, and the modules *board data updater* and *piece cover table updater* can be processed in pipeline.

Position Evaluator A block diagram of *Position Evaluator* is shown in the upper half of Figure 5. *Position Evaluator* has many sub-modules. These modules can be divided into the following three groups.

Position Updater

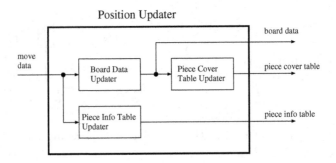

Fig. 4. A block diagram of *Position Updater*

1. Piece square table evaluator.
2. Modules for evaluating pins, discovered attacks, piece mobility and king distance.
3. Modules for evaluating the soma value, piece shape, promotion threats, castle shape and king position.

These three groups can be processed in pipeline, while the modules in each group can be processed in parallel. A position is evaluated by these modules, and the results of this evaluation are stored in the data structures for the game conditions such as *weak piece table*, *strong piece table* and *king in danger*.

Plausible Move Generator Before generating plausible moves, each position is evaluated and only promising moves are generated. The hardware modules shown in the lower half of the Figure 5 are used for this process. All these modules can be processed in parallel.

Game Tree Builder The game tree is stored in the memory of the FPGA. A node is represented by the move data and a relative address from its parent. This module builds nodes by adding this address to the move data, and then stores the node information in memory.

4.4   Architecture of Endgame Circuit

As we pointed out earlier, the weights of the features of the evaluation function change in the endgame. Despite these changes, the endgame circuit still should perform a tree search, so the basic structure is similar to that of the circuit used for the middle game. There is one exception, as there is a need for a circuit performing specialized mating search. This module is a vital part of strong shogi programs. Details of a tsume shogi circuit will be given later in Section 7.

5   Implementation of the Shogi Processor

Unfortunately, reconfigurable hardware is not a magic box where a shogi program in a common programming language like C can be entered, a button can be pushed and the FPGA chip rolls out at the other end. We already pointed at the importance of good

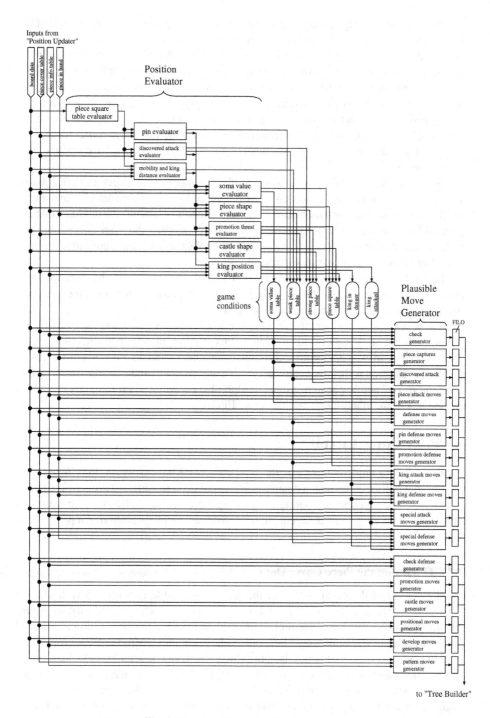

Fig. 5. The structure of the shogi processor.

hardware design, taking advantage of the parallel and pipeline processing abilities of the FPGA chip. In the next two sections we will see how the special features of the FPGA make it necessary to rewrite some of the basic building blocks of our shogi program.

As a result of these FPGA specific design issues, implementing a full shogi processor is a time consuming task. However, if we wait until the shogi processor is finished, we miss an opportunity to take advantage of another appealing feature of FPGAs: they are relatively cheap and can be used by many different researchers. By combining the efforts of different people, we can reduce the work on the full shogi processor and we can also take advantage of improvements by others on modules that have already been implemented.

As a first step, we have made two partial implementations to get an indication of how a shogi program can be speeded up by the use of FPGAs. There are of course many candidates for partial implementation, but we chose to implement two modules that are general for all strong shogi programs: a circuit for computing the *piece cover table* and a circuit for mating search, the *tsume shogi solver*. As these modules are stable, there is probably no need for important modifications later.

In Section 6, we describe the *Piece Cover Circuit* in detail. From this description, it will be clear that the method for acquiring piece cover data used in this circuit is general and can be used in most other modules as well.

The *Tsume Shogi Circuit*, described in Section 7 is not as general as the *Piece Cover Circuit*, but the advantage of this module is that it is completely independent from the other modules. Move generation is different (only checks and defenses against checks), position evaluation is different (estimate of distance to mate) and even the tree search is different (proof number search instead of alpha-beta search). Therefore, this module is a good test of the stand-alone abilities of an FPGA.

6 Piece Cover Circuit

As a first step in implementing our shogi processor, we have designed the *Piece Cover Circuit* that will be described in detail below. As pointed out earlier, it is necessary to use parallelism and pipeline computing for good speed-up results. The implementation on an FPGA is therefore very different from the software implementation of a piece cover table.

6.1 Obtaining the Piece Cover Data

Parallelization of the computation of the piece cover table is achieved by loading 9 squares (= one rank) of board data simultaneously at each clock cycle. Ideally, we would like to scan all 81 squares in parallel. However, the required memory bandwidth would be too wide and the circuit would become too large to fit on a single FPGA chip.

By loading one rank of board data, the whole board can be scanned in 9 clock cycles (as the dimension of a shogi board is 9x9). In our method, we use two scans of the board: a top-down scan and a bottom-up scan. To balance the size of the circuits the top-down scan and the bottom-up scan, each give piece cover data of four of the eight adjacent squares. An example of the top-down scan is given in Figure 6(a), while the bottom-up scan is given in Figure 6(b).

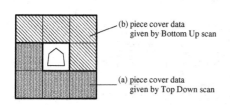

(b) piece cover data given by Bottom Up scan

(a) piece cover data given by Top Down scan

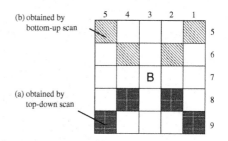

(b) obtained by bottom-up scan

(a) obtained by top-down scan

Fig. 6. The piece cover data given by each scan

Fig. 7. An example of obtaining the piece covering of long range pieces.

Two separate scans are necessary, because the problem with a single scan is that piece cover data for *long range pieces* (rook, bishop and lance) can not be obtained. These pieces can not only move to an adjacent square, but also to squares that are further away. An example of how the piece cover data for the bishop is obtained by a top-down scan and a bottom-up scan is given in Figure 7. Let's assume that the fifth rank is currently scanned. The board data of this rank is loaded, but there will be no piece cover data as there is no piece on this rank. We will get the same result after scanning the sixth rank, as there is still no piece found. Only after loading the board data of the seventh rank, the bishop is found and the piece cover data of the squares $(2, 8)$ and $(4, 8)$ is obtained. The information that there is a bishop on $(3,7)$ is stored in a register. The information in this register can be used at the next clock cycle, as the board data for the eight rank is loaded and the piece covering of square $(1, 9)$ and $(5, 9)$ by the bishop is obtained. The top-down scan is now finished, and the bottom-up scan is performed next. The piece covering of the squares in Figure 7(b) can now be obtained in the same way as for the top-down scan.

After both the top-down and the bottom-up scan have finished, the piece cover table is simply the logical OR of the piece data obtained from the top-down scan and the bottom-up scan.

Getting the piece cover data of the rook is a little more difficult, as the horizontal piece cover data of the rook is opposite to the scan direction. In this case an extra left-to-right scan and a right-to-left scan is necessary on the rank where the rook is positioned. The board data of this rank is stored in a register and the two horizontal scans are performed on this register.

Here is a summary of the computation of the piece cover table:

1. Carry out the top-down scan and the computation of the rook rank covering in parallel:
   - A top-down scan of the board is executed to compute piece cover data shown in Figure 7(b). The top-down piece cover data is stored in memory on the FPGA.
   - Compute the rook rank cover data by a left-to-right and a right-to-left scan. The rook rank cover data is stored to a register.
2. Carry out a bottom-up scan to obtain the piece cover data shown in Figure 7(a). The bottom-up piece cover data is stored in memory on the FPGA.

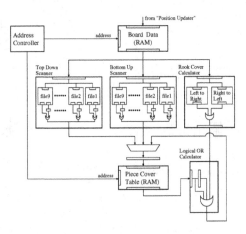

Fig. 8. A block diagram of the circuit for computing the piece cover data.

3. Compute the logical OR of the top-down piece cover data, the bottom-up piece cover data and the rook rank cover data.

6.2   Structure of the Hardware

Figure 8 shows the design of the circuit. The circuit has the following modules:

Board Data   One piece is represented by 10 bits and one rank has 9 squares, so the board data of one rank is 90 bits. The board has 9 ranks, therefore the complete description of the board requires 810 bits. This module is a 9×90 bit size RAM in which the board is stored. 90 bits of board data are loaded at one clock cycle.

Top Down Scanner/Bottom Up Scanner   The *Top Down Scanner* does a top-down scan. The input of this module is one 9×10 bit data block from the *Board Data*. This data block is split into 10 bit blocks that are processed in parallel by the sub-modules *file1*, *file2*, ..., *file9*. Each of the sub-modules checks the data of one square and gives the piece cover data on that square as output. This data is stored in *Piece Cover Data*. One rank of piece cover data is $360\ (= 9 \times 40)$ bits wide.

The *Bottom Up Scanner* does a bottom-up scan. The construction of the *Bottom Up Scanner* is similar to that of *Top Down Scanner*. Both the parallel processing and the pipeline processing is performed in the same way.

It takes eight clock cycles to compute one rank of piece cover data. Obtaining the piece cover data of the whole board takes 25 clock cycles. For this it is necessary to use pipeline processing, where the computation of the piece cover data of one rank does not need to finish before computing the piece cover data of the next rank. This is shown in Figure 9.

Rook Cover Calculator   The *Rook Cover Calculator* performs a left-to-right scan and a right-to-left scan to compute the rook rank piece cover data. The piece cover data obtained by these scans is stored to registers in the module.

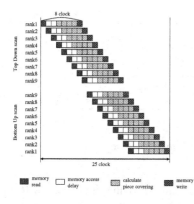

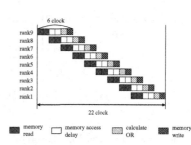

Fig. 9. Pipeline processing of the piece cover table computation

Fig. 10. Pipeline processing of the calculation of the logical OR

Table 1. Piece cover computation time per position

	Frequency [MHz]	Computation time [μsec]	Performance ratio
SPEAR	400	74.89	1
FPGA	38.46	1.20	62

Logical OR Calculator This module calculates the logical OR of the top-down piece cover data, the bottom-up piece cover data and the rook rank cover data. It takes 6 clock cycles to calculate a logical OR and write it to RAM. The computation of the final piece cover table finishes in 22 clock cycles by pipeline processing as shown in Figure 10.

Piece Cover Table One rank of piece cover data requires 360 bits because each square has 40 bits of piece cover data. Thus 3240 (= 9 ranks × 360) bits are needed to represent the whole board. The top-down piece cover data, the bottom-up piece cover data and the logical OR each require 9 × 360 bits, therefore the *Piece Cover Table* can be stored in 27 × 360 bits of RAM. 360 bit of the piece cover table is stored to or loaded from RAM at one clock cycle.

6.3 Performance

To test the performance of the piece cover circuit, the computation time was compared to that of SPEAR. The result of this test is shown in Table 1.

The piece cover table computation of SPEAR was done on a 400MHz Pentium II. The test consisted of 100 shogi positions. The average computation time of the piece cover table in these positions was 74.89μsec.

Figure 11 shows that in hardware the number of clock cycles required to compute the piece cover table is always 46. The frequency of the circuit was calculated with the simulator in MAX+PLUS II, which is a circuit design tool. MAX+PLUS II can find the critical path in the circuit because of the delay-predictable connections in an FPGA [1].

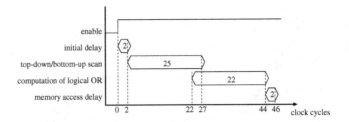

Fig. 11. A timing chart of the circuit for piece cover table

The longest delay was 26.0 ns and the maximum frequency of the circuit was 38.46MHz. Therefore, the computation time of the cover table is

$$46 \times \frac{1}{38.46} = 1.20 \, [\mu\text{sec}].$$

Therefore, the computation speed of the circuit is 62 times faster than that of SPEAR.

Note that the scanning method described in this section is not limited to the piece cover circuit. It will be the same for most of the modules of the position evaluator and move generator. Therefore, the performance of the *Piece Cover Circuit* is a good indication of the overall performance that can be expected of our shogi circuit.

7   The Tsume Shogi Circuit

In this section we describe the circuit for finding mate in the endgame. Building a program for solving mating problems in shogi (the *tsume shogi solver*) has been an independent research area in shogi for years with impressive results. The most important difference between mating problems in chess and tsume shogi is that in tsume shogi every move by the attacker has to be a check. As a result, the search space of tsume shogi is much smaller than that of normal shogi play, as the branching factor of the search tree is only about 5 [7][12]. Tsume shogi problems are used for endgame training or just as puzzles and there are tsume shogi problems that have solutions of several hundred ply. The longest tsume shogi problem has a solution of 1525 ply. The quest for solving this problem called *Microcosmos* inspired researchers for years until Masahiro Seo build a program that could solve *Microcosmos* in 1997. Tsume shogi is currently the only area where programs outperform the human experts.

A partial module for tsume shogi was our first test for using FPGAs in shogi. It seems clear that the use of FPGAs will have more impact in a search domain with a high branching factor, so it is interesting to compare the results for tsume shogi with the speed-ups given in the previous section. Even though the tsume shogi circuit was not completely finished, the performance of a complete tsume shogi circuit can be estimated by the intermediate results we have collected so far.

7.1   Structure of the Circuit

The tsume shogi circuit finds mate by processing the following three stage in pipeline:

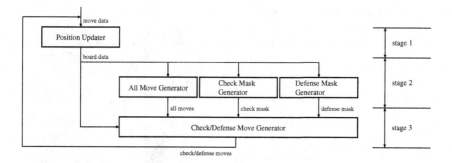

Fig. 12. A block diagram of the circuit for shogi mating problem

1. Update position after a move.
2. Generate all moves, generate check mask and generate defense mask.
3. Distinguish check/defense moves from all moves, and store them in memory.

To implement this procedure, we designed the circuits *Position Updater, All Move Generator, Check Mask Generator, Defense Mask Generator* and *Check/Defense Move Generator*. The function of each circuit is as follows.

Position Updater This module updates the position. The board data is updated based on the move data.

All Move Generator The structure of *All Move Generator* is similar to the *Piece Cover Circuit* in Section 6. The piece cover table contains information about which piece can move to which square, so the data collected by this circuit is the same as the data that is needed to generate all legal moves.

Check Mask Generator Generating a *check mask* makes it possible to discard the non-checking moves in an efficient way. This module scans the board twice (top-down and bottom-up) and checks if it is possible for a piece to check the king from a square. Figure 13 shows an example of a board position and the generated check mask. A white square means that there is a piece that can check the king by moving or dropping there. A black square means that no piece can check the king from that square.

Defense Mask Generator Generates a defense mask in the same way as a check mask. This module also scans the board twice and tests if there is a piece that can defend against the opponent check from a square.

Check/Defense Move Generator All legal moves are generated in *All Move Generator* and *Check/Defense Move Generator* distinguishes the check and defense moves from the other moves. This module generates check moves using all moves and the check mask, and defense moves using all moves and the defense mask.

7.2 Performance

Although we have not implemented the module *Defense Mask Generator* yet, we can estimate the performance of the full tsume shogi circuit by looking at the results of the

K: King   R: Rook   B: Bishop

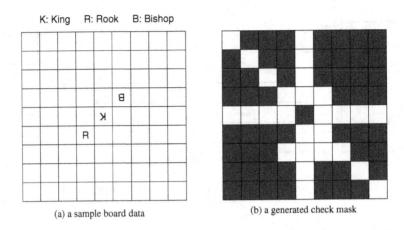

(a) a sample board data   (b) a generated check mask

Fig. 13. An example of board data and the check mask.

Table 2. The frequency of each modules

module	frequency [MHz]
Position Updater	80.06
All Move Generator	54.87
Check Mask Generator	90.07
Check/Defense Move Generator	22.32

partial implementation. Table 2 shows the frequency of the modules in the circuit. The frequency of each module was calculated with MAX+PLUSII.

The performance of the whole tsume shogi circuit can now be estimated as follows:

- The *Defense Mask Generator* runs at almost the same frequency of *Check Mask Generator* because the structures of these two modules are similar. Thus, the frequency of the whole circuit does not change if *Defense Mask Generator* is added to the circuit.
- The required clock cycles do not change if *Defense Mask Generator* is added to the circuit because it runs in parallel to *Check Mask Generator*.

Therefore, the frequency of the whole circuit is 22.32 MHz even if the *Defense Mask Generator* is added to the circuit. The required number of clock cycles is 48 (obtained with MAX+PLUSII), thus the computation time is

$$48 \times \frac{1}{22.32} = 2.15 \ [\mu sec].$$

We implemented a tsume shogi solver in C++ and compared its computation time for 100 tsume shogi problems to that of the circuit. The software ran on a Pentium II 700MHz. The result in Table 3 shows that the computation time of the hardware is about 9 times faster than that of the software.

In our current implementation, the operation speed of *Check/Defense Move Generator* is relatively slow. It is possible to improve the results by further pipelining the

Table 3. The computation time of tsume shogi

	Frequency[MHz]	Computation time[μsec]	Performance ratio
Software	700	18.9	1
FPGA	22.32	2.15	8.8

circuit, improving the operation speed to more than 54.87MHz. This is the frequency of *All Move Generator*, which is then the new bottle-neck of the tsume shogi circuit. If we can speed up *Check/Defense Move Generator* in this way, we can expect a speed-up of 23 instead of a speed-up of 8.8 as given in Table 3.

8   Conclusions

In this paper, we have described the architecture of a shogi processor based on reconfigurable hardware. Reconfigurable hardware promises significant speed-ups with a short development time. Also, reconfigurable hardware is relatively cheap, so a cooperative effort in the hardware design of a shogi processor is possible.

To evaluate the use of *Field Programmable Gate Arrays* (FPGAs) as the reconfigurable hardware for our shogi processor, we have implemented two modules of the shogi processor. The speed-up of the module *Piece Cover Circuit* was 62 times, while the speed-up of the *Tsume Shogi Circuit* was 9 times compared to an equivalent software program. We will continue our work on the processor by finishing the tsume shogi circuit and designing the other modules of the shogi processor that we have described in this paper.

FPGAs are a rather new technology that are still being improved fast and the speed gain is almost proportional to the hardware size. In the near future one PCI board with several FPGAs can be expected to be a thousand times faster than our shogi software by improvements of the hardware technology alone.

Another area that will be improved is how the FPGA chips are being reconfigured. At the moment it is still necessary to write circuits for each module in a hardware description language. This requires more time than programming in a traditional programming language like C. There have been many proposals for generating hardware circuits automatically from software programs, but so far it has been difficult to generate efficient circuits. However, by preparing hardware libraries for efficient data management in shogi, we expect that it is possible to write most of the modules of a shogi processor in C without negative effects on the performance. Preparing these libraries is also a future work.

References

1. Altera Corporation, San Jose, CA, USA, *Embedded Programmable Logic Data Sheet*, August 1999, ver 2.02.
2. D. Beal. A Generalised Quiescence Search Algorithm. *Artificial Intelligence*, 43:85–98, 1990.
3. R. Grimbergen, "Candidate Relevance Analysis for Selective Search in Shogi", in *Advances in Computer Chess Conference*, Paderborn, Germany, 1999.

4. R. Grimbergen, "A Plausible Move Generator for Shogi Using Static Evaluation", in *Game Programming Workshop*, pp. 9-15, Kanagawa, Japan, 1999.
5. S.Hamilton and L.Garber, "Deep Blue's Hardware-Software Synergy", *IEEE Computer*, Oct., 1997, pp.29-35.
6. G.Kakinoki, "The Search Algorithm of the Shogi Program K3.0", In H.Matsubara, editor, *Computer Shogi Progress*, pp.1-23. Tokyo: Kyoritsu Shuppan Co, 1996. ISBN 4-320-02799-X. (in Japanese).
7. H.Matsubara and K.Handa, "Some properties of shogi as a game", *Proceedings of Artificial Intelligence*, 96(3):21-30, 1994. (in Japanese).
8. H.Matsubara, H.ida and R. Grimbergen, "Natural developments in game research", *ICCA Journal*, 19(2):103-112, June 1996.
9. Monty Newborn, "Kasparov versus Deep Blue: computer chess comes of age", Springer, 1997, ISBN 0-387-94820-1.
10. J. Pearl. *Heuristics: Intelligent Search Strategies for Computer Problem Solving*. Addison Wesley Publishing Company: Reading, Massachusetts, 1984. ISBN 0-201-05594-5.
11. J. Schaeffer. The History Heuristic and Alpha-Beta Search Enhancements in Practice. *IEEE Transactions on Pattern Analysis and Machine Intelligence*, 11(11):1203–1212, 1989.
12. M. Seo, "A tsume shogi solver using conspiracy numbers", in H. Matsubara, editor, *Computer Shogi Progress 2*, pp.1-21, Tokyo:Kyoritsu Shuppan Co, 1998, ISBN 4-320-02799-X. (in Japanese).
13. C.E.Shannon, "Programming a computer for playing chess," *Philosophical Magazine* 41(1950): 256-75.
14. H.Yamashita, "YSS: About its Datastructures and Algorithm", In H.Matsubara, editor, *Computer Shogi Progress 2*, pp.112-142. Tokyo: Kyoritsu Shupppan Co, 1998. ISBN 4-320-02799-X. (in Japanese).
15. Xilinx, Inc., San Jose, CA, USA, *The Programmable Logic Data Book*, 1999.
16. URL: http://www.fccm.org
17. URL: http://xputers.informatik.uni-kl.de/FPL/FPL99/index.html
18. URL: http://www.ece.cmu.edu/˜fpga2000

Plausible Move Generation Using Move Merit Analysis with Cut-Off Thresholds in Shogi

Reijer Grimbergen

Department of Information Science, Saga University
1 Honjo-machi, Saga-shi, Saga-ken, Japan 840-8502
grimbergen@fu.is.saga-u.ac.jp

Abstract. In games where the number of legal moves is too high, it is not possible to do full-width search to a depth sufficient for good play. Plausible move generation (PMG) is an important search alternative in such domains. In this paper we propose a new method for plausible move generation in shogi. During move generation, Move Merit Analysis (MMA) gives a value to each move based on the plausible move generator(s) that generated the move. These values can be used for different cut-off schemes. We investigate the following alternatives: 1) Keep all moves with a positive MMA value; 2) Order the moves according to their MMA value and use cut-off thresholds to keep the best N moves. PMG with MMA and cut-off thresholds can save between 46% and 68% of the total number of legal moves with an accuracy between 99% and 93%. Tests show that all versions of shogi programs using PMG with MMA outperform an equivalent shogi program using full-width search. It is also shown that MMA is vital for our approach. Plausible move generation with MMA performs much better than plausible move generation without MMA. Cut-off thresholds improve the performance for $N = 20$ or $N = 30$.

Keywords: Plausible move generation, move merit analysis, cut-off thresholds, shogi.

1 Introduction

Full-width search has been very successful in two-player complete information games. DEEP BLUE in chess [20], CHINOOK in checkers [19] and LOGISTELLO in Othello [5] are examples of well-tuned full-width search programs that perform at the level of the human world champions.

In full-width search all legal moves in any given game position are generated. However, this does not mean that all legal moves are searched to the same depth. Based on domain-dependent heuristics, selectivity is added: some moves will be searched deeper than other moves. Examples of methods to add selectivity to the full-width search are *quiescence search* [3], *singular extensions* [1] and *futility pruning* [8].

Full-width search has not always been the main approach. *Plausible Move Generation* (PMG) was very important in the early days of chess research. A plausible move generator would select a small number of moves using domain-specific knowledge [15,4,7]. The remaining candidates were then searched as deep as possible with

T.A. Marsland and I. Frank (Eds.): CG 2000, LNCS 2063, pp. 315–332, 2001.

alpha-beta search. For example, Bernstein's chess program [4] generated only 7 plausible moves in any position. Plausible move generation is the ultimate form of selectivity: discarding moves without any search. In chess, the risk of discarding a good search candidate was too high and full-width search has been the dominant approach since the CHESS 4.5 program in the early seventies [23].

However, there are games in which it is impossible with current technology to search deep enough with standard full-width search to get a high performance program. Examples are games with a large average number of legal moves like *Go* and *shogi* [14] and single agent search problems with extremely long solution sequences such as *sokoban* [10]. To make a high performance program in these domains, some method for plausible move generation is needed [6,24,12,26,11]. Especially in Go, most of the available time per move is spent on generating promising looking moves, leaving little time for search [6].

In this paper we propose a new method of plausible move generation. Even though the method has been designed for shogi, we will present a framework for plausible move generation that will be applicable to other two-player complete information games as well. We think that one of the problems of plausible move generation in current programs is the lack of a combined effort to develop a general and satisfactory method. Shogi is a good example of this, and we will see in Section 7 that there are big differences between the plausible move generation methods used in some of the top programs. With the plausible move generation method that we present in this paper, we hope to make it easier to compare the different methods that are currently used and also make it easier for others to implement and improve their own plausible move generation method.

In section 2 we will explain why plausible move generation is a good alternative to full-width search in shogi. Then, a set of plausible move generators for shogi will be defined in Section 3. In Section 4 we will explain how analysing the merit of a move using these plausible move generators can improve move ordering and reduce the set of candidate moves generated by the set of plausible move generators. With this analysis of move merit additional cuts in the number of candidate moves can be made. These additional cuts are explained in Section 5. In Section 6 we will show that shogi programs based on plausible move generation with move merit analysis outperform an equivalent program using full-width search. In Section 7 we will compare our method with other plausible move generation methods used in shogi. We will end with some conclusions and ideas for future work in Section 8.

2   Why Is Plausible Move Generation Necessary in Shogi?

The main difference between chess and shogi is the possibility of re-using pieces. A piece captured from the opponent becomes a *piece in hand* and at any move a player can *drop* a piece he captured earlier on a vacant square instead of moving a piece on the board. As a result of these drop moves, the number of legal moves in shogi is on average much larger than in chess. The average branching factor of the search tree in chess is about 35, while in shogi the average branching factor is about 80 [13].

In shogi the average branching factor does not tell the whole story. In chess the branching factor rapidly decreases towards the endgame and finally gets to a point

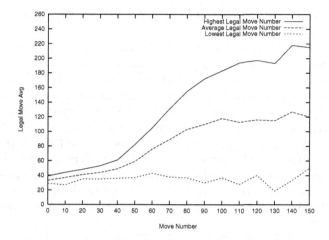

Fig. 1. Highest number of legal moves, average number of legal moves and lowest number of legal moves by move number in 100 test games.

where the exact theoretical game value can be retrieved from endgame databases [25]. This is not the case in shogi, where the branching factor of the search tree increases as the game progresses. To illustrate this behaviour, we have analysed the number of legal moves in 100 expert shogi games. The games have been selected to give a good coverage of the different types of positions that occur in shogi. The games therefore involve many different expert players (112) and have many different opening strategies (15).

The number of legal moves in the test games is given in Fig. 1. This figure shows that the number of legal moves in shogi increases as the game progresses. As more pieces get captured, the number of possible drop moves increases, leading to an average number of legal moves that is higher than 100 in the endgame. The top line in Fig. 1 shows that peaks of more than 200 can also be expected. The result of a shogi game is often decided in the endgame, so being able to deal with search trees that have such a high branching factor can mean the difference between winning and losing.

Not only the high average number of legal moves is a problem for building a strong shogi program. There is also the problem of strict time constraints. In shogi, the available time for finding the next move under tournament conditions is much less than in chess. There are two reasons for this. First, the average game length of shogi is about 115 ply [9], while the average game length of chess is about 80 ply. Therefore, even under the same conditions, a shogi program will have 30% less available time per move. Second, the tournament conditions for shogi programs are much stricter than in chess. In the annual CSA tournament, the computer shogi world championships, the available time per game is only 20 minutes. Therefore, even with the help of an opening book, on average only about 30 seconds per move are available for search.

To deal with large search trees under strict time constraints it is necessary to make good decisions about which moves the available search time should be spent on. Plausible move generation is a method to make these decisions and therefore can be an important alternative to full-width search in shogi.

3   A Set of Plausible Move Generators for Shogi

As the general game-playing system METAGAMER [17] shows, a wide range of games have the notion of goal, threat and positional improvement in common. The goal of a game can be to win material (e.g. chess, checkers), to occupy the largest territory (e.g. Go, Othello), or to reach a certain board configuration (e.g. five in a row, sokoban). Some goals are more important than other goals. In chess, for example, mating (winning the king) is more important than winning a queen, which in turn is more important than winning a pawn. A threat is a move that, if not defended against, will reach a goal on the next move or after a forced sequence of moves. Finally, there are moves to improve the player's position without actually threatening to reach a goal. An example is too improve the mobility of a piece. It is also possible to defend against this type of move by playing a move that makes such a positional improvement impossible. An example in chess is pinning a piece.

For each of these move classes a plausible move generator can be build which generates the moves in this class:

1. **PMG-Goal**:
   Moves that reach a goal.
2. **PMG-Th**:
   Moves that threaten to reach a goal.
3. **PMG-DefTh**:
   Moves that defend against a threat.
4. **PMG-PIm**:
   Moves that improve the position.
5. **PMG-DefPIm**:
   Moves that make it impossible for the opponent to improve the position.

For each game in which this basic set of plausible move generators is used, the PMGs have to be refined to reflect the features of that specific game. In shogi, the goal of the game is different than in Go and *PMG-Goal* is therefore different as well. For shogi, we have split the five basic PMGs above in 21 shogi specific PMGs:

1. **PMG-Goal:**
   - *Capture material.*
   - *Promote piece.*
2. **PMG-Th:**
   - *Check.*
   - *Attack king.*
   - *Attack material.*
   - *Discovered attack.*
     Moving a blocking piece leads to check or to a material attack.
   - *Threaten promotion.*
3. **PMG-DefTh:**
   - *Defend against checks.*
   - *Defend king.*
   - *Defend material.*

- *Defend discovered attacks.*
- *Defend against promotion threat.*

4. **PMG-PIm:**
   - *Defend pins.*
   - *Tie improvement.*
     If a piece can not move because it is tied to the defence of another piece P, defend piece P.
   - *Defend undefended pieces.*
   - *Defend against the exchange of pieces.*
   - *Cover squares in own camp.*
     Moves that gain control over a square in one's own camp.
   - *Develop pieces.*
     Patterns and move sequences taken from expert games for 1) standard opening sequences, 2) building castles, and 3) positional pattern moves.

5. **PMG-DefPIm:**
   - *Pin piece.*
     Moving the pinned piece puts the king in check or loses material.
   - *Cover squares in opponent camp.*
     Moves that gain control over a square in the opponent camp.
   - *Avoid development.*
     Moves that do not allow the opponent to develop its position.

4   Move Merit Analysis

Generating each plausible move only once is faster than having the same move generated several times. Therefore, having multiple plausible move generators with the possibility of generating moves more than once might not be efficient. However, the possibility of generating a move by more than one PMG is vital for our approach. If a move is generated by more than one PMG, it is often better than a move that is generated only once. For example, moving a piece away from an attack is in general more powerful if it is attacking an opponent piece at the same time.

Knowledge about which PMGs generated a move can be used to analyse the merit of the move. In our method, each PMG assigns a value to the generated moves based on the importance of the PMG. We will now give a detailed description of the values that the *Move Merit Analysis* (MMA) gives to the moves generated by the PMGs in shogi. As a reference, we have given the values of the shogi pieces that are used as the basic units of calculation of the MMA values in Table 1. It should be noted that because of the dynamic nature of shogi, there is no agreement on the values of the pieces as there is in chess.

1. **PMG-Goal:**
   - *Capture material*:
     Estimate of the material gain after a capture sequence on a square. For this estimate, MMA uses a static evaluator for sequences of piece captures on the same square.

Table 1. Piece values of shogi pieces as used for MMA value calculation.

Piece	Value
King	10000
Promoted rook	1300
Promoted bishop	1200
Rook	900
Bishop	800
Gold/Silver	500
Knight/Lance	300
Pawn	100

– *Promote piece*:
Promotion value of the piece. Table 1 shows that this is the equivalent of 4 pawns for rook, bishop and pawn and 2 pawns for knights and lances (knights and lances promote to gold).

2. PMG-Th:

– *Check*:
A constant based on whether the check is on a safe square or a sacrifice:
- Safe square: $V_{MMA} = 3 \times V_{Pawn}$.
- Sacrifice: $V_{MMA} = 1\frac{1}{2} \times V_{Pawn}$.

where V_{Pawn} is the value of a pawn.

– *Attack king*:
A constant based on whether the attack is on a safe square or a sacrifice:
- Safe square: $V_{MMA} = 3 \times V_{Pawn}$.
- Sacrifice: $V_{MMA} = 1\frac{1}{2} \times V_{Pawn}$.

– *Attack material*:
The MMA value is an estimate of how strong the attack is. An attack on a high value piece like a rook has a higher MMA value than an attack on a low value piece like a pawn. The basic piece attack value is:

$$AttackVal = \frac{1}{10} \times V_{AtteckedPiece}$$

where $V_{AttackedPiece}$ corresponds to the piece value as given in Table 1.

For pinned pieces, a bonus is given based on the value of the piece that would be lost if the pinned piece would move:
- King (check): $AttackValBonus = \frac{1}{6} \times V_{PinnedPiece}$.
- Major piece: $AttackValBonus = \frac{1}{7} \times V_{PinnedPiece}$.
- Gold and silver: $AttackValBonus = \frac{1}{8} \times V_{PinnedPiece}$.
- Knight and lance: $AttackValBonus = \frac{1}{9} \times V_{PinnedPiece}$.
- Pawn: $AttackValBonus = \frac{1}{10} \times V_{PinnedPiece}$.

where $V_{PinnedPiece}$ is the piece value of the pinned piece.

In case of attacking a dead piece the piece attack value is doubled.

– *Discovered attack*:
  A constant based on which piece the discovered attack is attacking.
  - King: $V_{MMA} = 1\frac{1}{2} \times V_{Pawn}$.
  - Other piece: $V_{MMA} = \frac{1}{10} \times V_{AttackedPiece}$.

– *Threaten promotion*:
  - $V_{MMA} = \frac{1}{10} \times V_{Prom}$.
  where V_{Prom} is the expected value of the promotion.

3. PMG-DefTh:

– *Defend against checks*:
  No MMA value is given to a defence against a check, as there are no other moves generated, so any MMA value would be given to each move anyway.

– *Defend king*:
  The MMA value is an estimate of how strong the attack of the opponent on a square SQ adjacent to the king is:
  - SQ controlled by the opponent: $V_{MMA} = 3 \times V_{Pawn}$.
  - SQ controlled by the player to move: $V_{MMA} = \frac{1}{2} \times V_{Pawn}$.
  - SQ controlled by neither player: $V_{MMA} = 1\frac{1}{2} \times V_{Pawn}$.

– *Defend material*:
  - $V_{MMA} = \frac{1}{2} \times V_{ExpectedLoss}$.
  where $V_{ExpectedLoss}$ is the value of the expected material loss after the capture sequence.

– *Defend discovered attacks*:
  Estimated improvement of the discovered attack. These values are the same as for playing the discovered attack.

– *Defend against promotion threat*:
  A constant based on which piece was threatening to promote. These values are the same as for promotion threats.

4. PMG-PIm:

– *Defend pins*:
  The MMA value is based on the value of the piece that is being pinned. These values are the same as the bonus values in *Attack material*.

– *Tie improvement*:
  The MMA value is based on the value of the piece that is tied and the material loss that would be the result if the tied piece should move:

$$V_{MMA} = \frac{V_{TiedPiece} + V_{TieLoss}}{40}$$

where $V_{TieLoss}$ is the material loss resulting from moving the tied piece.

- *Defend undefended pieces*:
  - $V_{MMA} = \frac{1}{8} \times V_{UndefendedPiece}$.

- *Defend against exchange of pieces*:
  - $V_{MMA} = \frac{1}{10} \times V_{ExchangePiece}$
  where $V_{ExchangePiece}$ is the value of the piece that can be exchanged.

- *Cover squares in own camp*:
  - $V_{MMA} = \frac{1}{10} \times V_{Pawn}$.

- *Develop pieces*:
  - Piece development moves: $V_{MMA} = \frac{1}{2} \times V_{Pawn}$.
  - Castle moves: $V_{MMA} = 1\frac{1}{2} \times V_{Pawn}$.
  - Pattern moves: $V_{MMA} = \frac{1}{4} \times V_{Pawn}$.

5. **PMG-DefPIm:**
   - *Pin piece*:
   The MMA value is based on the value of the piece that is being pinned. These values are the same as the bonus values in *Attack material*.

   - *Cover squares in opponent camp*:
     - $V_{MMA} = \frac{1}{10} \times V_{Pawn}$.

   - *Avoid development*:
     - $V_{MMA} = \frac{1}{8} \times V_{Pawn}$.

The values of the constants and weights used in the calculation of the MMA values have all been tuned by hand. An interesting future work is to investigate if these values can be learned automatically.

Negative MMA values are also possible because MMA can give a negative value to three types of moves:

- Material sacrifices: The penalty for sacrifices is the value given by the static exchange evaluator.
- Drops far from the kings: A drop move is given a penalty of $5 \times \min(BlackKingDistance, WhiteKingDistance)$
- Passive moves, i.e. moves that are not a threat: These moves are given a penalty of half a pawn.

If the penalty is higher than the expected merit of the move, this will result in a negative MMA value.

After move merit analysis, the plausible moves can be ordered according to their MMA value. If this MMA based ordering is a good estimate of the importance of a move, this will improve the performance of alpha-beta search. However, the main advantage of using MMA is to make additional cuts in the number of candidate moves generated by the PMGs. A natural cut-off is to discard all candidate moves with a negative MMA value. In this paper, plausible move generation without MMA will be called *PMG-All* and plausible move generation that cuts all moves with a negative MMA value will be called *PMG-MMA*.

5 Cut-Off Thresholds

The MMA values can be used for further cuts in the number of candidate moves. Assuming that the move ordering based on MMA values is a good indication of the quality of a move, the number of moves to search can be further reduced by taking only the best N moves after MMA based move ordering. We have investigated the *cut-off thresholds* $N = 20, 30, 40, 50$. The PMGs with cut-off thresholds will be called *PMG-N*.

Some modifications are needed in the application of these cut-off thresholds. First, moves with a high MMA value should not be discarded, even if they fall outside the first N moves. Second, in shogi the merit of checks and captures is hard to judge statically. A captured piece can be dropped back on almost any vacant square, and static evaluation of good drop points is difficult. Therefore, checks and captures of pieces higher than pawns are included regardless of their MMA value. These two modifications can increase the number of candidate moves above the basic threshold N.

The third modification helps to improve the number of cut-offs. If moves have an MMA value that is much smaller than the top moves, they are discarded even if they fall in the range N.

6 Results

We have analysed the behaviour and performance of our method of plausible move generation with four tests:

1. Plausible move generation test.
2. Move ordering test.
3. Search comparison test in tactical shogi problems.
4. Self play experiment.

With the plausible move generation test and the move ordering test the accuracy and savings of *PMG-All*, *PMG-MMA* and *PMG-N* were analysed by comparing the moves generated by the PMGs with moves played by expert players. Of course, expert performance is not equivalent to perfect performance, but there are several reasons why the results of these tests are interesting. First, they give a first indication about the performance of each PMG method. Second, the differences between the PMG methods will be clearer. We will see that the balance between accuracy and savings varies considerably between *PMG-All*, *PMG-MMA* and *PMG-N*. Third, the results will make it easier to compare our PMG method with other methods.

The search comparison test and the self play experiment compared the search performance of shogi programs using *PMG-All*, *PMG-MMA* and *PMG-N* with an equivalent

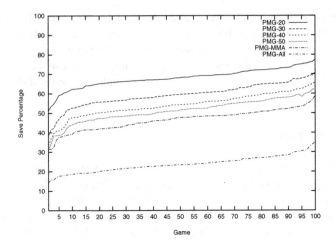

Fig. 2. Savings of different versions of plausible move generation in 100 test games, ordered by percentage of moves saved.

program using full-width search. This is necessary to answer the most important question: is our PMG method cost-effective, i.e. is the time spent on plausible move generation worthwhile?

6.1   Plausible Move Generation Test

First, we compared the savings and move generation accuracy of *PMG-All*, *PMG-MMA* and *PMG-N*. For this comparison, we used the 100 test games described in Section 2. These 100 test games have a total of 12097 positions. We tested the accuracy of *PMG-All*, *PMG-MMA* and *PMG-N* by checking if the move played by the expert was generated by the plausible move generators. We also calculated the savings of our approach, i.e. the difference between the total number of legal moves and the total number of moves generated by the PMG versions. The savings for *PMG-All*, *PMG-MMA* and *PMG-N* are given in Fig. 2. This figure shows that there are small areas of good and bad results, but that the majority of the savings are close to the average. It is also clear from the figure that the savings of *PMG-MMA* are much better than the savings of *PMG-All*. *PMG-N* gives a gradual increase in savings for smaller N. *PMG-20* and *PMG-30* give much more savings than *PMG-MMA*, but the difference between *PMG-40*, *PMG-50* and *PMG-MMA* is much less prominent.

Vital is the balance between the savings of the plausible move generation and the accuracy. The savings and accuracy results of *PMG-All*, *PMG-MMA* and *PMG-N* are summarised in Table 2. *PMG-All*, the basic plausible move generation without any additional cuts, can generate 99.4% of all expert moves in the test games. This version on average reduces the number of moves with 23.7% compared to the total number of legal moves. The savings can be almost doubled if MMA is used. *PMG-MMA* saves 46.5% of all moves at the cost of 0.5% accuracy compared to *PMG-All*.

Table 2. Results of the different versions of plausible move generation on 12097 positions. *NG* is the number of moves played by the expert, but not generated by the PMG; *Ac* is the accuracy of the PMG; *Sv* are the savings of the PMG.

Version	NG	Ac(%)	Sv(%)
PMG-All	81	99.4%	23.7%
PMG-MMA	144	98.9%	46.5%
PMG-50	178	98.6%	51.2%
PMG-40	231	98.1%	54.2%
PMG-30	401	96.7%	59.2%
PMG-20	892	92.7%	68.0%

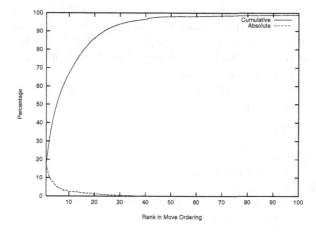

Fig. 3. Absolute and cumulative move ordering results in 100 test games.

The results indicate that *PMG-50* is a good improvement over *PMG-MMA* as it improves the savings with 4.7%, while the accuracy is only reduced by 0.3%. There is also a good improvement of savings between *PMG-40* and *PMG-30* at 5.0%. However, the loss of accuracy compared with *PMG-40* is 1.4%. Finally, savings of 68% can be achieved with *PMG-20*, but the accuracy will then drop to 92.7%. Other tests are needed to show if the relatively low accuracy or the high savings are the deciding factor in the performance of *PMG-20*.

6.2   Move Ordering Test

We have also looked at the move ordering results of MMA for each position in the set of test games. The results are given in Fig. 3. MMA orders the expert move first in 17.3% of the positions. Almost half of the expert moves (49.6%) are ordered among the best five moves. If the first ten moves in the ordering are considered, 66.9% of the expert moves are produced by the PMGs.

The absolute move ordering curve in Fig. 3 shows that there are only very few expert moves ordered lower than 30. These results support our assumption that MMA leads to a reasonable move ordering. The results also make it highly unlikely that a *PMG-N* for *N* higher than 50 will give better results than the *PMG-N* we investigate here.

6.3 Search Comparison Test in Tactical Shogi Problems

As a first test to compare the search performance of shogi programs using *PMG-All*, *PMG-MMA* and *PMG-N* with an equivalent program using full-width search, we compared the performance of different versions of the same shogi program on a set of tactical shogi problems. The PMG based programs all use plausible move generation at every node in the search tree.

The shogi problems in the test were taken from the weekly magazine *Shukan Shogi*. The test set consists of 300 problems published in issues 762 (November 4th 1998) to 811 (October 20th 1999). The problems in each issue are divided into six classes, ranging from starting level to expert level. It should be noted that the starting level is already quite advanced and is too hard for beginners. Two of the problems in the test set are incorrect and have been removed from the test set.

The basic program that we used for all versions has the following features, which are common in most shogi programs:

- Iterative alpha-beta search.
- Principal variation search [16].
- Quiescence search [3].
- History heuristic and killer moves [18].
- Null-move pruning [2].
- Hashtables for transposition and domination [22].
- Specialised mating search [21].

All three programs have the same evaluation function. The evaluation function has the following features:

- Material.
- King safety.
- Piece mobility.
- Pinned pieces.
- Discovered attacks.
- Promotion threats.
- King distance of pieces.
- A number of piece patterns to evaluate good and bad piece formations.

All versions of the program were given 30 seconds per problem on a 700 MHz Pentium II, which is about the same time as can be expected to be available under tournament conditions.

The results of this test are given in Table 3. The categories in the table correspond to the categories in Shukan Shogi. The table shows that MMA is vital to our approach. There is almost no difference between the number of solved problems by full-width search and *PMG-All*. However, *PMG-MMA* and *PMG-N* solve significantly more problems than full-width search and *PMG-All*.

This test shows no significant improvement of the performance when cut-off thresholds are used. *PMG-N* solves about the same number of problems as *PMG-MMA*.

Table 3. Results of full width search and different versions of plausible move generation on 298 tactical shogi problems.

Cat	Tot Pos	Full width	PMG All	MMA	50	40	30	20
1	50	17	21	22	21	23	23	22
2	50	9	8	13	12	12	13	10
3	50	10	10	11	11	12	10	11
4	50	9	7	8	8	9	9	8
5	50	5	4	4	4	4	5	7
6	48	4	5	6	6	6	6	6
Tot	298	54	55	64	62	66	66	64

6.4   Self Play Experiment

As a final experiment to compare full-width search, *PMG-All*, *PMG-MMA* and *PMG-N* we played different versions of the same shogi program against each other. One program was using full-width search and the other programs were using different versions of our plausible move generation method. Each program version played the other versions twenty times with a time limit of 20 minutes per side per game. This is the same time limit as used in the annual CSA tournament. The results of this tournament are given in Table 4.

The results show that a PMG based program is playing better than a program without plausible move generation. Full-width search won only 10 out of 120 games. The results also show how important move merit analysis is for our approach. *PMG-All* only managed to win against the full-width search program and lost all other matches by a wide margin, on average winning less than 4 games against any of the MMA based programs. Move merit analysis is clearly an improvement of the general plausible move generation approach.

From the results of this tournament, it is difficult to draw conclusions on the importance of cut-off thresholds. *PMG-MMA* did surprisingly well, losing against *PMG-20* and *PMG-30* with only the narrowest possible margin of 11-9. Actually, *PMG-MMA* scored more points in the tournament than *PMG-20*, which won most matches. It seems that *PMG-50* and *PMG-40* are not an improvement over *PMG-MMA*, but *PMG-30* and *PMG-20* might be an improvement. Further tests with more games and against different opponents are needed to evaluate the importance of cut-off thresholds.

7   Related Work

In this section we will present different approaches for plausible move generation in shogi and compare them with the method presented in this paper. We will look at the plausible move generation in the shogi programs IS SHOGI, winner of the 1998 and 2000 CSA computer shogi championships; KANAZAWA SHOGI, winner in 1996 and 1999; YSS, winner in 1997; and KAKINOKI SHOGI, winner of the Computer Shogi Grand Prix in 1999.

Table 4. Results of a self play experiment between a full-width search shogi program and shogi programs using different plausible move generation versions.

No	Version	1	2	3	4	5	6	7	P	W	L
1	PMG-20	x	10-10	11-9	11-9	11-9	16-4	17-3	5.5	76	44
2	PMG-30	10-10	x	11-9	13-7	10-10	16-4	20-0	5	80	40
3	PMG-MMA	9-11	9-11	x	12-8	12-8	15-5	20-0	4	77	43
4	PMG-40	9-11	7-13	8-12	x	11-9	18-2	19-1	3	72	48
5	PMG-50	9-11	10-10	8-12	9-11	x	17-3	18-2	2.5	71	49
6	PMG-All	4-16	4-16	5-15	2-18	3-17	x	16-4	1	34	86
7	Full-width	3-17	0-20	0-20	1-19	2-18	4-16	x	0	10	110

7.1 IS Shogi

IS SHOGI [24] uses the following plausible move generators:

– Best move of the previous iteration.
– Capture opponent piece that just moved.
– Move piece that was attacked on the previous move to a safe square.
– Killer move.
– Null move.
– Attack king or attack material.
– Discovered attacks.
– Defend piece that was attacked on the previous move.
– Defence moves.
– Other special moves.

The categories are strictly ordered. If a move in a category leads to an alpha-beta cut-off or has a sufficiently high evaluation, none of the moves in the categories ordered below it will be generated. It is unclear which moves are generated by *Other special moves* as the description of this PMG is very short. There is only one example given of a special defence move to shut out pieces from attack.

The advantage of this plausible move generation method compared to our method is that no time is spent on generating moves that do not influence the search. The disadvantage is that it is not possible to make use of the extra information of moves that are generated by multiple plausible move generators.

7.2 Kanazawa Shogi

In the program KANAZAWA SHOGI, plausible move generation is used only in the following special cases [12]:

– Best move of the previous iteration.
– Take piece with the highest value.
– Move attacked piece with the highest value.
– Killer move.
– Null move.

If none of these moves are good enough to stop the search, all remaining legal moves are generated. After this, all moves are played and evaluated. Based on the evaluation, a decision is made on which moves to search and which moves to discard.

Strictly speaking, KANAZAWA SHOGI does not use plausible move generation, since in most cases all legal moves are generated. However, some legal moves are discarded without any search, so the method is very similar to plausible move generation. No detailed data on the proportion of moves that is being discarded without search is being given, so it is difficult to compare the savings of the method KANAZAWA SHOGI uses to MMA. The advantage of having all legal moves available is that it is easier to recover from a bad decision about the moves to search. The disadvantage is that all moves need to be evaluated. The number of legal moves in a position can be high and evaluation in shogi is expensive, so the extra number of evaluations can slow down the search.

7.3 YSS

Yamashita's YSS [26] uses 30 move categories. In YSS, the plausible move generation is strongly related to the search depth. Moves are only generated if the remaining search depth is enough to show that the move can actually reach the goal implied by the move category. For example, a move that attacks a piece is not generated at depth 1, because it is not possible to show that the attack will have a positive effect on the position. Also, some moves are only generated at the start of the search and based on the stage of the game (opening, middle game or endgame). Here are YSS's move categories in detail:

- **Remaining search depth is at least 1:**
  - Capture opponent piece that just moved.
  - Capture undefended piece.
  - Promote piece.
  - Checks that do not sacrifice material.
  - Move attacked piece with highest value.
- **Remaining search depth is at least 2:**
  - Defend against strong threat.
  - Attack material.
  - Discovered check.
  - Attack king from the front.
  - Discovered attack.
  - Attack pinned pieces.
  - Drops of bishop and rook in the camp of the opponent.
- **Remaining search depth is 3 or higher:**
  - Attack pieces around the opponent king.
  - Attack tied defending pieces.
  - Capture material that has a higher value than a pawn.
- **Moves only generated at the first ply of search:**
  - Develop inactive pieces.
  - Sacrifices with check.
  - Pawn drops far from the promotion zone.
- **Moves only generated at the first two ply of search:**
  - Pawn pushes in front of rook and lance.

- Material sacrifices that lead to a fork.
- Pawn promotion sacrifice.
- Dangling pawn.
- Block opponent rook or bishop.
- Move gold sideways in the camp of the opponent.
- Attack opponent piece that just moved with a pawn drop.
- Attack opponent piece that just moved.
- Move the king.
– **Moves only generated at the first two ply of search in the opening:**
  - Attack a pawn with a piece.
  - Drop a pawn to make an attacking base (covering the opponent camp).
  - Develop inactive pieces.

YSS uses more plausible move generators than our method (30 PMGs instead of 21 PMGs), which might improve the quality of the plausible move generation, but also takes more time. Also, control of the search seems difficult with a plausible move generation that depends heavily on the search depth.

Still, YSS's practical results are very good, so this method deserves further investigation. It is not difficult to use our plausible move generation method in the same way. Only minor modifications are needed to relate plausible move generation to the search depth. Further research is needed to investigate if this improves the performance of our method and this is a future work.

7.4   Kakinoki Shogi

Kakinoki Shogi uses 8 basic move categories [11]:

– Capture material.
– Defend material.
  - Move away from attack.
  - Cover attacked piece.
  - Take attacking piece.
  - Interpose piece between attacker and attacked piece.
– Promote piece.
– Defend against promotion threat.
– Attack king.
– Defend king.
– Other attacks.
– Other defences.

Most of Kakinoki's move categories correspond to the PMGs in Section 3. Although not clear from his description, we assume that moves like *Attack material* and *Pin piece* fit in the category *Other attacks*, while moves like *Defend undefended pieces* are part of the category *Other defences*. Absent from Kakinoki's move category description are non-tactical moves for piece development. Search in KAKINOKI SHOGI is a tactical search only and moves for piece development are handled differently. They become plans that can be played if there are no tactical problems detected by the search.

8   Conclusions

Full-width search has been the dominant approach in most game playing programs and has been the subject of much scientific research into two-player complete information games. In this paper we have argued that plausible move generation is an important alternative that deserves further investigation. There are games where the full-width search paradigm can not be successfully applied because of a large average number of search alternatives.

We have proposed a new plausible move generation method for shogi. This plausible move generation method uses five basic move categories that are general for a wide range of games. For each game these basic move categories need to be refined to match the specific features of the game under investigation. For shogi the five basic move categories resulted in 21 different plausible move generators. Each generated move is given a value based on the plausible move generator(s) that generated the move. This *Move Merit Analysis* (MMA) is an indication of the importance of a move. The MMA value of moves can be used for different cut-off schemes. In this paper we compared *PMG-All* (no cut-offs), *PMG-MMA* (cut moves with a negative MMA value) and *PMG-N* (cut moves that are ranked lower than N after MMA; $N = 20, 30, 40, 50$).

Results in shogi show that plausible move generation with move merit analysis gives important savings in the search candidates without compromising accuracy. Savings between 46% and 68% can be achieved, losing only between 1% and 7% of the moves chosen by expert players.

Using plausible move generation with MMA also significantly improves the performance of a shogi program. In tactical shogi problems, a shogi program using plausible move generation without MMA performs only slightly better than an equivalent program using full-width search, but PMG with MMA can solve significantly more problems. More importantly, programs based on *PMG-MMA* and *PMG-N* beat *PMG-All* and a full-width search program by a wide margin under tournament conditions. Our tests also suggest that further cutting the number of moves by using thresholds based on the MMA value can improve the playing strength of a shogi program. In this case rigorous extra cuts are needed, keeping no more than 30 moves. However, the results are not conclusive on this point and further testing is needed to support this.

References

1. T. Anantharaman, M.S. Campbell, and F. Hsu. Singular Extensions: Adding Selectivity to Brute-Force Searching. *Artificial Intelligence*, 43:99–109, 1990.
2. D. Beal. Experiments with the Null Move. In D.Beal, editor, *Advances in Computer Chess 5*, pages 65–79. Elsevier Science Publishers: The Netherlands, 1989.
3. D. Beal. A Generalised Quiescence Search Algorithm. *Artificial Intelligence*, 43:85–98, 1990.
4. A. Bernstein and M. de V. Roberts. Computer v Chess-Player. *Scientific American*, 198:96–105, 1958.
5. M. Buro. The Othello Match of the Year: Takeshi Murakami vs. Logistello. *ICCA Journal*, 20(3):189–193, September 1997.
6. K. Chen. Some Practical Techniques for Global Search in Go. *ICGA Journal*, 23(2):67–74, June 2000.

7. R. Greenblatt, D. Eastlake III, and S. Crocker. The Greenblatt Chess Program. In *Proceedings of the Fall Joint Computer Conference*, pages 801–810, 1967.

8. E. Heinz. Extended Futility Pruning. *ICCA Journal*, 21(2):75–83, June 1998.

9. Japanese Shogi Federation. *Heisei 10 Nenban Shogi Nenkan*. Nihon Shogi Renmei, 1999.

10. A. Junghanns and J. Schaeffer. Domain-Dependent Single-Agent Search Enhancements. In *Proceedings of the Sixteenth International Joint Conference on Artificial Intelligence (IJCAI-99)*, pages 570–575, 1999.

11. G. Kakinoki. The Search Algorithm of the Shogi Program K3.0. In H. Matsubara, editor, *Computer Shogi Progress*, pages 1–23. Tokyo: Kyoritsu Shuppan Co, 1996. ISBN 4-320-02799-X. (In Japanese).

12. S. Kanazawa. The Kanazawa Shogi Algorithm. In H. Matsubara, editor, *Computer Shogi Progress 3*, pages 15–26. Tokyo: Kyoritsu Shuppan Co, 2000. ISBN 4-320-02956-9. (In Japanese).

13. H. Matsubara and K. Handa. Some Properties of Shogi as a Game. *Proceedings of Artificial Intelligence*, 96(3):21–30, 1994. (In Japanese).

14. H. Matsubara, H. Iida, and R. Grimbergen. Natural developments in game research: From Chess to Shogi to Go. *ICCA Journal*, 19(2):103–112, June 1996.

15. A. Newell, C. Shaw, and H. Simon. Chess Playing Programs and the Problem of Complexity. *IBM Journal of Research and Development*, 2:320–335, 1958.

16. J. Pearl. *Heuristics: Intelligent Search Strategies for Computer Problem Solving*. Addison Wesley Publishing Company: Reading, Massachusetts, 1984. ISBN 0-201-05594-5.

17. B. Pell. A Strategic Metagame Player for General Chess-like Games. *Computational Intelligence*, 12(2):177–198, 1996.

18. J. Schaeffer. The History Heuristic and Alpha-Beta Search Enhancements in Practice. *IEEE Transactions on Pattern Analysis and Machine Intelligence*, 11(11):1203–1212, 1989.

19. J. Schaeffer. *One Jump Ahead: Challenging Human Supremacy in Checkers*. Springer-Verlag New York, Inc., 1997. ISBN 0-387-94930-5.

20. J. Schaeffer and A. Plaat. Kasparov Versus Deep Blue: The Rematch. *ICCA Journal*, 20(2):95–101, June 1997.

21. M. Seo. The C* Algorithm for AND/OR Tree Search and its Application to a Tsume-Shogi Program. Master's thesis, Faculty of Science, University of Tokyo, 1995.

22. M. Seo. On Effective Utilization of Dominance Relations in Tsume-Shogi Solving Algorithms. In *Game Programming Workshop in Japan '99*, pages 129–136, Kanagawa, Japan, 1999. (In Japanese).

23. D. Slate and L. Atkin. Chess 4.5: The Northwestern University Chess Program. In P. Rey, editor, *Chess Skill in Man and Machine*, pages 82–118. Springer Verlag, New York, 1977.

24. Y. Tanase. The IS Shogi Algorithm. In H. Matsubara, editor, *Computer Shogi Progress 3*, pages 1–14. Tokyo: Kyoritsu Shuppan Co, 2000. ISBN 4-320-02956-9. (In Japanese).

25. K. Thompson. 6-Piece Endgames. *ICCA Journal*, 19(4):215–226, December 1996.

26. H. Yamashita. YSS: About its Datastructures and Algorithm. In H. Matsubara, editor, *Computer Shogi Progress 2*, pages 112–142. Tokyo: Kyoritsu Shuppan Co, 1998. ISBN 4-320-02799-X. (In Japanese).

Abstraction Methods for Game Theoretic Poker

Jiefu Shi[1] and Michael L. Littman[2]

[1] Department of Computer Science
Duke University
Durham, NC 27708
jshi@cs.duke.edu
[2] AT&T Labs–Research
180 Park Ave. Room A275
Florham Park, NJ 07932-0971 USA
mlittman@research.att.com

Abstract. Abstraction is a method often applied to keep the combinatorial explosion under control and to solve problems of large complexity. Our work focuses on applying abstraction to solve large stochastic imperfect-information games, specifically variants of poker. We examine several different medium-size poker variants and give encouraging results for abstraction-based methods on these games.

Keywords: poker, game theory, imperfect information games, Texas Hold'em.

1   Introduction

One of the principle challenges in developing computer-based solutions to real-world problems is dealing with uncertainty. Four principle types of uncertainty are:

- *effect uncertainty*: In an unfamiliar environment, a decision maker might not know the possible effects of its decisions. We do not treat this type of uncertainty in this paper, although work in the area of reinforcement learning [2,6] has focused on this issue.
- *outcome uncertainty*: In games of chance, for example, although the decision maker may know the set of possible outcomes and their probabilities, there is no way to know exactly which outcome will occur. For example, dealing a random card from a deck could result in a red card or a black card being dealt, but we would not know which until the card is shown.
- *state uncertainty*: The decision maker might not know information that may affect the outcomes of its future decisions. For example, in the game of Scrabble[TM], a player cannot see the other player's tiles.
- *opponent uncertainty*: In multi-agent systems, and especially games, the decision maker does not know precisely how other agents in the system will respond.

Traditional work in game theory has a mathematical framework for reasoning about uncertainty. Outcome and state uncertainty are modeled by probability theory; decision makers maintain probability distributions over the current state and future events, and decisions are made to maximize expected utility. Because opponents need not behave

T.A. Marsland and I. Frank (Eds.): CG 2000, LNCS 2063, pp. 333–345, 2001.

according to any fixed probability distribution, however, opponent uncertainty is handled differently. In the *game-theoretic approach* [7], an agent makes decisions to maximize its expected utility assuming the opponent makes decisions to minimize the agent's utility. Thus, although the decision maker does not know how the opponent will actually behave, it does as well as possible in the worst case.

The game-theoretic approach has been applied to a large number of games, such as chess, backgammon, checkers, and many others. In fact, it is safe to say that this is the predominant approach in use in computer games research. Billings et al. [1] argue that this approach is not appropriate for games such as poker where (a) opponent play is probably not very strong, and (b) repeated encounters give the decision maker an opportunity to learn patterns in the opponent's play. Thus, opponent uncertainty can, and perhaps should, be treated as a type of state/effect uncertainty.

We have decided to attack poker using a game-theoretic approach, mainly because we believe this is still the best way to create an extremely high-quality player. Learning techniques run the risk of being fooled into low-quality play by a sufficiently clever adversary and the game-theoretic approach guards against this. In addition, finding an optimal or approximately optimal strategy using game theory can lead to unexpected insights into the structure of the game. We describe an instance of this from our poker player in Section 5. Note that, throughout the paper, we treat only 2-player zero-sum poker.

In Section 2 we introduce the game of poker and some of its unique challenges. Sections 3 and 4 describe two techniques we developed for attacking poker games using game theory. Section 5 presents some results and Section 6 concludes.

2   Poker Introduction

Consider the following mini-poker game. We start with a deck of 3 cards: J, Q, K. We deal one card to each of two players. Each player contributes one dollar to the pot (ante) and looks at his or her card. Next, a betting round commences. In the betting round, the players look at their cards, then alternate either betting (adding a dollar to the pot) or passing. If a player passes when the other player has contributed more money to the pot, that player "folds" and forfeits the pot. On the other hand, if there are two consecutive passes or both players have added the maximum number of dollars to the pot (1, for this example), the players reveal their cards and the one with the higher card wins the pot.

We seek a game-theoretic strategy for this game. Consider the game tree shown in Figure 1, where the leaves are labeled with player 1's winnings if the corresponding branches are followed. If the game were one of perfect information (players can see each others' card), an optimal strategy could be found by "minimaxing" up from the leaves. This computes, for each node in the tree, what player 1 would win on average if she bet optimally while player 2 made optimal responses.

Of course, a complete-information version of this game would be pointless, as whichever player was dealt the lower card would pass and the other player would win the ante. The challenge here is that, because cards are private, the game-tree nodes that share information sets with each other (ovals in the figure) are indistinguishable and the player must make the same choice from states in these sets. Using a simple bottom-up

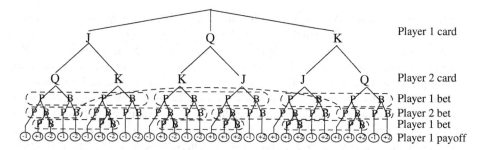

Fig. 1. Game tree for a very small poker game

approach, it is not at all evident how this constraint can be satisfied. Further, optimal play in imperfect information games is often randomized and simple game-tree search cannot reveal this.

Koller et al. [3] provide an algorithm that maps a game tree notated with information sets to an optimization problem, namely a linear program. When this optimization problem is solved, it reveals the optimal (possibly randomized) strategy for the game. Roughly speaking, variables in the linear program represent stochastic decisions, constraints represent making the same decision from all states in the same information set, and the objective function captures the optimal expected minimax winnings.

An optimal strategy for this type of game can have very rich and interesting structure. Consider what player 1 could do when she has the J. Player 2 might have either the Q or the K. In either case, player 1 will lose, but if player 2 has the Q and player 1 can convince him she has the K (perhaps by betting whenever possible), player 2 might fold and she would win. This is an example of "bluffing". Consider also what player 1 could do if she had the K. If she bets grudgingly (only matching player 2's bets when necessary), she might be able to get player 2 to bet before beating him. This is sometimes called "slow playing". These strategies are practiced by experienced poker players and emerge from game-theoretic solutions as well [4].

Thus, interesting and surprising strategies for complex imperfect information games can emerge from game-theoretic solutions. Furthermore, using the linear-programming approach, solution time is polynomial in the size of the game tree. Unfortunately, these game trees can get quite large. Consider a poker game in which c cards are dealt from a deck of d cards and players carry out b betting rounds each with a maximum of r raises per round[1]. The game tree for such a game contains at least $(d - c + 1)^c \, 2^{rb}$ leaves. The justification for this equation is that the branching factor for each of the c dealt cards is the number of cards not yet dealt (at least $d - c + 1$). Each of the b betting rounds consists of at most r raises, each of which can at most double the game-tree size (one for each order the two players carry out their raises).

Koller and Pfeffer [4] report solving games of size $(b, c, d, r) = (1, 2, 128, 1)$ (more than 32K leaves), $(1, 2, 3, 11)$ (more than 8K leaves), and $(1, 10, 11, 3)$ (more than 8K

---

[1] Throughout this paper, we use "raise" to mean putting money in the pot after the ante. This is different from the term "betting levels" used by some authors, which does not count the first bet as a raise.

leaves). Poker games played by people are much larger. The two-player version of Texas Hold'em, discussed in Section 4, has $(b, c, d, r) = (4, 9, 52, 3)$ or more than 3.7×10^{18} leaves. Note that, often, a more appropriate representation for a game is a directed acyclic graph, since the order cards are dealt is not important sometimes (player's hole cards or the flop in Texas Hold'em). We do not consider this optimization in our analysis, although it can substantially decrease the game representation. In either case, though, full games are well beyond what can be solved using the linear-programming approach. In this paper, we examine several abstraction-based approaches that can be used to generate approximations to large imperfect information games. Note that a related theoretical treatment of game-theoretic abstraction is underway [5].

Section 3 describes binning methods, which effectively reduce the number of cards in the deck d. Section 4 describes ways of treating betting rounds independently to effectively reduce the number of betting rounds r. Section 5 describes the results of applying these ideas in a game we invented called Rhode Island Hold'em.

3   Binning

The simple three-card game in Section 2 can be easily scaled up by considering larger decks. Even with a 52-card deck and 3 raises, the game is not very challenging to solve (around 21K leaves in the game tree). However, it is not a huge leap to imagine a poker variation where players are dealt 5-card hands and must bet on them. This is similar to, but not identical to, dealing each player one card from a deck of size 52 choose 5 (around 2.6M) cards with the highest card winning, since there is a well defined total order over all possible hands. The full game tree has more than 5.4×10^{13} leaves.

For games like this, with a single betting round based on the entire hand, we can use a grouping method to reduce the number of distinct hands considered. We rank all possible hands by their strength to obtain a ranking of each hand (a hand with higher value always beats a hand with lower value). For any hand, we can determine its ranking very fast either by looking it up in a database, or just calculating it in real-time.

Next, we group hands into equal-size bins. Each bin contains hands with similar rankings. The game is then solved at the level of bins: we imagine that players are randomly assigned to bins, with the highest ranking bin the winner and ties broken arbitrarily; betting strategies are computed for the resulting game. The number of bins used in the approximation controls the degree of abstraction and can be adjusted to accommodate space and time requirements.

At a high level, this may not be too different from how humans play poker. We do not really care whether we have two kings with a five of spades or whether we have two kings with a six of diamonds; we treat these situations fairly similarly in terms of our decision making. Also, note that if we produce a strategy using the same number of bins as the total number of hands possible, then what we have is the true optimal strategy. Of course, the accuracy of the approximation depends on how the groups are formed.

For our test-bed, we used a game with 200 possible hands. Both players initially ante one chip into the pot. Each player is then dealt a hand and the betting round begins. For this game, $(b, c, d, r) = (1, 2, 200, 1)$. Its game tree contains at least 79K leaves.

Game Without Potential

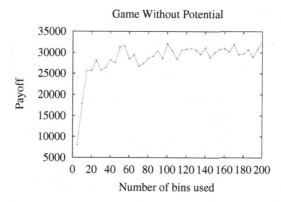

Fig. 2. Payoff vs. optimal for a game with betting only at the end

We first generated the optimal strategy for player 1. We then ran experiments dividing hands into from 4 to 200 bins and produced strategies for player 2 based on each of these groupings. So, for example, when we used 50 bins to solve our game, bin number 1 (the lowest bin) contained the lowest hands of 1, 2, 3 and 4. Similarly, bin number 50 contained the highest hands of 197, 198, 199, and 200.

Figure 2 shows how well player 1 fared against the optimal strategy for player 2 based on 1,000,000 games (we felt this number of games would be sufficient to see statistically significant results). Note that player 2 has an advantage in this and most poker games by virtue of gaining information from seeing the other player's initial selection. The results are quite encouraging; using as few bins as 10% of the number of hands, the resulting play is almost as good as that of the optimal strategy.

In popular poker games, players must bet based on partial hands and the principle concern is "hand potential". The ranking trick just described does not directly apply here because there is no direct linear ranking. For example, a partial hand consisting of four spades could turn into a flush if another spade is added (good hand) or nothing in particular if a non-spade is added (bad hand).

One way to score partial hands is to use the average rank of all possible complete hands that the partial hand can develop into. Partial hands can then be binned according to their assigned scores. Note that this is only an approximation and a convenient way of grouping hands into bins; we are not actually using these scores to judge hands. We play each bin against another to get the expected payoff, and this is used in defining the payoffs for the "abstracted" game.

We introduced another game to test our method for dealing with games with potential. For this game, we use a 52-card deck. Each player is dealt one private card. A third "public" card is dealt onto the table to be shared by both players. Players then bet (up to three raises). A final public card is then dealt and the player with the best 3-card hand (one private and two public cards) wins the pot[2]. For this game, $(b, c, d, r) = (1, 4, 52, 3)$, so the game tree has over 46M leaves.

---

[2] Three-card hands are ranked slightly differently from five-card hands: three of a kind beats a straight, and a straight beats a flush. See
http://conjelco.com/faq/poker.html#P15 for details.

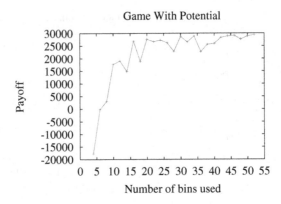

Fig. 3. Payoff vs. optimal for a game with hand potential

Although this game tree would be too large to solve with our available software, we can eliminate two of the cards from the computation. First, because one public card is dealt before any betting takes place, it is *public knowledge* and we can essentially treat this card as part of the problem statement (assuming we are willing to solve a linear program at game time). Second, because the other public card is dealt after betting is complete, we can compute the *expected value* of the winnings over all possible cards when we compute the values for the leaves. Therefore, we really only have to reason specifically about the two hidden cards, or a game tree with around 21K leaves. Figure 3 shows the result of player 2 versus player 1 (player 2 is using the optimal strategy). Similar to the game without potential, we can do quite well using a very small number of bins.

In a game with multiple betting rounds, a different binning scheme would be used at each round, with transition probabilities calculated between consecutive pairs of bins. These probabilities can be calculated by enumeration (costly) or by random sampling (risky). For a game with b betting rounds at r raises per round with B bins used between each round, the abstracted game tree has $B^b 2^{rb}$ leaves. This can be a substantial reduction if $b << c$ (many cards dealt between each betting round) or $B << d$ (far fewer bins than unique cards in the deck).

4   Independent Betting Rounds

Texas Hold'em is a poker variant used to decide the world champion. Each player is dealt two cards, followed by a betting round, then three public cards are dealt, followed by a betting round, then one public card, followed by a betting round, then one final public card and betting round. Thus, the game-tree size for Texas Hold'em is huge.

In the context of our current work, we introduce *Rhode Island Hold'em*, which is intended to be similar to Texas Hold'em in style, but much smaller. In Rhode Island Hold'em, each player is dealt one private card, followed by a betting round, followed by a public card, followed by a betting round, followed by a final public card and betting round. The three-card hands based on the private card and two public cards

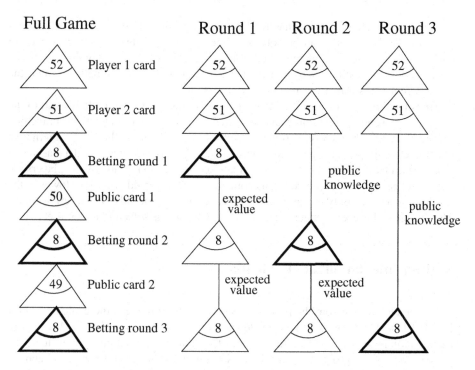

Fig. 4. Approximating Rhode Island Hold'em by handling betting rounds independently (using expected value for future cards, filling in public knowledge from previous rounds)

are compared. Although this is clearly a much smaller game than Texas Hold'em, at $(b, c, d, r) = (3, 4, 52, 3)$ the game tree still has more than 2.9B leaves.

Although the binning idea from the previous section could be applied to these games with perhaps between 5 and 10 bins, we decided to explore a different type of approximation. Whereas binning techniques are useful in controlling the combinatorial explosion due to the number of different hands that can be dealt, the techniques in this section address the explosion due to the large number of betting rounds. The main idea is to treat betting rounds quasi-independently instead of solving the entire game at once.

Figure 4 illustrates the process of handling the betting rounds independently. We replicate the full game tree three times, once for each betting round. To solve the first betting round, we make the assumption that our decisions in the first betting round are not very sensitive to whether future betting decisions come before or after additional cards are dealt. By moving the dealing of the public cards to the end of the game, we can then compute expected values over the dealt cards and remove the corresponding branches from the game tree.

The average expected payoff can be calculated by exhaustively going through each node. This works for a game such as Rhode Island Hold'em. But, for a larger game such as Texas Hold'em, a randomized simulation can be performed to estimate the average expected payoff.

Using this technique, the game tree for solving the first betting round now contains around 1.4M leaves and can be solved offline in approximately 15 minutes. We are currently uncertain regarding the accuracy of the strategy obtained using this technique. Because no information about past betting decisions is lost, we hypothesize that the strategy found this way is optimal or very close to it.

To solve the game tree for the second betting round, we need to build the part of the tree that represents the previous round as well. Again, we need to deal with the branching caused by future chance events (outcome uncertainty). We have chosen the alternative of completely ignoring the previous round, so we can keep our game tree at a reasonable size. The tradeoff here is the loss of information (the decisions made by the players are lost and those may reveal information about the opponent's hidden cards). However, the resulting game tree only has around 170K leaves and can be solved in a few seconds. The third round is even smaller (around 21K leaves) and is solved in under a second.

5   Implementation and Evaluation

Using the techniques from the previous section, we have reduced the game tree(s) for Rhode Island Hold'em to a manageable size. However, there is a danger that the approximation produced will lead to strategies far from the optimal play. To evaluate the performance of our program, we designed several opponents for it to compete against. First, we designed a rule-based player (RB), detailed in the Appendix,, that emulates how a good human player might play the game of Rhode Island Hold'em. The program generates randomized strategies depending on the state of the game. Next, we modified the rule-based program to model different player behaviors. Some players are willing to risk more (risk-seeking players) and pursue more aggressive strategies (AG), while others are more conservative (CON) and do not risk much (risk-averse players). We use appropriate rules to model both types.

We also have an opponent-modeling program (OPP) to play the game of Rhode Island Hold'em. This program plays by assuming its opponent is RB. It uses its knowledge of RB's behavior to compute a probability distribution over the hidden card given the observed betting behavior. It then makes choices to maximize its expected winnings (calculated by random sampling). This is a strategy used in the successful Loki program for playing Texas Hold'em [1].

Using the five opponents, we evaluated the effectiveness of our program. A thousand games were played between our program and each of the five opponents. During the competition, each program played 500 games as player 1 and the other 500 as player 2. This is done so that no one program has the advantage due to position. We felt 1,000 total games, which corresponds to roughly 30 hours of poker playing, would make for a meaningful comparison. The results, in terms of units gained, are summarized in Table 1.

From Table 1, we see that our program beat all of the opponents. The opponent modeling program was the toughest competitor, but the system was still able to win on average, to the tune of 5.4 units per 100 games. It is interesting to see that the aggressive player (AG) and the conservative player (CON) both did worse than the normal rule-based system (RB). AG lost more money probably due to its aggressiveness in betting

Table 1. Units gained per 100 hands by our system versus opponents

	RB	AG	CON	OPP
Result	+8.3	+12.7	+9.1	+5.4

RB = Rule-based (Normal)   AG = Rule-based (Aggressive)
CON = Rule-based (Conservative) OPP= Opponent-modeling

and bluffing in inappropriate situations. CON lost more because it did not take advantage of bluffing situations and also probably got bluffed out of many hands.

We analyzed the program's play and found that it was successfully using standard strategies such as bluffing and slow playing. One unexpected aspect was early round bluffing: human players are usually advised against bluffing in the early rounds of play with a low potential hand, whereas the program was bluffing in these situations. Because of the approximation due to computing independent betting rounds, it was possible that this behavior was a suboptimal artifact. To try to rule this out, we generated a new strategy that matched the computed strategy everywhere, but inhibited first and second round bluffing with low potential hands[3]. As player 2, this strategy scored +8.5 against RB, whereas the computed strategy scored +11.3, suggesting that this type of bluffing is a valuable strategy in this game.

6 Conclusion

Our program did an excellent job of playing the game of Rhode Island Hold'em. We would like to see it scale up to play a more complex game such as Texas Hold'em. Some of the main challenges that remain to be solved are the coarseness of the derived strategy from using a limited number of bins in the computation (probably between 5 and 10) and the fact that the system needs to either cache a large number of precomputed games or solve large linear programs during a match. Also, most Texas Hold'em games are played between 7 to 12 players, and currently, we are not aware of game-theoretic approaches for solving large multi-player games of imperfect information.

Games are excellent places to explore new ideas because of their clearly defined rules, specific goals, and the opportunity they present for comparison to human experts. In the past, game-playing research focused on deterministic games of perfect information. But, in computer science and emerging electronic commerce applications, many problems are made difficult by unreliable and imperfect information. It is our hope that by tackling a stochastic game with imperfect information such as poker, we can learn more about dealing with uncertainty and apply our abstraction ideas to other similar problems.

---

[3] We considered replacing these bluffs with an equal number of semi-bluffs from stronger hands, but since the program is already semi-bluffing, we thought additional semi-bluffs would only hurt its performance further.

A   Rule-Based System for Rhode Island Hold'em

For Player 1 (who goes first), B represents the probability of betting, PB represents the probability of betting after Player 1 passes and Player 2 bets. For Player 2, pB represents the probability of betting after seeing a pass and bB represents the probability of betting after seeing a bet.

Round 1

Player 1 Holding	Player 1 B	Player 1 PB
2,3,4,5	0.05	0
6,7,8,9	0.4	0.1
10,J	0.9	0.9
Q,K,A	1	1

Player 2 Holding	Player 2 pB	Player 2 bB
2,3,4	0.3	0
5,6,7,8	0.7	0.4
9,10	0.9	0.8
J,Q,K,A	1	1

Round 2

Player 1 Holding	Player 1 B	Player 1 PB
Pair	1	1
2 card straight flush	0.9	1
2 card straight	0.7	0.7
2 card flush (9-A)	0.8	0.8
2 card flush (6-8)	0.6	0.6
2 card flush (2-5)	0.4	0.4
high card (J-A)	0.5	0.5
high card (7-10)	0.5	0.4
high card (5-6)	0.3	0.1
high card (2-4)	0.1	0

Player 2 Holding	Player 2 pB	Player 2 bB
Pair	1	1
2 card straight flush	1	1
2 card straight	0.9	0.7
2 card flush (9-A)	1	1
2 card flush (6-8)	0.8	0.6
2 card flush (2-5)	0.8	0.4
high card (J-A)	0.8	0.5
high card (7-10)	0.6	0.4
high card (5-6)	0.3	0.1
high card (2-4)	0.0	0.0

Round 3

Player 1 Holding	Player 1 B	Player 1 PB
straight Flush	1	1
three of a kind	1	1
straight	1	1
flush (9-A)	1	1
flush (6-8)	0.8	0.8
flush (4-5)	0.7	0.8
flush (2-3)	0.4	0.4
pair	1	1
(we have the hole card that makes up the higher pair, no flush possible)		
pair	0.9	1
(we have the hole card that makes up the lower pair, no flush possible)		
pair	0.8	1
(we have the hole card that makes up the higher pair, flush possible)		
pair	0.5	0.9
(we have the hole card that makes up the lower pair, flush possible)		
pair (kicker J-A)	0.8	0.8
(pair on the board)		
pair (kicker 8-10)	0.8	0.6
pair (kicker 5-7)	0.4	0.2
pair (kicker 2-4)	0.2	0
high card (J-A)	0.8	0.8
(no flush possible)		
high card (8-10)	0.7	0.5
(no flush possible)		
high card (5-7)	0.3	0.2
(no flush possible)		
high card (2-4)	0.1	0
high card (A)	0.8	0.8
(flush possible)		
high card (J-K)	0.7	0.7
(flush possible)		
high card (8-10)	0.5	0.3
(flush possible)		
high card (2-7)	0.2	0
(flush possible)		

Player 2 Holding	Player 2 pB	Player 2 bB
straight Flush	1	1
three of a kind	1	1
straight	1	1
flush (9-A)	1	1
flush (6-8)	1	0.8
flush (4-5)	0.8	0.5
flush (2-3)	0.5	0.3
pair	1	1

(we have the hole card that makes up the higher pair, no flush possible)

pair	1	1

(we have the hole card that makes up the lower pair, no flush possible)

pair	1	1

(we have the hole card that makes up the higher pair, flush possible)

pair	1	0.8

(we have the hole card that makes up the lower pair, flush possible)

pair (kicker J-A)	1	0.8

(pair on the board)

pair (kicker 8-10)	0.9	0.5
pair (kicker 5-7)	0.2	0.1
pair (kicker 2-4)	0	0
high card (J-A)	0.9	0.8

(no flush possible)

high card (8-10)	0.7	0.3

(no flush possible)

high card (5-7)	0.3	0

(no flush possible)

high card (2-4)	0	0
high card (A)	0.9	0.8

(flush possible)

high card (J-K)	0.8	0.7

(flush possible)

high card (8-10)	0.5	0.3

(flush possible)

high card (2-7)	0.2	0

(flush possible)

References

1. Darse Billings, Denis Papp, Jonathan Schaeffer, and Duane Szafron. Opponent modeling in poker. In *Proceedings of the 15th National Conference on Artificial Intelligence*, pages 493–499, 1998.
2. Leslie Pack Kaelbling, Michael L. Littman, and Andrew W. Moore. Reinforcement learning: A survey. *Journal of Artificial Intelligence Research*, 4:237–285, 1996.
3. Daphne Koller, Nimrod Megiddo, and Bernhard von Stengel. Efficient computation of equilibria for extensive two-person games. *Games and Economic Behavior*, 14(2):247–259, 1996.
4. Daphne Koller and Avi Pfeffer. Representations and solutions for game-theoretic problems. *Artificial Intelligence*, 94(1–2):167–215, 1997.
5. Avi Pfeffer, Daphne Koller, and Ken T. Takusagawa. State-space approximations for extensive form games. Workshop paper at First World Congress on Game Theory, 2000.
6. Richard S. Sutton and Andrew G. Barto. *Reinforcement Learning: An Introduction.* The MIT Press, 1998.
7. J. von Neumann and O. Morgenstern. *Theory of Games and Economic Behavior.* Princeton University Press, Princeton, NJ, 1947.

Reasoning by Agents in Computer Bridge Bidding

Takahisa Ando and Takao Uehara

Tokyo University of Technology, 1401-1,Katakura-cho,Hachioji-shi,Tokyo,Japan
{and,uehara}@ue.it.teu.ac.jp

Abstract. The authors propose an agent oriented model for a bidder in the auctioning stage of bridge. Each agent selects a bid according to the criteria: Cooperate with the partner to get maximum profit and compete against opponents to minimize loss. Since bridge auction is a task of imperfect information, each agent has hypothetical reasoning ability and generates images of other players' hands by abduction from the observed bidding sequence. This paper shows a framework for reasoning about each others' knowledge and the details of analysis on typical examples. It is shown that the difference between one's own real hand and its image in a partner's knowledge motivates an agent to continue bidding. We also analyze an example of reasoning by an agent to select a sacrifice bid where the expected score of the bid is better than the score of an opponent's possible contract. Experimental results show that the reasoning by the agent is flexible enough to play with a human partner and other computer bridge programs.

Keywords: Computer bridge, Bidding, Imperfect information game, Agent, Hypothetical reasoning, Constraint logic programming

1 Introduction

Auction in the game of bridge is an interesting field for a case study of a multi-agent system. There are only 4 agents/players, but cooperation with a partner and competition against opponents happen in this small world. Language for communicating with each other is restricted (38 possible bids: 1 club to 7 no trump, pass, double, redouble) but a bid may have various meanings according to the context. The goal of each bidder agent is to reach a reasonable contract in the auction.

Wasserman wrote a bidding program in ALGOL with the collaboration of bridge experts [1]. It is reported that the operation was satisfactory so long as all bids were determined by the program, but the program was weak when the bidding took a course (as in a quiz) different from the choice made by the program. There will be similar problems when a human player bids as a partner of the program. Lindelöf proposed COBRA, a new bidding system, and showed through a series of computer experiments that the system has excellent ability[2]. Unfortunately, there are few players who can bid with his system as the partner. This paper aims at a more humanlike program than the program of Wasserman or Lindelöf. Each player is modeled as an agent shown in Section 2.

Gambäck and his colleagues claimed that it is important to estimate the hand based on the bid[3]. They described the inference rule to estimate the hand in Prolog. Uehara

T.A. Marsland and I. Frank (Eds.): CG 2000, LNCS 2063, pp. 346–364, 2001.
© Springer-Verlag Berlin Heidelberg 2001

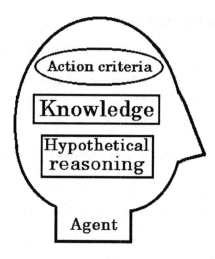

Fig. 1. Agent Model

described a bidding system in a constraint logic programming language, and used the same description for both selection of a bid and estimation of the hand [4].

Since bridge is a game of imperfect information, deductive inference is not so useful in this game. The hypothetical reasoning implemented in a constraint logic programming language is used in this paper. A framework and typical examples of reasoning by bidder agents are explained in Section 3. When a bid has several meanings, the agent may misunderstand the situation. Methods to prevent misunderstanding are discussed in Section 4. Some experimental results on partnership with a human player and competition against other computer bridge programs are shown in Section 5. Rule-based approaches to bidding are fairly common. Many implementations attempt to construct some kind of model of their own partner's hands, but using the same techniques to reason about the opponent's hand isn't so well advanced. We use the same technique to guess the opponent's possible contract in order to make a successful sacrifice bid.

2 Bidder Agent

Figure 1 shows the agent model proposed by the authors [5]. The agent has knowledge of a well known bidding system, that is, the summary of Charles Goren's book [6]. The bidding system is described as rules in the constraint logic programming language ECLiPSe [7]. In the case where the bidding system knowledge says nothing, general criteria are used for an agent to select a bid. This is what most human players do. The action criteria are often used because the bidding system summary says little about bids after the opener's rebid. The important feature of our agent is a hypothetical reasoning mechanism. The agent generates an image of the other agents' hidden hands by abduction from the observed bidding sequence. If the agent finds a consistent hypothetical image, it is used for selecting a bid according to the action criteria.

2.1   Knowledge of Bidding System

The summary in Charles Goren's book [6] is used as the rule-based knowledge of the bidding system for all agents. Examples of rules are shown below, where the list of variables [NS,NH,ND,NC] denotes the number of cards (in spade, heart, diamond and club respectively), P is the points, and LT is the losing tricks [8] in the bidder's hand.

```
% opening_bid(OpeningBid,Hand)
% Rule 1
    opening_bid([1,spade], [[NS,NH,ND,NC],P,LT]):-
        NS #>= 5,
        P #>= 13, P #<= 21,
        LT #<=7, LT #>= 5.
% response(OpeningBid,Response,Hand)
% Rule 2
    response([1,spade],[2,spade], [[NS,NH,ND,NC],P,LT]):-
        NS #>= 3,
        P #>= 8, P #<= 9,
        LT #= 9.
% Rule 3
    response([1,spade],[3,spade], [[NS,NH,ND,NC],P,LT]):-
        NS #>= 4,
        P #>= 10, P #<= 12,
        LT #= 8.
% Rule 4
    response([1,spade],[1,no_trump], [[NS,NH,ND,NC],P,LT]):-
        P #>= 6, P #<= 12,
        LT #<= 10, LT #>= 8.
% rebid(OpenningBid,Response,Rebid,Hand)
% Rule 5
    rebid([1,spade],[1,no_trump],[2,spade],
                                [[NS,NH,ND,NC],P,LT]):-
        NS #>= 6,
        P #>= 13, P #<= 17,
        LT #<= 7, LT #>= 6.
```

These rules are described in the constraint logic programming language ECLiPSe [7]. Operators with # have the same meaning with Prolog operators without # when variables are grounded (instantiated). In Prolog it fails when a variable is not grounded, but ECLiPSe keeps it as the delayed goal. So the bidding rules described in ECLiPSe can be used two ways. One way is to select a bid for the given hand (when Hand is grounded and Bid is not grounded). The other way is to guess the bidder's hand from the observed bid (when Bid is grounded and Hand is not grounded. Constraints on the Hand are kept as delayed goals). Now our program has 400 rules.

2.2   Action Criteria

The goal of auction for each agent is to reach the contract with the maximum gain (or minimum loss). The bonus score is given if a game, a small slam or a grand slam contract is achieved. So it is important to bid a game, a small slam or a grand slam when it is possible. The action criteria for partnership bidding are conditions to select such a bid. The general idea is as follows.

The number of winning tricks in a no-trump contract is estimated by using the high card point[6]. 26, 33 and 37 high card points are enough for 3 no trump, 6 no trump and 7 no trump, respectively. An eight card suit is good for a trump. The number of winning tricks in a trump contract is counted by using the losing trick count[8], that is, 'winning tricks of a pair = 24 - losing tricks of the pair'.

We show examples of criteria for partnership bidding, where the partner's image (the point, the losing tricks and the number of trumps) has to be guessed from the observed bidding sequence.

```
% Rule 6.1
    select_bid([4,spade],OwnNS,MinPartnersNS,
                        OwnLT,MaxPartnersLT):-
    OwnNS + MinPartnersNS #>= 8,
    OwnLT + MaxPartnersLT #>= 13,
    OwnLT + MaxPartnersLT #<= 14.
% Rule 6.2
    select_bid([4,heart],OwnNS,MinPartnersNS,
                        OwnLT,MaxPartnersLT):-
    OwnNH + MinPartnersNH #>= 8,
    OwnLT + MaxPartnersLT #>= 13,
    OwnLT + MaxPartnersLT #<= 14.
% Rule 7
    select_bid([6,spade],OwnNS,MinPartnersNS,
                        OwnLT,MaxPartnersLT):-
    OwnNS + MinPartnersNS #>= 8,
    OwnLT + MaxPartnersLT #= 12.
% Rule 8
    select_bid([7,spade],OwnNS,MinPartnersNS,
                        OwnLT,MaxPartnersLT):-
    OwnNS + MinPartnersNS #>= 8,
    OwnLT + MaxPartnersLT #<= 11.
% Rule 9
    select_bid([3,no_trump],OwnP,PartnersP):-
    OwnP + PartnersP #>= 26,
    OwnP + PartnersP #<= 32.
% Rule 10
    select_bid([6,no_trump],OwnP,PartnersP):-
    OwnP + PartnersP #>= 33,
    OwnP + PartnersP #<= 36.
% Rule 11
    select_bid([7,no_trump],OwnP,PartnersP):-
    OwnP + PartnersP #>= 37,
    OwnP + PartnersP #<= 40.
```

The next example is the invitational bid to game, where the game seems to be possible, but it is not certain.

```
% Rule 12
   select_bid([3,spade],OwnNS,MinPartnersNS,
                        OwnLT,MaxPartnersLT,MinPartnersLT):-
       OwnNS + MinPartnersNS #>= 8,
       OwnLT + MaxPartnersLT #= 15,
       OwnLT + MinPartnersLT #<= 14.
```

Again these rules can be used in two ways. One way is to select a bid for the given hand. The other way is to guess the bidder's hand from the observed bid. Since the partner's hand is hidden, the partner's point and losing tricks are not known exactly. We show how to guess the partner's hand in Section 3.2.

In competitive bidding, the action criteria are more complicated.

```
Step 1)   Guess the possible contract of one's own side.
Step 2)   Guess the possible contract of the opponent's side.
Step 3)   Find the contract which gives the maximum gain or
          minimum loss,and bid the contract immediately.
```

Suppose that the possible contracts of one's own side and the opponents' side are [M,S] and [N,T], respectively. When an opponent bids [N1,T], a player selects a bid according to the following rules. [M,S]>[N,T] shows that the bid [M,S] is callable after the bid [N,T]. s([M,S],R) is the score of making [R,S] for the contract [M,S]. If R<S, s([M,S],R) has minus value. [M,S]* means doubled [M,S].

```
Case 1)   If [M,S]>[N,T]>=[N1,T],
          then bid the lowest [X,S] where [X,S]>[N,T] and
               s([X,S],M)=s([M,S],M).
Case 2)   If [M,S]>[N1,T]>[N,T] and s([M,S],M)<-s([N1,T]*,N),
          then double [N1,T].
Case 3)   If [M,S]>[N1,T]>[N,T] and s([M,S],M)>=-s([N1,T]*,N),
          then bid the lowest [X,S] where [X,S]>[N1,T] and
               s([X,S],M)=s([M,S],M).
Case 4)   If [N1,T]>[M,S] and [N1,T]>[N,T],
          then double [N1,T].
Case 5)   If [N,T]>[M,S] and [N,T]>=[N1,T],
          then bid the lowest [Y,S] where [Y,S]>[N,T] and
               s([Y,S]*,M)>-s([N,T],N).
          If such [Y,S] does not exist,
          then bid the highest [Y1,S] where [Y1,S]>[N1,T] and
               s([Y1,S]*,M)>-s([N,T],N).
```

The difficult part in implementing the criteria is how to guess the possible contracts on each side. This is discussed in Section 3.3.

2.3   Facts

Some of the most essential information for reasoning in competitive bidding consists of the facts of the game.

Fact 1) The total number of cards in each suit is 13.
That is,

```
NorthNS + EastNS + SouthNS + WestNS #= 13.
NorthNH + EastNH + SouthNH + WestNH #= 13.
```

so on.
Fact 2) The total high card point is 40.
That is,

```
NorthP + EastP + SouthP + WestP #= 40.
```

These are examples of facts. Each fact is described in ECLiPSe as a relation among variables with domains. The use of these constraints is shown in Section 3.3.

2.4 Hypothetical Reasoning

Each agent has a hypothetical reasoning mechanism. When an agent observes a fact G and G cannot be proved logically from the agent's knowledge F, the agent generates a hypothesis h (by abduction based on F), so that G is proved from F∪h (where h is consistent with F). We implement this mechanism in ECLiPSe where a set of possible hypotheses are represented by variables with domains.

When an agent observes a bidding sequence, the agent searches rules in the bidding system knowledge (or in the action criteria) which match with bids in the sequence. Constraints on variables such as NS,NH,ND,NC,P and LT in those rules are kept in the system as delayed goals. The assignment of values to the variables which satisfies all delayed goals corresponds to a hypothesis. There may be several rules which match with each bid in the bidding sequence and this is a potential cause of misunderstandings. The agent checks them one by one and selects a set of consistent rules which can explain the observed bidding sequence. The agent revises its belief when the old hypothesis does not explain a newly observed bid. A misunderstanding during the early stage of an auction may be corrected later by selecting an alternative set of rules.

3 Reasoning by Bidder Agent

3.1 Framework for Bidding and Reasoning

The method described up to this stage is implemented using the constraint logic programming language. It is outlined in the following:

```
select_bid(Bid,BidSeq,Player,Hands):-
        bidding_system(Bid,BidSeq,Player,Hands);
        action_criteria(Bid,BidSeq,Player,Hands).
```

The program searches for the rule in the bidding system that meets the condition of the player's hand. If no such rule exists, the decision is made by the action criteria.

```
action_criteria(Bid,BidSeq,Player,Hands):-
       abduction(BidSeq,Player,Hands,Images),
       partnership(OurLevel,TheirLevel,Player,Hands,Images),
       competitive(Bid,BidSeq,OurLevel,TheirLevel).
```

As the first step of the action criteria, the hypothetical images (Images) of hands are generated by abduction, so that the observed bidding sequence (BidSeq) can be accounted for. As the next step, the program for partnership bidding estimates the possible contract of each pair (OurLevel,TheirLevel) based on the aggregate parameters like points, losing trick count, and length of suit in the hypothetical images (see Rule 6 to 10 in Section 2.2). The last step is the decision of a competitive bid (Bid) to get the best score (see Case 1 to 5 in Section 2.2).

```
abduction( [ ],Player,Hands,Images).
abduction( [LastBid|SubBidSeq ],Player,Hands,Images):-
       next_player(LastPlayer,Player),
       select_bid (LastBid,SubBidSeq,LastPlayer,Images),
       abduction( SubBidSeq,LastPlayer,Hands,Images).
```

When the bidding sequence is given, the program for abduction recursively generates hypothetical images of all players. As shown in Figure 2, the image of the partner owned by the player includes the image of his own that is probably owned by the partner. This is a nested structure since that image of his own includes past images of the partner. All multiple nested images of each player's hand should be consistent. If they are not consistent, the program searches an alternative set of rules to be used for abduction. The image need not be consistent with the real hand although it is the aim of partnership bidding to make them as close as possible.

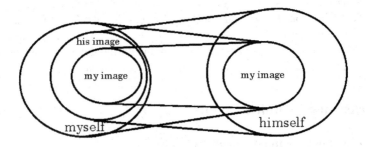

Fig. 2. A basic model for partnership bidding

3.2   Partnership Bidding by Agents

It is assumed that opponents always pass in partnership bidding. The assumption is not realistic in an actual game, but it is used in Section 3.2 to keep the explanation of reasoning simple. An example of reasoning in competitive bidding is shown in Section

3.3. It is an easy task for an agent to select a bid if its hand matches with a rule in the bidding system knowledge shown in Section 2.1. Suppose South and North have hands shown below. The agent South bids 1 spade as a dealer by Rule 1, and the agent North bids 1 no trump by Rule 4.

	South		North
S	AQxxxx	S	Kxx
H	xxx	H	xx
D	Ax	D	QJx
C	Kx	C	Axxxx

Suppose the auction by agents is as follows. What is North's next bid?

South	West	North	East
1 spade	pass	1 no trump	pass
2 spades	pass	?	

Since the responder's rebid is not included in the bidding system knowledge, the agent uses the general action criteria shown in Section 2.3 for selecting the bid. When the agent North observes South's opening 1 spade bid, North generates a hypothetical hand of South by abduction using Rule 1 in Section 2.1 as delayed goals. That is,

```
SouthNS #>= 5,
SouthP  #>= 13, SouthP  #<= 21,     (1)
SouthLT #<= 7,  SouthLT #>= 5.
```

Then North observes South's rebid, and generates the following constraints by abduction using Rule 5.

```
SouthNS #>= 6,
SouthP  #>= 13, SouthP #<= 17,      (2)
SouthLT #<= 7, SouthLT #>= 6.
```

From (1) and (2), North believes that

```
MinSouthNS = 6,
MaxSouthLT = 7,
MinSouthLT = 6,                     (3)
MaxSouthP  = 17,
MinSouthP  = 13.
```

In other words, (3) is the image of South in North. Since the agent North knows that

```
NorthNS = 3,
NorthLT = 8,                        (4)
```

North bids 3 spades by Rule 10 in the action criteria.

South	West	North	East
1 spade	pass	1 no trump	pass
2 spades	pass	3 spades	pass
?			

What is South's next bid? When the agent South observes North's 1 no trump response, South generates a hypothetical hand of North by abduction using Rule 4 in Section 2.1. That is,

```
NorthP   #>= 6,   NorthP   #<= 12,
NorthLT #<= 10, NorthLT #>= 8.          (5)
```

Then South observes North's 3 spades rebid, and generates the following constraints by abduction using Rule 10 in the action criteria.

```
NorthNS + MinSouthNS #>= 8,
NorthLT + MaxSouthLT #=  15,          (6)
NorthLT + MinSouthLT #<= 14.
```

Since the agent South can trace the reasoning by North, South knows its own image in North is (3). So, South gets (7) by substituting (3) for (6).

```
NorthNS + 6 #>= 8,
NorthLT + 7 #= 15,          (7)
NorthLT + 6 #<= 14.
```

Now South solves (5) and (7), and South believes that

```
MinNorthNS = 2,
NorthLT= MaxNorthLT=MinNorthLT=8.        (8)
```

Since South knows that

```
SouthNS = 6,
SouthLT = 6,          (9)
```

South bids 4 spades by Rule 6, because

```
SouthNS + MinNorthNS = 8,
SouthLT + MaxNorthLT = 14.          (10)
```

According to similar analysis North believes that MinSouthLT is 6 and North passes 4 spades.

```
        South       West        North         East
     1 spade       pass      1 no trump       pass
     2 spades      pass       3 spades        pass
     4 spades      pass         pass          pass
```

This example shows a step by step approach to the best contract by the cooperation of two agents in partnership bidding. It is important for an agent to know the image of its own hand in the partner as shown in Figure 3. If the image is different from the real one, the agent selects the bid to make the partner aware of the difference.

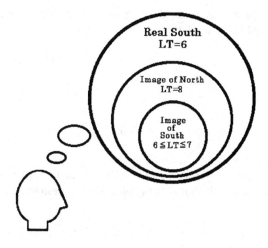

Fig. 3. Images in the agent South

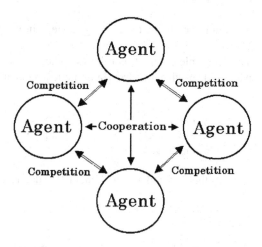

Fig. 4. Model of competitive bidding

3.3 Competitive Bidding by Agents

Figure 4 shows 4 agents who cooperate with a partner and compete against opponents.

We assume that 4 agents have the same knowledge. Since the eight-page handy summary in Charles Goren's book [6] does not include competitive bidding, some basic rules such as 'takeout double' and 'overcalls' are added into the agent's bidding system knowledge. Response and rebid rules are modified to make them competitive according to the "Commonsense Bidding" by W. Root[9]. However, still in many cases an agent has to select a bid by using its general criteria. The next example is a quiz from R. Klinger's book [8]. Suppose South has the hand shown below and the auction is as follows. What is the best response bid by South?

```
S   Kxxxx
H   xx
D   x
C   Jxxxx


North       East    South   West
2 spades    pass    ?
```

This weak two opening bid is described as Rule 11 in the bidding system knowledge. This is a competitive bid which makes the opponents' communication difficult.

```
% Rule 11
    opening_bid([2,spade],  [[NS,NH,ND,NC],P,LT]):-
       NS #>= 6,  NH #<= 3,
       P  #>= 6,  P  #<= 10,
       LT #<= 8,  LT #>= 6.
```

The reasoning by South is explained according to steps in the general criteria described in Section 2.2.

Step 1) Guess one's possible contract as follows. When the agent South observed North's opening 2 spades bid, South generated a hypothetical hand of North by abduction using Rule 11. That is,

```
NorthNS #>= 6,  NorthNH #<= 3,
NorthP  #>= 6,  NorthP  #<= 10,        (11)
NorthLT #<= 8,  NorthLT #>= 6.
```

Since South knows SouthNS=5 and SouthLT=8, (12) is deduced from (11)

```
SouthNS + MinNorthNS #>= 11,
SouthLT + MaxNorthLT #<= 16.           (12)
```

That is, South can make 2 spades (8 winning tricks = 24 -16) even if North has 8 losing tricks. It is the worst case of (11).

If the opponent's possible contract cannot be known or it is lower than 2 spades, South will pass the partner's 2 spades.

Step 2) Guess the opponent's possible contract as follows. South knows

```
SouthNH = 2,
SouthP  = 4.                           (13)
```

South believes that from (11)

```
MaxNorthNH = 3,
MaxNorthP  = 10,                       (14)
```

in North's hypothetical hand. On the other hand, Fact 1 and Fact 2 in Section 2.2 are

```
NorthNH + EastNH + SouthNH + WestNH #= 13,
NorthP  + EastP  + SouthP  + WestP  #= 40.   (15)
```

Now South can calculate from (13),(14) and (15) as follows;

```
EastNH + WestNH #>= 8.
EastP  + WestP  #>= 26.        (16)
```

Then South converts 26 high card points to 14 losing tricks.

```
EastNH + WestNH #>= 8.
EastLT + WestLT #<= 14.        (17)
```

As a result South believes that the opponents can make at least 4 hearts (see Rule 6).

Step 3) Select the bid which gives the maximum gain or minimum loss. The agent South compares the score of the opponent's 4 hearts with the score of 4 spades by North. The 4 spades are expected to be doubled and it may go two down according to the analysis in Step 1. The agent selects a bid according to the criteria shown in Section 2.2 (see Case 5, where [M,S]=[2,spade] and [N,T]=[4,heart]). If North and South are vulnerable and East and West are not, South selects 3 spades and hopes that the opponent does not bid 4 hearts. At any other vulnerability, South bids 4 spades, which will give the minimum loss.

This conclusion is a little bit different from the answer in the R. Klinger's book [8] which says that South should bid 4 spades at any vulnerability because the opponent's game is certain and a slam may be possible. The general criteria we have prepared is not bad, but we should improve the analysis of the opponents possible contract in the future. Klinger's book also suggests a psychic bid such as 4 no trump Blackwood (signing off in 5 spades later). Other players will misunderstand South's hand, but North will pass the 5 spades by South anyway. Our agent does not have such a strategy at present, but it is not difficult to build it in the action criteria.

4 Misunderstanding

We use delayed goals in ECLiPSe to implement the hypothetical reasoning. It is difficult to model such goals in a disjunctive way. This seems to be a quite severe limitation against bidding systems with multiple interpretations (multi-2D, etc). In practice, it is not so severe because each bidding system prepares bidding sequences to show which interpretation is right. Our agent may have the wrong interpretation the first time, but as soon as an agent finds inconsistency in the delayed goals, the agent searches for an alternative interpretation and puts it in the new set of delayed goals. The belief revision is performed by backtracking instead of searching for disjunctive goals.

Sometimes a player intentionally gives a wrong or an ambiguous image to other players, including his partner. It is difficult for our agent to get a correct interpretation. The problem cannot be solved even if the disjoint possibilities are maintained as the belief sets. The following example explains how to prevent trouble in such cases.

Suppose South has the following hand.

```
S   Kxxxx
H   AK
D   x
C   Jxxxx
```

Auction is the same as the last example.

```
North       East     South     West
2 spades    pass       ?
```

Then South bids 4 spades according to Rule 6 as the possible contract. When the auction proceeds as follows, the agent North may misunderstand South's hand because South, with the hand in Section 3.2, also bids 4 spades.

```
North       East     South      West
2 spades    pass     4 spades   pass
   ?
```

The misunderstanding cannot be avoided as far as the hypothetical reasoning based on abduction is used. So it is important to analyze what happens as the result of a misunderstanding and to prevent trouble. The step by step approach to the best contract described in Section 3.1 may lead to an overbid because of the misunderstanding. Suppose that North has 6 losing tricks. North will bid (or explore) a slam . If South's actual hand is the hand with 8 losing tricks shown in Section 3.3, they lose because of their overbid. This situation is prevented by modification of Rule 11 as follows. The width of the losing tricks for the hand qualification is made no more than the difference of the losing tricks between a game and a slam. It is safer if the width is 1 (for example LT #= 8), but it reduces chances to use this powerful bid which makes the opponents' communication difficult.

```
% Rule 11m
     opening_bid([2,spade],  [[NS,NH,ND,NC],P,LT]):-
        NS #>= 6, NH #<= 3,
         P #>= 6,  P #<= 10,
        LT #<= 8, LT #>= 7.
```

Now, suppose that North has 7 losing tricks. Even after this modification North may bid 5 spades as an invitation to a slam because North believes that South may have 5 (= 13- 8) losing tricks by Rule 6. If South's 4 spades is a sacrifice bid, North's 5 spades causes one more down. This situation can be prevented by modification of Rule 6 to the following three rules. 2 no trump in Rule 6m2 is the convention asking the opener's hand and the rule is built into the bidding system knowledge.

```
% Rule 6m1
    select_bid([4,spade],[OwnNS,MinPartnersNS,
                       OwnLT,MaxPartnersLT,MinPartnersLT]):-
      OwnNS + MinPartnersNS #>= 8,
      OwnLT + MaxPartnersLT #<= 14,
      OwnLT + MinPartnersLT #>= 13.
```

```
% Rule 6m2(Ogust convention)
    response([2,spade],[2,no_trump], [[NS,NH,ND,NC],P,LT]):-
        NS #>= 2,
        LT #<= 5.

% Rule 6m3
    select_bid([5,spade],[OwnNS,MinPartnersNS,
                        OwnLT,MaxPartnersLT,MinPartnersLT]):-
        OwnNS + MinPartnersNS #>= 8,
        OwnLT + MaxPartnersLT #=  13,
        OwnLT + MinPartnersLT #<= 12.
```

After this modification North does not bid 5 spades as the invitation to a slam even if North has 7 losing tricks, because North believes that South's losing trick count is 6. The reasoning is as follows.

When North observed South's 4 spades, North generated (18) by abduction using Rule 6m1.

```
SouthLT + MaxNorthLT #<= 14,
SouthLT + MinNorthLT #>= 13.            (18)
```

North knows that South observed North's 2 spades and got the following values by abduction using Rule 11m;

```
MinNorthLT = 7,
MaxNorthLT = 8.                         (19)
```

Values of (19) is substituted to (18).

```
SouthLT + 8 #<= 14,
SouthLT + 7 #>= 13.                     (20)
```

Now North solves (20) and believes that

```
SouthLT = 6.                            (21)
```

North understands that the slam contract is impossible even if North has 7 losing trick hand. So, North does not bid 5 spades as an invitation to a slam. North will bid 5 spades if the opponents bid 5 hearts, and it is a reasonable contract even if South's hand is misunderstood.

Suppose that South bids 5 spades as follows.

```
North       East      South      West
2 spades    pass      5 spades   pass
    ?
```

South's 5 spades must be a sacrifice bid according to the modified rules, because South with a 5 losing tricks hand is supposed to bid 2 no trump by Rule 6m2. This is an example that shows how to prevent trouble caused by misunderstandings. Human players are careful if a bid has several meanings, but our agent is not, and it selects the first one so far as it is consistent with other observations. So we set the most general meaning for the bid at the first position in the knowledge of the same bid. One way to remove the chance of misunderstandings is to keep tweaking the preconditions of the rules. But this gets harder as the size of rule sets increases. So we are interested in keeping the size of the rule set compact.

5   Experimental Results

An agent is implemented as shown above. The experimental results on partnership with a human player and competitive bidding against commercial computer bridge programs are shown in this section.

5.1   Partnership with Human Player

Suppose South and North have the hands shown below.

```
     South              North
S    AQxxxx        S    Kxx
H    xxx           H    xx
D    Ax            D    QJx
C    Kx            C    Axxxx
```

Partnership bidding by two agents in Case 1 is shown in Section 3.1. In Case 2 to 6, South is the agent and North is a human player.

```
Case 1:     South        North
            1 spade      1 no trump
            2 spades     3 spades
            4 spades     pass

Case 2:     South        North
            1 spade      2 clubs
            2 spades     3 spades
            4 spades     pass

Case 3:     South        North
            1 spade      2 spades
            3 spades     4 spades
            pass

Case 4:     South        North
            1 spade      3 spades
            4 spades     pass

Case 5:     South        North
            1 spade      2 no trump
            4 spades     pass
```

```
Case 6:      South        North
             1 spade      2 heart
             2 no trump   3 no trump
             4 spades     pass
```

The agent South may have some misunderstanding but the contract is reasonable. In Case 6 the human player North tried to cheat every other player, but the agent South prefers 4 spades rather than 3 no trump. If South has only 5 spades, the agent South will pass 3 no trump. We performed many experimental biddings to check the flexibility and the robustness of our agents in partnership bidding[5]. We deleted some conventions/rules from the bidding system knowledge of an agent. In most cases, the agent behaves like a smart beginner and reaches a reasonable contract. This is mainly because the selection of a bid is based on the hypothetical reasoning and the action criteria shown above.

5.2   Competition against Computer

We took two commercial computer bridge programs, Bridge Baron (version 9.00.01) [10] and GIB(version 2.5.8) [11], and adjusted their bidding system options (Weak Two, Limit Raise, Two over One, Forcing 1 No Trump etc.) to match the system used by our agent. Experimental auctions were tried on 14 randomly generated hands with constraints that South opens 2 spades and North has the hand described in Section 3.2. It was assumed that East and West were vulnerable. The next example is one of the best results we obtained.

```
                        S   K8752
                        H   75
                        D   8
                        C   J9843
West                    North               East
S   A3                                       S
H   A943                                     H   KQ1086
D   AJ52                                     D   KQ10943
C   AQ2                                      C   76
                        South
                        S   QJ10964
                        H   J2
                        D   76
                        C   K105
```

Auctions and estimated results are as follows, where BB stands for Bridge Baron and TUT is the team of our agents. The result is estimated by a human player not playing by computers.

Details of the experiment are reported in [12]. Sometimes TUT lost more than 10 IMP in one board. With the average of 14 boards, TUT won about 2 IMP per board in auction. We do not claim that our bidding program is better than Bridge Baron or GIB, however we do claim that our simple-minded agent was able to reach a reasonable contract even in competitive bidding with sophisticated commercial programs.

Experiment 1(TUT vs. BB): TUT won 1500(17 IMP)
Room A: 4 hearts by East(BB): made 7:

South(TUT)	West(BB)	North(TUT)	East(BB)
2 spades	2 no trump	pass	3 hearts
pass	4 hearts	pass	pass
pass			

Note: TUT pass BB's 2 no trump because TUT could not
understand the bid. Our agent should ask what
the bid means.

Room B: 7 hearts by East(TUT): made 7:

South(BB)	West(TUT)	North(BB)	East(TUT)
2 spades	double	3 spades	4 no trump
pass	5 clubs	5 spades	7 hearts
pass	pass	pass	

Experiment 2 (TUT vs. GIB): TUT won 1410(16IMP)
Room A: 5 spades double by South(TUT): down 4

South(TUT)	West(GIB)	North(TUT)	East(GIB)
2 spades	double	4 spades	5 hearts
pass	pass	5 spades	double
pass	pass	pass	

Room B: 7 hearts by East(TUT): made 7:

South(GIB)	West(TUT)	North(GIB)	East(TUT)
2 spades	double	4 spades	4 no trump
pass	5 clubs	pass	7 hearts
pass	pass	pass	

Fig. 5. Bidding sequences produced by TUT and GIB

We performed another test against GIB(version 3.4.2) on 156 boards played at two tournaments of the Japan Contract Bridge League in 1999. The result is shown in the Appendix, where all hands were played by GIB. In general GIB is better in lower level competition and TUT is strong in higher level trump contract. GIB selects a bid using Monte-Carlo simulation of possible hands. So GIB often changes its bid in tests of the same deal, although TUT always keeps the same bid. We prefer the method which can explain the reason why the bid is selected.

6   Conclusions

Several examples of reasoning by bidder agents have been analyzed. The action criteria were set so that the difference between the agent's hand in the partner's image and the real hand motivated the agent to continue bidding to reach a reasonable contract. Experiments on partnership bidding showed that this step-by-step approach to the best contract gives the agent flexibility and robustness. Examples of reasoning in competitive bidding were also analyzed. The action criteria were a simple selection of the best score contract. The difficult task was to guess the opponents' possible contract. Our agent guessed the opponents' hands by subtracting its own hand and the partner's hand from the total cards. Sacrifice bids by our agents worked fine in experiments on competitive bidding against other computer bridge programs.

Now our program has about 400 rules. We think that about 500 rules are enough for implementing a chosen bidding system fully. If an expert bridge player puts in more and more rules, our program will work better as an expert system; however, we are interested in making a program which works well without so many rules. Some aspects of this program still require revision. Since we have used the naive losing trick count, sometimes agents overbid at the slam level. Reasoning by using the cover card count [8] should be built into the agent in the future. No theoretical problem exists in a game against players who use other bidding systems, but the practical task of preparing those systems is not easy. For example, partial modification of competitive bids (overcall etc.) of our bidding system is necessary.

Hypothetical hands generated by reasoning from bidding is also useful for selecting a card to play in the game. Actually we use 208 variables which represent possession of 52 cards by four players. Constraints on these variables generated by reasoning on aggregate parameters are sometimes useful to guess the location of honor cards in the play of the game. We are now going to develop a player agent who can guess other players' strategies.

Acknowledgments

The authors thank the faculty, staff and students of Tokyo University of Technology for the opportunity to perform this research, especially Professors M. Shioya, W. Miyao, K. Ibuki and K.Fuchi, and students who participated in the project as graduate thesis research. The authors are grateful for advice from Mr. F. Nishino, Fujitsu Laboratories Limited, and Dr. K.Sato, Hokkaido University, as well as the help in using CHIP provided by Dr. W. O'Riordan and Mr. M. Rigg of ICL, and for assistance with ECLiPSe provided by PARC of Imperial College.

References

1. Wasserman A.:"Realization of skillful bridge bidding program", Proc. FJCC(1970)433-444
2. Lindelöf E.:"COBRA:The computer-designed bidding system", Victor Gollancz Ltd., London(1983)
3. Gambäck B., Rayner M. and Barney P.:"Pragmatic reasoning in bridge", Technical report No.299, University of Cambridge, Computer Laboratory(1993)
4. Uehara T.'F"Application of abduction to computer bridge", Transactions of the institute of electronics, information and communication engineers(D-II), J77-D-II,No.11(1994-11)2255-2264
5. Ando T.,Sekiya Y. and Uehara T.:"Partnership Bidding for Computer Bridge", Transactions of the institute of electronics, information and communication engineers,Vol.J81-D-II,No.10(1998)2366-2375

6. Goren C.H.:"Goren's new bridge complete", Doubleday(1985)
7. URL:http://www.icparc.ic.ac.uk/eclipse/
8. Klinger R. :"The modern losing trick count; Bidding to win at bridge", Victor Collancz Ltd(1991)
9. Root W.S. :"Commonsense bidding",Crown Publishers,Inc.(1986)
10. URL:http://www.bridgebaron.com/
11. Ginsberg M.L. :"GIB: Steps toward an expert-level bridge-playing program", University of Oregon Eugene(1997)
12. Ando T., Kobayashi N. and Uehara T. :"Cooperation and competition of agents at auction of computer bridge",Transactions of the institute of electronics, information and communication engineers, Vol.J83-D-I,No.7(2000)759-769

Appendix: Test against GIB(version 3.4.2)

```
Nintendo Cup 1(26 boards)    Total IMPs  TUT 69 : GIB 43
        TUT won 10 boards    GIB won 7 boards   9 boards were tied

Nintendo Cup 2(26 boards)    Total IMPs  TUT 67 : GIB 50
        TUT won 10 boards    GIB won 8 boards   8 boards were tied

Nintendo Cup 3(26 boards)    Total IMPs  TUT 53 : GIB 35
        TUT won 12 boards    GIB won 8 boards   6 boards were tied

Nintendo Cup 4(26 boards)    Total IMPs  TUT 41 : GIB 56
        TUT won 5 boards     GIB won 10 boards  11 boards were tied

Princes Takamatsu Cup1(26 boards)  Total IMPs  TUT 78 : GIB 47
        TUT won 11 boards    GIB won 8 boards   7 boards were tied

Princes Takamatsu Cup2(26 boards)       Total IMPs  TUT 47 : GIB 47
        TUT won 9 boards    GIB won 7 boards   10 boards were tied

Total    boards 156     Total IMPs  TUT 355 : GIB 278
        TUT won 57 boards    GIB won 48 boards  54 boards were tied
```

Linguistic Geometry for Solving War Games

Boris Stilman

University of Colorado at Denver, USA
bstilman@carbon.cudenver.edu
and
STILMAN Advanced Strategies, USA
boris@stilman-strategies.com

Abstract. The purpose of this paper is to introduce various types of war games and a theory for construction of winning strategies for these games, Linguistic Geometry. The paper includes examples of application of LG to planning of Suppression of Enemy Air Defenses (SEAD mission).

Keywords: war games, abstract board games, symmetric/asymmetric war games, linguistic geometry, strategies, search problems, reduced search.

1   War Games

When thinking about modern or future military operations, the game metaphor comes to mind right away. Indeed, the air space, together with the ground, may be viewed as a gigantic three-dimensional game board. The groups of aircraft performing a single task may be viewed as friendly pieces whereas the enemy manned and unmanned aircraft together with the ground targets may be viewed as the opponent's pieces. The mission commanders on various levels have a place in this picture as game players. These are called Abstract Board Games (ABG) [9]. To help the commander to achieve the mission objectives, we would need algorithms finding winning and near-winning strategies for the respective games. A game strategy describes behavior in terms of *moves*. A move represents the smallest activity of pieces discernable from the game point of view. It may include physical motion of pieces as well as their actions such as shooting a missile, turning on radar, or employing a camera to make aerial photos. A move can be associated with a time interval. Depending on the task level (tactical, operational, or strategic) the interval could be measured in seconds, minutes, hours, or days. A strategy may offer a significant "look ahead" so that various intended or unintended consequences of immediate actions could be seen.

Application of game strategies does not exclude a human commander out of the loop. Indeed, whatever problems would be solved by the intelligent computer strategist, they are eventually human problems and, as such, they may not be completely solvable by a machine. Instead of replacing the mission commander, an application of game strategies will help the commander to conduct and monitor the operations with less effort and with greater chance of success. However, without an ability to find a winning strategy, games would serve mostly to display the situation, rather than provide models from which solutions could be derived. [1,2, 9, 11-14].

T.A. Marsland and I. Frank (Eds.): CG 2000, LNCS 2063, pp. 365–383, 2001.
© Springer-Verlag Berlin Heidelberg 2001

2   Asymmetric War Games

Asymmetric warfare is a new type of conflict where the participants do not initiate full-scale war [1, 12]. Instead, the sides may be engaged in a limited open conflict (e.g., operation "Desert Storm"), or one or several sides may covertly engage another side using unconventional or less conventional methods of engagement. Unconventional engagements may include information systems war ("cyber war"), biological weapons, terrorism, public-relations warfare etc. [1, 12].

The most critical challenges are as follows:

- Each type of an asymmetric conflict is very specific. It requires special modeling effort to adequately capture the conflict type in order to support its simulation and/or its control. In contrast, in conventional warfare generic models are more widely applied;
- Participants of an asymmetric conflict act concurrently within multi-dimensional space;
- Asymmetric conflict participants demonstrate highly personalized behavior. However, classic war-gaming models are intended for the analysis of generic behaviors which is not acceptable;
- Multiple sides of an asymmetric conflict should be represented by the intelligent stakeholders versus scripted adversaries usually modeled by the classic war-gaming tools.

The need to find innovative approaches that are timely and effective across the full range of military operations is well recognized by the US military forces. This includes the area of asymmetric conflicts. Frequently, war-gaming is used for planning and control of conventional military operations. Modern war-gaming uses software tools that include components similar to software for computer games. These tools provide display for analysis, as well as control of possible courses of actions within the conflict environment. The major shortcoming of this approach is that it does not adequately model the intelligent strategists that may be employed by the sides at conflict. The corresponding modeling capability of war-gaming is usually limited to playing fixed scripts for the adversary strategists. (Below, we will interchange the term "stakeholders" for the term "strategists.")

The modeling deficiency is exacerbated for asymmetric conflicts [1, 12]. Individual differences between the intelligent stakeholders are more important, than those in conventional conflicts.

There may be special constraints on the actions of the participating sides for the asymmetric conflict (e.g., ideology, personality, multi-sided influences on the intelligent stakeholders, goals influenced and formulated by means of ideology), that are not usually considered with respect to conventional conflicts.

Large amount of intelligence data about the stakeholders, (e.g., Milosevich, Saddam Hussein, others) their traits and behaviors, is frequently available from the CIA/FBI archives, open literature, etc. However, war-gaming software for conventional conflicts typically does not utilize this information.

The influence of the concurrent actions for an asymmetric conflict can be more significant than that of a conventional conflict because of a number of hidden concurrent actions accomplished by at least one of the conflict sides.

Evaluation of the current status of an asymmetrical conflict is more difficult than that of a conventional conflict. Part of this problem is rooted in a challenge of adequately modeling the enemy. Another part is in quantifying the value of a current state of the conflict.

3   Conventional War Gaming Approaches

Game-based approaches have frequently been applied to military command and control (C^2). The games used by many game-based approaches are *normalized* games or one-step games. They were introduced and investigated by Von Neumann and Morgenstern [13] half a century ago and later developed by multiple followers. This approach allows analyzing full game strategies, representing entire games. It does not allow breaking a game into separate moves and comparing them. Only full strategies, the entire courses of behavior of players can be compared. This significant limitation makes this approach inadequate for real world C^2 problems. Von Neumann-Morgenstern games and respective strategies have been represented in the discrete or continuous (differential) form. For both types of games, discrete and differential, advanced theoretical results have been received. However, these advanced theories lack scalability. The classic approaches based on the conventional theory of differential games are insufficient, especially in case of dynamic, multi-agent models. It is well known that there exist a small number of differential games for which exact analytical solutions are available. There are a few more differential games for which numerical solutions can be computed in a reasonable amount of time, albeit under rather restrictive conditions. However, each of these games must be one-to-one, which is very far from the real world combat scenarios, especially from the asymmetric war games. They are also of the "zero-sum type" which does not allow the enemy to have goals other than diametrically opposing to those of the friend. Other difficulties arise from the requirements of the 3D modeling, limitation of the lifetime of the agents, or simultaneous participation of the heterogeneous agents such as on-surface and aerospace vehicles.

Another class of games is called *extended* games. LG-based approach (Section 4 and [9]) utilizes this class of games. Extended games are usually represented as trees, which include every alternative move of every strategy of every player. Application of this class of games to real world problems requires discretization of the domain, which can be done with various levels of granularity. In addition, in the real world problems, moves of all the pieces (aircraft) and players (Red and Blue) are concurrent, and this can be represented within extended, but not within normalized, games. Thus, the extended games allow us to adequately represent numerous problem domains including military C^2. The main difficulty for any game approach is the "curse of dimension." Even for a small-scale combat, an extended game is represented by a game tree of astronomic size, which makes this game intractable employing conventional (non-LG-based) approaches.

Consider, for example, a small concurrent game with 10 pieces total so that each can make 10 distinct moves at a time. If the game lasts for at least 20 moves (not unusual for battlefield examples), the size of the game tree would be about 10^{200} nodes. To be more specific, the JFACC Game (Sections 5, 10) includes 8 mobile pieces with 18 legal moves each, while the game lasts 70 moves.

No computer can search such tree in a lifetime. Even the most presently promising search algorithms on the game trees, those that utilize alpha-beta pruning, would result in insufficient search reduction. Even in the best case the number of moves to be searched employing alpha-beta algorithms grows exponentially with the power of this exponent divided by two with respect to the original game tree [3]. In the above example the reduced tree would have 10^{100} nodes, which is just as impossible to search as the unreduced tree. Moreover, the alpha-beta pruning method is applicable to sequential alternating games only (Blue-Red-Blue-... moves), whereas most of the real world games, including air and other military operations are concurrent. For the games with concurrent actions the number of moves to be searched "explodes" even more dramatically than for the sequential games. This is because of all the possible combinations of moves for different pieces can be included in one concurrent move. With conventional non-LG approaches, the question of scalability of extended concurrent games cannot be even raised.

4   LG Games

Following [9] we define a "discrete universe" by observing "the laws of discrete physics." The problems in such universe are very close to the 2D board games like chess, checkers, etc. An abstract board, an area of the discrete universe, is represented by an arbitrary finite set X. Abstract pieces represent the agents standing or moving with a constant or variable speed. We introduce concurrent movement, explosion of agents (removal from the system), collision, and collision avoidance.

DEFINITION 4.1
Abstract Board Game (ABG) is the following eight-tuple:

$$< X, P, R_p, SPACE, val, S_i, S_t, TR>,$$

where
- $X = \{x_i\}$ is a finite set of *points*, LG cells, which represent locations of pieces;
- $P = \{p_i\}$ is a finite set of *pieces*; P is a union of two disjoint subsets P_1 and P_2 called the *opposing sides*;
- $R_p(x, y)$ is a set of binary relations of *reachability* in X (x and y are from X, p is from P);
- *val* is a function on P with positive integer values describing the *values* of pieces.
- SPACE is the state space. A state $S \in$ SPACE consists of a partial function of *placement* ON: $P \rightarrow X$ and additional parameters.
- The value ON(p) = x means that piece p occupies location x at state S. Thus, to describe function ON at state S, we write equalities ON(p) = x for all pieces p, which are present at S. We use the same symbol ON for each such partial function, though the interpretation of ON may be different at different states. Every state S from SPACE is described by a list of formulas $\{ON(p_j) = x_k\}$ in the language of the first order predicate calculus, which matches with each relation a certain Well-Formed Formula (WFF).
- S_0 and S_t are the sets of *start* and *target* states. Thus, each state from S_0 and S_t is described by a certain list of WFF $\{ON(p_j) = x_k\}$. S_t is a union of three disjoint

subsets S_t^1, S_t^2, and S_t^3. S_t^1, S_t^2 are the subsets of target states for the opposing sides P_1 and P_2, respectively. S_t^3 is the subset of target draw states.

- TR is a set of transitions (moves), TRANSITION, of the ABG from one state to another. These transitions are described in terms of the lists of WFF (to be removed from and added to the description of the state and a list of WFF of applicability of the transition. These three lists for state S $\in$ SPACE are as follows:

    Applicability list: $(ON(p) = x) \wedge R_p(x, y)$;

    Remove list: $ON(p) = x$;

    Add list: $ON(p) = y$,

where $p \in P$. The transitions are defined and carried out by means of a number of pieces p from P_1, P_2, or both. This means that each of the **lists** may include a number of items shown above. Transitions may be of *two types*.

A transition of the *first type* (shown above) occurs when piece p moves from x to y without removing an opposing piece. In this case, point y is not occupied by an opposing piece.

A transition of the *second type* occurs if piece q does not move at all: it is removed from the board. Typically, the opposing piece q, OPPOSE (p, q), has to occupy y before the move of p has commenced. In the latter case, the **Applicability list** and the **Remove list** include additional formula $ON(q) = y$. For concurrent systems (DEF. 4.2), this is not necessary: pieces p and q may be removed from any locations.

The *goal of each side* is to reach a state from its subset of target states, S_t^1 or S_t^2, respectively, or, at least, a draw state from S_t^3. The problem of the optimal operation of the ABG is considered as a problem of search for a sequence of transitions leading from the Start State of S_0 to a target state of S_t assuming that each side makes only the *best* moves, i.e., such moves (transitions) that could lead the ABG to the respective subset of target states.

To *solve an ABG* means to find a *strategy* (an algorithm to select moves) for one side, if it exists, that guarantees that the proper subset of target states, S_t^1, S_t^2, or S_t^3, will be reached assuming that the other side makes arbitrary moves.

DEFINITION 4.2

A *Totally Concurrent* (TC) game is the ABG where all, some, or none of the pieces of both sides can move simultaneously (plus some of the pieces located at various cells can be destroyed).

Various technical and human society systems including military combat systems, systems of economic competition, positional games, etc., can be represented as ABGs [4-10, 15]. A development of such representation is the key issue in the application of LG to various problem domains. To apply the LG algorithm we have to reflect all of the components of the ABG: the operational district X, the mobile units, i.e., the set of pieces P broken in two subsets – opposing sides, P_1 and P_2, and the moving abilities of the pieces, relations of reachability $R_p(x, y)$, etc.

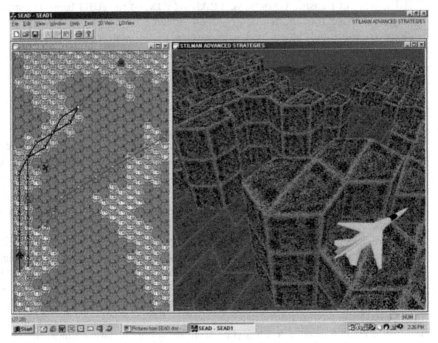

Fig. 1. Operational District for the JFACC Game

5   LG War Gaming

Combat control problems with explicit mobile entities organized in adversarial teams can be represented as totally concurrent ABGs. The LG approach is based on the ABGs, where each side participating in a conflict can control its pieces so that their actions will bring about the fulfillment of the side's goals. The pieces are placed on the Abstract Board. The game description provides rules describing legal motions for each piece, as well as its interactions with other pieces, including the enemy pieces. The game board can be viewed as analogous to a 2D/3D military map that shows current dispositions of military forces. The key difference is in the dynamic nature of the game in a sense that it can be watched like an interactive movie controlled by the intelligent adversaries.

Combat planning and control problems can be represented as ABGs as follows [8, 10]:

- X represents the operational district, which is the area of combat operation, broken into smaller 2D/3D areas, LG cells, e.g., in the form of the 2D/3D grid. It could be a land operation, where X represents the set of 2D cells, an Air Force or Navy combat with 2D/3D cells. A space mission can be represented as an ABG by introducing X as a set of orbital segments. The specific shape of a cell is unimportant from the LG point of view and can be chosen as the most convenient for the specific domain.

JFACC combat model is an example of the conflict to which the LG tools may be applied. The prototype software JEC (JFACC Experiment Commander) developed for DARPA by STILMAN Advanced Strategies is capable of conducting multiple experiments (Section 10 and [8, 10]). The pieces within the prototype may be moved to different locations within the space, which consists of 25,000 3D cells (50x50x10), Fig. 1. Every cell is a hexagonal prism, which represents a real world 3D hexagonal prism with the diameter of the inscribed circle of 2 nautical miles and 1-mile height. The LG tools provide strategic planning for the mission described below. A mission for Suppression of Enemy Air-Defenses (SEAD) is modeled employing ABG modeling techniques. Such a mission may include an aircraft strike package moving into a canyon to a target area with the opposing side having ground-to-air missiles, anti-aircraft batteries, fighter wings, and radars. The corresponding Abstract Board captures 3D air space, terrain, the aircraft trajectories, positions of the batteries, strategic features of the terrain, such as bridges, and their status (destroyed, not destroyed), radars and illuminated space, etc. Two different views are provided by JEC: a 3D animated view for realistic representation of the "Cyberland" (right window, Fig. 1) and a 2D map view for ease of analysis and control (left window, Fig. 1). The JEC allows us to control the Blue and Red forces by generating optimal (or near-optimal) strategies for both sides. In Fig. 1, 4, 5, left, the trajectories connecting red and blue pieces with various locations represent combat Zones (Section 5). The right window (Fig. 1, 4, 5) shows animated movement and actions of the aircraft along the trajectories following the strategy generated by JEC.

• P is the set of autonomous vehicles, e.g., tanks, subs, or aircraft. For the space war games it represents a fleet of spacecraft. It is broken into two subsets P_1 and P_2 that represent adversarial teams of vehicles with opposing interests.

For the JFACC project, an ABG can fully represent a SEAD mission (Fig. 1). Such mission may include an aircraft strike package (Blue forces – P_1) moving into a canyon to a target area with the opposing side having ground-to-air missiles, anti-aircraft batteries, fighter wings, and radars (Red forces – P_2).

• $R_p(x,y)$ represents moving abilities of various vehicles for different problem domains: piece p can move from point x to point y if $R_p(x, y)$ holds. Some of the vehicles can crawl, others can jump or ride, sail and fly, or even move from one orbit to another. Some of them move fast and can reach point y (from x) in "one step", i.e., $R_p(x, y)$ holds; others can do that in k steps only, and many of them cannot reach certain points at all. For example, a spacecraft with small fuel tanks cannot reach certain cells (orbits' segments) significantly different from its current orbit, especially, the orbits in a different plane. During one time increment, some non-sophisticated vehicles, tanks, fighters, space interceptors, can move to the *adjacent* cell only. However, vehicles with more advanced moving abilities, i.e., with different R_p, can "jump over" and reach remote cells during the same time increment. Moreover, during the operation, a vehicle can *change the speed, replicate*, or *convert itself* to another level of sophistication by changing representation from p_1 to p_2 with the new set of relations of reachability

$R_{p_2}(x, y)$. In this way we can represent a start of a missile from the aircraft or a change of the spacecraft mobility after the fuel recharge.

In the JFACC project relations of reachability R_p for an aircraft allow it to move to the hexes adjacent to the current location of an aircraft. This reflects cruising speed. The maximum speed allows an aircraft to "jump over" the adjacent cells and reach next cells (two hexes away) within the same time increment. The duration of the time increment is about 30 seconds, therefore these two reachabilities reflect two values of speed, 4 miles/min and 8 miles/min. Additionally, these relations reflect the direction of movement, the velocity of an aircraft.

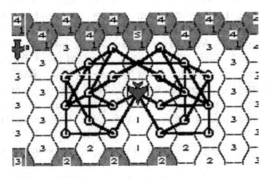

Fig. 2. Trajectories of the aircraft changing direction (making a circle)

In particular, an aircraft flying with certain velocity cannot turn around 180° and move in the opposite direction. This maneuver will require a number of steps. Relations of reachability allow the change of velocity during one step (one time increment) within the range of ±60° from the current velocity. A bundle of trajectories of an aircraft making a circle is shown in Fig. 2. The orientation of the aircraft (down, Fig. 2) shows the direction of the current velocity. To make a circle (360°) and return to the same cell with different velocity it has to make at least 8 steps. Every link of a trajectory (a time stamp) is marked by a small circle in the center of the respective hex.

Another type of reachability in the JFACC project reflects missile shooting range. Two types of air-to-air missiles are involved: long-range with the range of up to 20 miles (10 hexes away) and short-range with the range up to 3 miles (adjacent hex). Ground-to-air and air-to-ground missiles' ranges are reflected analogously. The main difference of the shooting reachabilities from the flying ones is the requirement of one-step direct visibility. An object to be shot must be acquired by the missile radar, which means that it should be visible from the aircraft, i.e., a straight line connecting the aircraft and the target should not cross any obstacles (hexes filled as part of terrain). This condition requires having a copy of the 3D operational district as 3D Euclidian space. Then, a straight line connecting the centers of the corresponding hexes (of the aircraft and the target) can be checked if it crosses obstacles. Another difference is that the missile shot including missile's flight time and explosion take place in zero time without spending a time increment.

Computation of both types of reachabilities for generating trajectories of aircraft is very time consuming. These procedures are optimized by computing reachabilities for all the district locations in advance and storing them in RAM in large hash tables for future use.

- Equation $ON(p) = x$ holds, if vehicle p is at the location x. For the land or navy operations, this represents the fact that vehicle p is located within the area x. In case of the space combat, spacecraft p is in certain orbit (within the orbit's segment x).

  In the JFACC project location x is the identifier of the respective hex. This means that an aircraft is located within the 3D cell limited by the hexagonal prism. It can be airborne or landed. Similarly, other objects like bridges, military headquarters, missile batteries, radars, etc., are located within the respective hexes. A number of aircraft and other objects can be located in the same hex simultaneously.

- Value $val(p)$ for a vehicle p might be determined by the technical parameters of the vehicle, the striking power of a missile or a maximum speed of a submarine. It can also include an immediate value of this vehicle for the given combat operation.

  In the JFACC project $val(p)$ reflects the value of the aircraft p as well as the value of other mobile and immobile objects.

- S_o is an arbitrary Start State of operation for analysis; S_t is the set of target states. These might be the states where vehicles of each side reached specified cells. On the other hand, S_t can specify states where opposing vehicles of the highest value are destroyed or trapped. The list of formulas $\{ON(p_j) = x_k\}$ corresponds to the list of objects with their coordinates in each state.

  In the JFACC project at the Start State S_o some of the aircraft are airborne, others are on red alert, mobile missile launchers are at their respective cells. The set S_t^1 of target states for Blue forces includes all the states after accomplishing a SEAD mission, i.e., Blue forces have destroyed Red headquarters and returned to the Blue base. This mission requires involvement of the Blue strike package, a number of bombers, strikers and radar jammers. Set S_t^2 includes the set of states after SEAD mission has failed.

- TRANSITION(p,x,y) represents the move of the vehicle p from location x to y; it can include *concurrent movements* of several vehicles. If at the moment of a move an opposing vehicle q is at location y, it is destroyed and removed. In this way we represent that the target was destroyed. A removal of q might happen even without movement of p, e.g., in case of a laser beam attack or a land mine explosion. For the space vehicles concurrent moves include constant movement within the orbit with respect to the rest of the vehicles and the Earth as well as orbit changes employing boosters.

In the JFACC project concurrent move may include movement of aircraft (Blue and Red), missiles launches and target hits.

A preliminary planning of operation with LG models can be done by running multiple experiments in order to select the best Start State, i.e., the best initial configuration of all the friendly agents involved in the operation and the best allocation of resources. After the engagement the control is conducted in real time by multiple re-planning by taking into account actual advancement of agents, actual losses, and changes of mobility. Similar planning and real-time control of operation can be conducted on a smaller scale by each team and each military unit. Planning of operation for the JFACC Game is shown in Section 10.

6   LG Tools

The LG approach is applied as follows [9]. First, the problem is defined in ABG terms, i.e., the players, the Board, the pieces, the game rules, etc., are identified. Then some methods are utilized to generate strategies that would guide the behavior of the designated players (i.e., the Friend and its allies) so that their goals would be fulfilled.

LG dramatically reduces the size of the search trees, thus making the problems previously considered unsolvable, computationally tractable. In addition, unlike most of conventional approaches, LG is ideally suited for totally concurrent games. To achieve the above, LG provides a formalization and abstraction of search heuristics of advanced experts in the form of the game strategies. Essentially, these heuristics replace the search by the construction of strategies. The formalized expert strategies yield efficient algorithms for problem settings whose dimensions may be significantly greater than the ones for which the experts developed their strategies. Moreover, these formal strategies proved to be able to solve problems for different problem domains far beyond the areas envisioned by the experts. These strategies are not intended to provide solutions that are always optimal, but they are intended to provide "good enough" solutions. Although for some classes of problems, these formalized expert strategies yield provably optimal solutions [9], for the rest of the problems the LG strategies are the best-known solutions. To formalize the heuristics, LG employs the theory of formal languages (i.e., formal linguistics), as well as certain geometric structures over the abstract board. Since both the linguistics and the geometry were involved, this approach was named Linguistic Geometry.

The basic building block for the LG strategies is the LG Zone structure. Intuitively, an LG Zone is a network of trajectories drawn in the game Board. There are several kinds of LG Zones, e.g., *attack*, *retreat*, *unblock*, etc., see [9]. Here we'll concentrate on the attack Zones but will add a new notion of a bundle of attack Zones. In addition, in contrast with [4-7, 9], our Zones will employ the *strike* trajectories and their bundles in addition to the regular trajectories.

Roughly speaking, an attack "Zone" (Fig. 3) has a main (friendly or adversary) piece (p_0 in Fig. 3) and a main strike trajectory (e.g., 1, 2, 3, 4, (5)) that is a path that the main agent needs to attain a local goal. A Zone includes a number of opposing pieces (q_0, q_1, q_2, q_3 – blue force) and their strike trajectories (e.g., 6, 7, 8, (3)) capable of preventing the main agent from achieving the goal. It also includes auxiliary friendly pieces counteracting the above actions of the enemy, counter-counteractions of the opponents, etc. A Zone is strict if the length of any negation trajectory t is equal to the number of moves that the acting piece on the negated by t trajectory has to make for reaching the target location of t, Fig. 3. There, black lines indicate the directions of physical moves, whereas the gray lines indicate the action, that is the weapon release as in the example on Fig. 3.

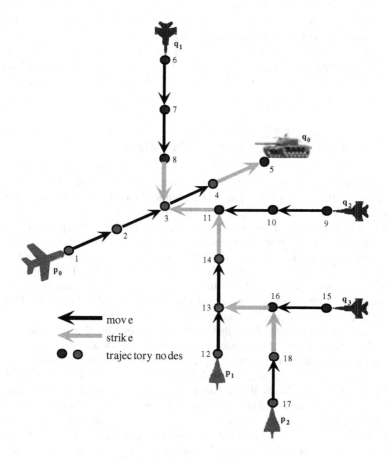

Fig. 3. A simple Attack Zone

The Zone shown in Fig. 3 has 3 aircraft for the Red side (p_i) and 3 aircraft and a tank for the Blue side (q_j). Therefore, the pieces are the 6 aircraft and the tank. With respect to this Zone, the aircraft are intended to move along the indicated trajectories. The small circles along the trajectories indicate the possible moves (trajectory nodes). For example, the red aircraft, p_0, will have the Blue Tank q_0 in shooting range in three moves, once the aircraft moves along its trajectory 1, 2, 3, 4. However, the blue aircraft q_1, and q_2 can get the position into their shooting range once they reach locations 8 and 11, respectively. The blue aircraft q_2 and q_3 can be prevented from reaching their respective destinations by the red aircraft p_1 and p_2. Finally, the blue aircraft q_3 can effectively prevent p_0 from achieving its goal. Thus, the red side would not win this local combat. Therefore, with respect to this Zone the piece p_0 is at disadvantage.

Assuming that the Blue pieces represent unmanned aerial vehicles, the Red Bomber is manned, and the Red fighters are either manned or unmanned, the intuitive meaning of the strategy represented by a single Zone in Fig. 3 is the following: pieces

q_2 and q_3 distract the Red fighters whereas q_1 finishes off the Red Bomber preventing destruction of q_0 which is the goal of the Blues.

In general, an LG strategy is derived from one or several *Zone bundles*, [8, 10]. To think of a bundle of Zones, it is enough to replace each strike trajectory t in a singleton Zone by a bundle of trajectories with the same source and target as those of t.

7   Asymmetric Games and LG Paradigm

The LG approach to war-gaming directly addresses the modeling challenges with respect to both the conventional and asymmetric conflicts. LG uses multi-agent Abstract Board Games (ABG) to model all the sides of a conflict, the conflict environments, and the constraints. Intelligent stakeholders are represented by means of LG generated intelligent strategies. To generate those strategies, in addition to its own "mathematical" intelligence, the LG algorithm may incorporate the intelligence of the top-level experts in the respective problem domains. The LG approach makes it possible to incorporate most of the known traits of a stakeholder, i.e., real personalities involved in the conflict with their specific individual style. Thus, all the sides of the conflict are modeled on the same level of sophistication.

LG models the individual differences between intelligent stakeholders. In particular, LG provides classification of the Zones that allows us to tell which kind of a Zone the stakeholder prefers to play. Currently LG differentiates among the following types of Zones: attack, block/relocation, domination, retreat, unblock. For example, given two Zones that achieve the same local goal, the stakeholder may prefer a relocation Zone. This may mean that the stakeholder would like to move a supportive piece, such as a military unit, equipment/weapons, etc., providing a strategic advantage over the previous position of the stakeholder. The relocation Zone preference may capture a personal trait, such as "patient pragmatist" (as are Milosevich, Saddam Hussein). By extending the classification of Zones to reflect a greater variety of personal traits of the stakeholders/strategists and by attaching preference rules to the stakeholders as players, the LG would model the intelligence data about stakeholders' personalities and thereby would represent their playing style;

LG allows us to capture the constraints on an asymmetric conflict by selecting the geometry of the Abstract Board and the rules of the Game. The geometry may reflect the real world geometry of the 2D, 3D, or 6D (coordinates and velocity) space, as well as abstract relationships among the agents, e.g., p may destroy q, p_1 must coordinate its moves with p_2, etc. The game rules may reflect actions of agents including motion, rules of engagement, shooting modes, probabilities of kill, etc.

LG allows modeling of intelligence data for all the stakeholders. Such data are incorporated in the geometry of the Abstract Board, in the rules of the game, in the algorithms for selection of the Zones, and in the construction of strategies.

The LG approach covers sequential, partially concurrent, and totally concurrent games. The key concept representing concurrent moves is a multi-move (TRANSITION). Each multi-move encompasses many concurrent actions. In addition, hidden concurrent actions can be analyzed by LG at the planning stage. Hidden actions can also be responded to by LG during conflict in real time. The basis for the responses is the visible side effects of the hidden actions that may be uncovered by LG at the planning stage;

LG evaluates a current state of a conflict with respect to a stakeholder. Specifically, for each state, LG determines a quantitative measure of a positional advantage/disadvantage for every agent (piece) of the stakeholder. On the basis of that, together with the LG worth function for the agents, LG determines the current positional advantage/disadvantage for the stakeholder.

The algorithm for LG strategy construction reduces search dramatically and, thus, is much more promising than conventional war-gaming approaches. The run time of the conventional gaming algorithms is more than exponential with respect to the size of the input. Theoretical results and experiments [8-10] with the LG models indicate that the LG run time is at most polynomial which makes it applicable to real world conflicts.

LG is well suited for applications to asymmetric warfare or asymmetric conflict, because it directly addresses the main challenges of this problem domain.

8   Hierarchy of Games

LG can be utilized to model and assist the operations at various levels of resolution. At the top (strategic) level, the lowest resolution model controls the global campaign-size operations, as well as the largest possible teams of military mobile, manned or unmanned, units. The full spectrum of mobility of those teams is employed. In the LG terms, the Abstract Board would be determined via a low-resolution grid covering the physical domain of the campaign (i.e., oceans, land, air, and, possibly, even near-planet space). The pieces would be battle units intended to fulfill uniform goals, e.g., either friendly or opposing teams of submarines, ships, land and air force units, etc. At this level, large, long range, possibly stealth, vehicles would be utilized for either long-range reconnaissance or for delivering weapons payload far behind the enemy lines. The LG motion reachability and weapon reachability relations would permit us to encapsulate the mobility and military strength of the battle groups and single vehicles into the ABG.

At the lowest levels of the hierarchy, high-resolution grids covering relatively small areas called tactical control elements (TCE) are employed. High-resolution LG models control swarms of manned and unmanned vehicles corresponding to a single task, as well as other agents of interest.

The LG approach is highly flexible with respect to treatment of the mapping between the game pieces and the real world agents such as swarms of vehicles. On the one hand, it is not necessary to represent each vehicle in a swarm as a game piece. The whole swarm may be represented as a single piece, as long as the individual vehicles are moving together. This saves computational time and simplifies the display of the game board for the mission commander. On the other hand, as the games progresses, this piece may be split into smaller pieces if an independent action by a swarm component would be required. Also, the same swarm may be represented as a single piece at the higher level of the game hierarchy and as several pieces at the lower level. For example, a group of aircraft, a strike package, may be viewed as a single piece at the operational level and as three pieces, Ground Suppression Piece, Striker Piece, and Air Suppression Piece at the tactical level.

A mission planning with LG models could be conducted by running multiple experiments (Section 10), so that the mission commander may select the best Initial State, i.e., the best initial configuration of all the friendly agents to be involved in the operation. After the Initial State is selected, LG application would generate an initial strategy for the mission. After the actual engagement starts, the mission execution control would be conducted in real time as follows. In the beginning, the initial LG strategy would be utilized to provide advice for the commander. As the mission progresses, the LG strategy would be updated by taking into account the actual advancement of agents, actual losses/gains, and changes of mobility, as well as the actual enemy actions. Similar mission planning and real-time control of mission execution may be conducted on a smaller scale by each team and each military unit reflecting their autonomy or subordination.

A variety of computers at the battlefield may be linked over the network. This would permit us to coordinate several LG battlefield assistants to the mission commanders at various levels of abstraction. Finally, a number of LG strategies may take advantage of strategic patterns developed beforehand by the military experts (either LG-assisted or not) and stored in a database. These retrieved strategies and patterns would allow us to utilize the historical experts knowledge by identifying strategies leading to familiar patterns of successful operations and by avoiding strategies leading to known failures. If they would not be available, the LG would provide advice based on the current situation only.

9   Games with Incomplete/False Information

Within the LG approach there is a distinction between *constructing strategies* for the players (say, for two players, Reds and Blues) and *simulation* where Reds and Blues fight it out using these strategies. In our game formalization of the military engagement, we call the snapshot of the battlefield the "game state." This is a static notion. A more important notion is that of the Worldview. It represents a space where the game state is permitted to evolve according to the game rules. If the sides of the conflict (in this case Reds and Blues) share the same Worldview, it is difficult to model combat with incomplete/false information. Thus, LG employs separate Worldviews for the strategy computation, as well as a "true" Worldview for the simulation. There are two approaches for modeling incomplete/false information.

Reactive Approach. Consider the Blue side. In order to develop the strategy (before each simulation step), the Blues have Reds in their Worldview. This Worldview is permitted to be slightly different from the Red Worldview, which, symmetrically, has the Blues inside it. The Blues create Zones (within the Blue Worldview), which contain both blue and red pieces. Assume for the moment that the Reds and Blues in these Zones share the same Blue Worldview. Assume the same for the Red Worldview. If the discrepancies with the True Worldview were uncovered during the simulation, the Blue would **react** by reconciling the Worldviews and re-computing their strategy. The Red would do likewise. However, if the discrepancy was the result of the deception engineered by the Reds, the Reds would be unable to prepare their strategy taking into account this deception. This would be because the Blues imagined by the Reds in the Red Worldview share the same Red Worldview.

Therefore, this approach deals with incomplete/false information only in part, although that is a reasonable step in this direction.

Proactive Approach. Within the Blue Worldview it would be beneficial to reflect a limited Red Worldview in the form of "what the Blues think it ought to be." Thus, when computing the Zones to figure out the Blue strategy, the imaginary Reds would not see what the Blues intend to conceal or would see the decoys created by the Blues. This would cause the Zone algorithm not to draw certain trajectories for the Reds that would otherwise appear as, say, first or third negation trajectories. Thus, the strategy for the Blues would reflect the expectation of the Blues that the reds are going to be deceived. This, for instance would allow them to shift forces to where they otherwise would not dare to move them. Of course, the Reds may not be actually deceived. We would be able to pursue experiments researching ways to evaluate the danger for the Blues of not deceiving the Reds. This may further influence the strategy for the Blues. Of course, the situation of Reds deceiving the Blues would be played in the same manner.

The LG approach permits us to provide both the *reactive* and *proactive* treatment of the incomplete/false information.

10   JFACC Games

Consider application of JEC to mission planning as part of the analysis conducted by the mission Commander of the Blue forces. All three experiments described in this Section required 10 minutes total on the 800 MHz PC.

The land shown in Fig. 1, 4, 5 is a remote island that contains enemy installations on its North side – the Red base (far from the viewer). Our Blue forces have been deployed to the South end of the island (close to the viewer) with a mission to destroy the Red headquarters. This island is largely covered by mountains, with two canyons leading from the South to the North. One of the canyons contains several enemy surface-to-air missile sites, or SAMs. In addition to these anti-aircraft defenses, the Red side also has two fighters: one is next to its headquarters, and the other one is located on the southeastern side of the island. Our Blue strike package consists of one striker, which is capable of destroying ground targets, as well as several escort fighters to defend the striker from enemy aircraft. We will demonstrate how JEC assists the Blue Force Commander in planning the SEAD mission by conducting the "what-if" analysis under different initial conditions and requirements. In the first two experiments, the Commander put the priority on the speed of the mission, and in the third, he controls the analysis by putting successful strategies ahead of speed. The flexibility of the LG tools allows us to accomplish analysis of different goals in real time. One of the benefits of JEC is that it allows the user to graphically observe the strategies that LG tools generate. The map in the left window (Fig. 1, 4, 5) demonstrates how the strategies are generated. Initially the shortest trajectories from the striker to the enemy base are generated – they are displayed in black. They go through the canyons because the JEC knows that it is always safer to hide than to fly above the mountains.

Then JEC generates negation trajectories, or intercept paths, which are shown in red. These negation paths show how the enemy aircraft and SAM sites can attempt to destroy the striker along its trajectory. JEC also generates second level negation trajectories, which instruct our Blue fighters on how they can destroy the enemy objects before they become a threat to the striker. These trajectories are shown in green. Although these trajectories are displayed on a 2-dimensional map they are actually calculated in 3 dimensions.

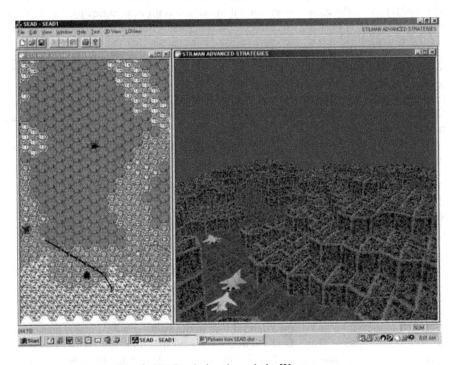

Fig. 4. SEAD mission through the West canyon

This collection of trajectories called Zone was discussed in Section 5. A typical Zone includes thousands of trajectories. The Zone is analyzed to see which of the trajectories for the striker yields the highest probability of success. Only such trajectories are displayed. In our example, the trajectory through the right (or East) canyon can be attacked by the SAM sites as well as the enemy fighters. This is represented graphically with red lines. However, the canyon on the left (or West) is "safer" – it does not contain as many enemy objects (Fig. 4). In the first experiment the striker has chosen the safer route (Fig. 4).

The action commences with an intercepting fighter coming from the East. Two of our blue fighters take out the interceptors, but two of our escorts are also destroyed in the melee. Our remaining striker and escort now join forces and fly into the westward canyon – since, as mentioned, this canyon contains fewer enemy units. The Zones and strategies (in the left window) are updated on each step to show the change in the battlefield dynamics. Shortly, the Blue aircraft are met by the remaining enemy fighter. The Red fighter is destroyed, but we also lose the third fighter. The remaining Blue striker continues the mission by destroying the now visible SAM unit.

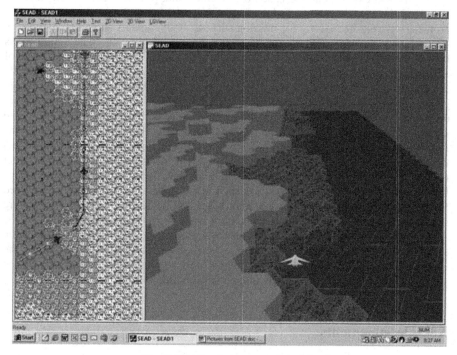

Fig. 5. SEAD mission around the island

After this, a clear path to the Red base is available, and the lone fighter destroys it. Each weapon has a probability of kill which allows us to model the situation more realistically. In this case, several missiles are needed because the probability of successfully destroying the enemy base is low for each attack. Once successful, the striker then turns around and flies toward the south end of the island where its hangar is located. The aircraft is safely guided back to the base by JEC, and the experiment is over with a victory for the Blue side.

The Commander decided to consider a different set-up of Blue forces. He altered the location of the Blue striker, and also changed the positions of our escort fighters to simulate a situation with less support for the striker. The requirement for a quick, unconditional destruction of the Red base forces the Blue striker to fly through the East canyon, despite its lower chances for success. As before, the action starts with a Red interceptor. We are able to destroy it, but in the process our only supporting fighter is lost. As a result, the striker enters the canyon alone, although the remaining fighters follow it, trying to catch up. Half way through the canyon, the striker is met by the remaining enemy fighter. Fortunately, this fight ends in our favor and the striker continues on its mission. Next, it has to destroy an enemy SAM site. With a more powerful missile range, the striker is able to destroy the SAM before coming into its radar. The next SAM site, however, is hidden better, and it locks onto our striker and destroys it. The failure predicted by chance has finally caught up to us. Without the striker, we cannot complete the mission, so this simulation ends with a Red victory.

For the next experiment, the Blue Force Commander put the success of the mission ahead of the speed, and JEC shows what happens when both canyons are well protected. Given the priority for mission success, both canyons should be avoided at all costs. As a result, JEC would have to come up with an alternate solution. As before, the action starts with a short dog fight with a Red fighter. However, instead of entering a canyon, our aircraft head East (Fig. 5). The strike package attempts to reach the North side of the island by flying along the East shore. Again, the strategy generation process is displayed in the left window. However, the chosen route is not entirely safe either – we encounter an enemy fighter coming from the North – the same fighter, which we encountered before inside a canyon.   This shows the flexibility of JEC, as the negation trajectory of the Red fighter was updated to reflect the changes the Blue force made. After that fighter is destroyed, the path to the target is clear. The strike package continues North, and then turns northwest to approach the Red headquarters. After the target is destroyed, our aircraft turn around and proceed South along the ocean towards the Blue base. As in the first experiment they are safely guided back to the base, and the simulation is over with a Blue victory.

The analysis is finished. Now it is up to the Commander to make the final decision, to choose the right strategy. Even after that, when the actual SEAD mission starts, the Commander will be assisted by JEC, which will automatically update the chosen strategy in real time employing new information coming through sensors.

Acknowledgements

This paper would be impossible without extensive R&D at STILMAN Advanced Strategies, LLC (Denver, CO, USA), founded in September of 1999. STILMAN served as a key subcontractor to Rockwell Science Center (Thousand Oaks, CA, USA) for the project "Agile Symbolic Mission Control and Hostile Counteraction Strategies" funded by DARPA within the 1999 Joint Force Air Component Commander (JFACC) program.

References

1. Battlefield of the Future, e-Journal,
   http://www.airpower.maxwell.af.mil.airchronicles/battle/bftoc.html
2. Grecu, D., Gonsalves, P. Agent-Based Simulation Environment for UCAV Mission Planning and Execution, AIAA Guidance, Navigation, and Control Conference, Denver, CO, Aug. 14-17, 2000.
3. Knuth, D.E. and Moore, R.W., An Analysis of Alpha-Beta Pruning, Artificial Intelligence, 293-326, 6(4), 1975.
4. Stilman, B., Linguistic Geometry for Control Systems Design. *Int. J. of Computers and Their Applications*, 1(2): 89-110, 1994.
5. Stilman, B., Managing Search Complexity in Linguistic Geometry. *IEEE Transactions on Systems, Man, and Cybernetics*, 27(6): 978-998, 1997.
6. Stilman, B., Network Languages for Concurrent Multi-agent Systems, *Intl. J. of Computers & Mathematics with Applications*, 34 (1): 103-136, 1997.

7. Stilman, B. and Fletcher, C., Systems Modeling in Linguistic Geometry: Natural and Artificial Conflicts. *Intl. J. of Systems Analysis, Modeling, Simulation*, 33:57-97, 1998.
8. Stilman, B. and Yakhnis, V., Solving Adversarial Control Problems with Abstract Board Games and Linguistic Geometry (LG) Strategies, *First DARPA-JFACC symposium on Advances in Enterprise Control* (AEC), San Diego, CA, November 15-16, 1999.
9. Stilman, B., *Linguistic Geometry: From Search to Construction*. Kluwer Academic Publishers, 416 pp, 2000.
10. Stilman, B. and Yakhnis, V., Adapting the Linguistic Geometry-Abstract Board Games Approach to the Air Operations, *Second DARPA-JFACC symposium on Advances in Enterprise Control* (AEC), Minneapolis, MN, July 10 & 11, 2000.
11. Szafranski, R., Parallel war and Hyper war: Is Every Want a Weakness? in *Battlefield of The Future,* January 4, 2000
12. Tucker, J., Asymmetric Warfare, FORUM for Applied Research and Public Policy, Internet.
13. Von Neumann, J. and O. Morgenstern, *Theory of Games and Economic Behavior,* Princeton University Press, 1947.
14. Warden III, J., Air Theory for the Twenty First Century, in *Battlefield of The Future,* January 4, 2000.
15. Yakhnis, V. and Stilman, B., A Multi-Agent Graph-Game Approach to Theoretical Foundations of Linguistic Geometry, in *Proc. of the Second World Conference on the Fundamentals of Artificial Intelligence* (WOCFAI 95), Paris, France, July, 1995.

Physics and Ecology of Rock-Paper-Scissors Game

Kei-ichi Tainaka

Department of Systems Engineering,Shizuoka University,
Hamamatsu 432-8561, Japan
tainaka@nsa.sys.eng.shizuoka.ac.jp

Abstract. From physical and ecological aspects, we review an interacting particle system which follows a rule of the Rock-Paper-Scissors (RPS) game. This rule symbolically represents a food chain in ecosystems. It also represents nonequilibrium systems which have a feedback mechanism. We describe the spatial pattern dynamics in lattice RPS system: the time dependence of each species is not fully understood, especially on two-dimensional lattice. Moreover, we modify and apply RPS rule to voter and biological systems. Computer simulation for both voter model and ecosystems exhibits counter-intuitive results in phase transition. Such results can be seen in many cyclic systems, and they may be related to the unpredictability in nonequilibrium systems.

1 Introduction

In most cases, large interactive systems organize themselves to stationary states. From the term of *stationary*, one may imagine the equilibrium. However, the stationary state does not always mean the equilibrium. When the principle of detailed balance is broken, the stationary state is in nonequilibrium. Such a state is called as the"cyclic balance". A typical example of nonequilibrium systems is an ecosystem where the cyclic balance holds by a food chain (web). In the present paper, we deal with a basic cyclic system whose reaction rule is represented by the so-called "Rock-Paper-Scissors"(RPS) game. The RPS system is one of the simplest ecological models.

In the present paper, we study the spatial pattern in lattice systems which contain three species, Rock, Paper and Scissors. Each lattice site occupies a single particle (individual) of three species. A fundamental interaction is represented by

$$R + S \longrightarrow 2R, \tag{1a}$$

$$P + R \longrightarrow 2P, \tag{1b}$$

$$S + P \longrightarrow 2S, \tag{1c}$$

where R, P and S denote a particle of Rock, Paper and Scissors, respectively. In the case of (1a), the species R beats (eats) S, and it reproduces an offspring. The strength relation among three species is cyclic. The relation of RPS symbolically represents the interaction of three species: producer (plant), herbivore and carnivore. The herbivore eats plant, and the carnivore eats herbivore; moreover, if the carnivore dies, its chemical components are served for the producer (plant). A more concrete example of the interaction (1) is the relation among three species of bees and wasps in Japan; that is, Japanese honeybee,

T.A. Marsland and I. Frank (Eds.): CG 2000, LNCS 2063, pp. 384–395, 2001.

(a) Random collision model

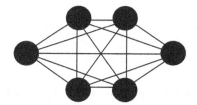

(b) Lattice model

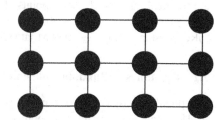

Fig. 1. Two different methods. (a) Random collision model which is called mean-field theory; collision occurs between any pair of particles. (b) lattice model; only adjacent particles can collide.

Japanese hornet and European honeybee. The Japanese honeybee is invaded by the introduced European honeybee, whereas the colony of the latter honeybee is quickly destroyed by attacks of the giant hornet. On the other hand, the Japanese honeybees can kill the hornet: more than 500 honeybees engulf and kill the hornet [1]. The strength relation among three species corresponds to the Rock-Paper-Scissors (RPS) rule. Another example of (1) is a side-blotched lizard [2]; there are three morpho-species of males distinguished by colors of throat. Males with orange throats are dominant to males with blue on their throat; males which have yellow throats and resemble females in morphology prevail the males with orange throats. These males also have the relation of RPS rule.

The RPS rule can be also seen in the Prisoner's Dilemma game on lattices. We assume that 1) the system contains "noise" or error, 2) the game is infinitely repeated, and 3) there are three strategies, All Defect (abbreviated as AD), Tit-For-Tat (TFT) [3, 4] and Pavlov [5, 6]. Each lattice site occupies a single particle (player) of three strategies. In this case, it is known that Pavlov completely beats TFT, but TFT beats AD. On the other hand, Pavlov is beaten by AD. The strength relation among these strategies completely corresponds to the Rock-Paper-Scissors game.

2   Dynamics of RPS Systems

2.1   Methods

We deal RPS system (1) by two different methods (Fig. 1): random collision and lattice models. In the former case, particles are regarded as gas molecules, and the interaction (1) occurs between any pair of particles (mean-field limit). In contrast, in the latter case, reactions (1) occur between adjacent particles. Simulation is carried out on one ($d = 1$) and two ($d = 2$) dimensional lattices. The method for lattice model is as follows [7];

1)Distribute three kinds of species, R, P and S in such a way that each lattice point is occupied by only one individual (particle).

2)Reaction processes are performed as follows: Choose one square-lattice point randomly, and then specify one of adjacent points. Let them react according to (1). Here we employ periodic boundary conditions: the next to the right edge is the site of left edge.

3)Repeat step 2) by L times for $d = 1$ and $L \times L$ times for $d = 2$, where L and $L \times L$ are the total number of lattice points. This step is called as Monte Carlo step. We set $L = 10^4$ for $d = 1$, and $L = 100$ or $L = 160$ for $d = 2$.
4)Repeat step 3) for 500–1000 Monte Carlo steps.

2.2   Dynamics of Random Collision Model

The population dynamics of random collision model was in detail studied by Itoh [8, 9]. He obtained the result that population dynamics reveals neutrally stable ("center"). In Fig. 2 (a), a typical attractor of population dynamics is shown; the population size of each species oscillates around the fixed point where three species exist with equal densities. In this figure, each vertex denotes that all particles become a single species, and the fixed point locates at the center of triangle. The oscillation profile depends on initial conditions. Itoh found two variables which are unchanged with time: one is the total number of particles, and the other is the product of three densities. He also proved that if the total number of particles is finite, the product always decreases. Namely, extinction of two species occurs eventually.

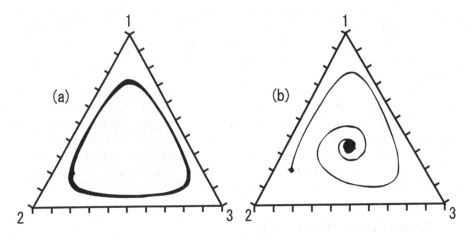

Fig. 2. Population dynamics of three species, where the star means initial densities. (a) random collision model, (b) lattice model.

2.3   Dynamics of Lattice Model

Lattice version of RPS system has been studied by many authors. I first reported the dynamics of lattice model in 1988 [7]. Dynamics for one dimensional lattice ($d = 1$) is relatively simple [7, 10, 11]; namely, "clusters" of each species slowly grow with time, where the cluster is the region (connected sites) of an identical species. The average size of clusters continues to increase. We can define "kinks" [12] which locate at the boundary of two different clusters. There are two kinds of kinks; K_L and K_R. The kink

K_L always moves to the left, but K_R moves to the opposite direction. If a couple of defects happen to meet, then they interact as follows:

$$K_L + K_R \longrightarrow \phi, \tag{2a}$$
$$K_L + K_L \longrightarrow K_R, \tag{2b}$$
$$K_R + K_R \longrightarrow K_L, \tag{2c}$$

where ϕ means the disappearance of kinks. From the above reactions, we find that the total number of kinks decreases. The number of kinks is proportional to $t^{-\alpha}$ according to (2a), where t means time and α is the constant. We have $\alpha = 1$ for infinitely large size of lattice. If the lattice size is finite, we also have $\alpha = 1$ in early stage. However, later, this value becomes $\alpha = 1/2$ originated in (2b) and (2c). Eventually, one of three species occupies the whole system. On two dimensional lattice ($d = 2$), the system evolves

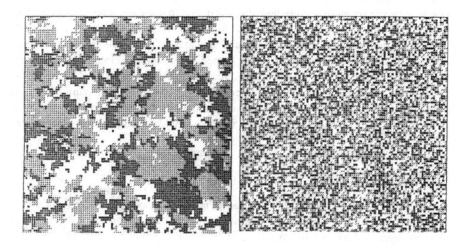

Fig. 3. Typical example of stationary pattern (left). For the sake of comparison, random distribution is displayed in the right [7].

into a stationary state as illustrated in Fig. 2 (b). Irrespective of initial conditions, three species coexist with equal density ("asymptotically stable"). Such a population dynamics cannot be explained by the first (mean-field) and second (pair) approximations. The latter gives a worse prediction compared to the former. It is still unknown which order of approximation explains the asymptotical stability. Note that the stable behavior is qualitatively explained by the introduction of "vortices" [12, 13].

Spatial pattern is also self-organized irrespective of initial patterns. Photographs of typical patterns are illustrated in Fig. 3, where the left and right represent the final stationary pattern and random distribution, respectively. It is found from Fig. 3 that the left is more realistic than the right. We can show that the stationary pattern (left) is fractal; strictly speaking, this is slightly different from true fractal [14].

3 Effects of Mass Media on an Election

3.1 Voter Model

The Rock-Paper-Scissors (RPS) rule is applied to real problems. First, we study the relation between mass media and an election [15, 16]. Suppose that only one person can win for an election. Consider the following model:

$$V_1 + V_2 \longrightarrow 2V_1, \tag{3a}$$

$$V_2 + V_3 \longrightarrow 2V_2, \tag{3b}$$

$$V_3 + V_1 \longrightarrow 2V_3, \tag{3c}$$

$$V_3 \xrightarrow{b} V_1, \tag{3d}$$

where V_i denotes the voter who supports candidate (color) i ($i = 1, 2, 3$), and b is the probability that reaction (3d) occurs. Provided that the reaction (3d) is neglected ($b = 0$), then the rule (3) is identical with the RPS system (1). In this case, three candidates have the cyclic strength; there is no dominant color in the hierarchy. On the other hand, the reaction (3d) is considered as an perturbation (*external field*). An example of such a field is the effect of mass media: By the news of mass media, the candidate 3 suffers the damages, while the candidate 1 receives advantages. An example is the scandal of the candidate 3. Thus, the parameter b measures the intensity of effect of mass communication. Note that the system (3) is an extension of a familiar model: if the candidate 2 was absent, or the reactions (3a) and (3b) were ignored, then (3) would be called the contact process [17, 18, 19].

3.2 Phase Transition

Simulations are carried out for various values of b. The two-body [(3a)–(3c)] and single-body [(3d)] reactions are performed alternately. The pattern dynamics of this system is also self-organized into a stationary state, irrespective of initial patterns. The density of each color in stationary state does not depend on the size of lattice. In Fig. 4 (left), the steady-state density of each candidate (color) is plotted against b, where the straight lines are the prediction of random collision model. If $b = 0$ (RPS), three colors coexist with equal densities. With increasing b, the population of color 3 increases, while the colors 1 and 2 decrease. Thus, Fig. 4 (left) exhibits a kind of paradox that the color 3 prevails the system and wins the election. In spite of reaction (3d), the color 1 is decreased. Moreover, it is found from this figure that the color 2 disappears (phase transition), when

$$b > b_1; \quad b_1 \sim 0.41.$$

In the case $b > b_1$, the color 2 disappears regardless of initial patterns; in this case, (3) becomes equivalent to the contact process. The steady-state densities for contact process are represented in Fig. 4 (right). It is found from this figure that with the increase of b, the color 3 conversely decreases. In particular, when

$$b > b_2, \quad b_2 \sim 0.83,$$

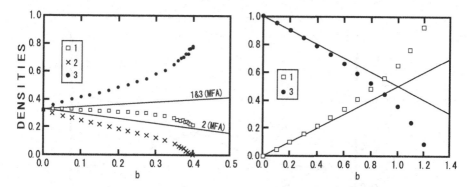

Fig. 4. The density of each candidate (color) in stationary state is plotted against the intensity b of mass communication [15]. Left: three colors coexist. The color 2 disappears by the scandal of the candidate 3. Nobody believes such a cause. **Right**: the color 2 is absent. The theoretical results of the random collision model (mean-field approximation: MFA) is also shown.

the density of color 1 exceeds that of color 3. Hence, when the effect of news is too great ($b > b_2$), then the candidate 1 wins the election. If b is very large ($1.22 \leq b$), the color 1 occupies the total system. The result in Fig. 4 (right) is reasonable, since the color 1 becomes stronger with the increase of b.

3.3 Direct Effect

Heretofore, we investigated the final stationary state, and exhibit the indirect effect of mass communication. In contrast, let us study the dynamic process (transient state), when the value of b is suddenly jumped up at a certain time t_0 (we set $t_0 = 0$). The experiment of jumping process from $b = 0$ to 0.3 is performed. This procedure is called press perturbation in ecology; the simulation is carried out with $b = 0.3$, where the initial condition is the stationary pattern for $b = 0$. In Fig. 5, the time-dependence of population of each color is depicted. This figure reveals that at first, the color 1 increases, but finally the population of color 3 exceeds that of color 1. Figure 5 therefore represents the direct effect of mass media: when mass media report the news just before the voting day, the candidate 1 wins the election. Since the direct effect is predictable, we can easily control the result of election in the country, where the democracy is sufficiently developed.

4 Paradoxical Effect of Habitat Destruction

4.1 Prey-Predator Model

Next, we deal with the problem of habitat destruction [20]. Consider a two-dimensional lattice consisting of two species of prey (X) and predator (Y). Each lattice site is labeled by X, Y, or O, where X (or Y) is the site occupied by prey (or predator), and O represents the vacant site. We assume the following interaction [21-25]:

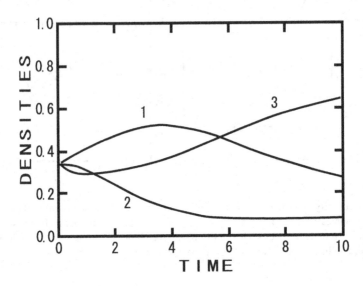

Fig. 5. The time-dependence of population of each color [15]. At time $t = 0$, b is suddenly jumped up from 0 to 0.3. Immediately after mass media report the news, the news gives a straightforward effect: the color 3 decreases, while color 1 increases. However, later, the color 3 increases.

$$X + Y \longrightarrow 2Y, \tag{4a}$$

$$X + O \xrightarrow{r} 2X, \tag{4b}$$

$$Y \xrightarrow{d} O, \tag{4c}$$

The above reactions respectively represent the predation, reproduction of prey (r) and the death (d) of predator.

The destroyed sites (*walls* or *barriers*) are put on the boundary (link) between neighboring lattice sites, where the barrier means the local destruction of habitat. For simplicity, we randomly put barriers in such a way that each link has a barrier by the probability p. Thus, p measures the intensity of habitat destruction. We assume that the interactions (4a) and (4b) occurs between adjacent lattice points, and that the barrier prohibits only (4b). Namely, the destruction only disturbs the reproduction of prey (X); in contrast predators (Y) receive no direct damage. It is well known in the field of physics that the barrier distribution shows *percolation transition* [26, 27]. When p takes an extremely small value, no barriers may connect with each other. On the contrary, when p takes a large value (near unity), almost all barriers are connected. Below, we call *cluster* for a clump of connected barriers, and *percolation* in the case that the largest cluster reaches the whole size of system. The probability of percolation takes a nonzero value, when p exceeds a critical point p_c; this value is given by $p_c = 0.5$ in our case (*link percolation in a square lattice*). Percolation ecologically means that the habitat region of species X may be fragmented into small segments for $p > p_c$.

We carry out a perturbation experiment by computer simulation of a lattice model. Before the perturbation, the system is assumed to stay in a stationary state of $p = 0$. At time $t = 0$ the barrier density is jumped from zero to a nonzero value of p as schematically illustrated in Fig. 1. We record the population sizes of both species X and Y for $t > 0$.

Each reaction process is performed in the following two steps:

*(i)*We perform a single particle reaction (4c). Choose one square-lattice point randomly; if the point is occupied by a Y particle, it will become O by a probability d.

(ii) Next, we perform two-body reaction, that is, the reactions (4a) and (4b). Select one square-lattice point randomly, and then specify one of the nearest-neighbor points. The number of these points is called the coordinate number (z); for square-lattice, this is given by $z = 4$. When the pair of selected points are X and O, and when there is no barrier (wall) between them, then the latter point will become X by a probability r. On the other hand, the barrier never effects the predation of Y: when the selected points are X and Y, the former point becomes Y. Here we employ periodic boundary conditions.

4.2 Result of Predator-Prey Model

We describe the result of perturbation experiments in the lattice model. Before the perturbation, the system is assumed to be in a stationary state. After the perturbation, the system changes into the other stationary state. Figure 6 shows the plots of densities of both species X and Y in the stationary states for various value of p, where the results of random collision model (mean-field theory) MFT and pair approximation (PA) are also depicted. The lattice model in Fig. 6 reveals the following results:

i)With the increase of barrier density p, the density P_Y of predator decreases. Especially, when $p > p_0$, the predator becomes extinct.

ii)The prey density P_X increases with p, and it takes the maximum value at $p = p_0$. When $p > p_0$, the prey density conversely decreases with p.

The species Y goes extinct, even though it suffers no direct damage by barriers, and there exist a lot of prey. Moreover, we find from Fig. 6 that the density P_X (or P_Y) for the lattice model is much larger (or smaller) than that predicted by MFT.

When the species Y becomes extinct ($p > p_0$), our system (1) is represented only by reaction (1b). It is therefore thought that the prey (X) occupies the whole lattice points. Nevertheless, this argument is not true: X cannot increases, since the fragmentation of habitat of X becomes severe for a large value of p. In particular, when p exceeds the percolation transition p_c ($p_c = 1/2$), the prey X is enclosed in small segments. Hence, the prey density decreases with increasing p (Fig. 6). In both theories (MFT and PA), the effect of fragmentation is not taken into account.

The extinction of predator (Y) observed in simulation thought to be understood by the following argument: The only way that species Y may reproduce is by consuming X. A domain containing only Y is unstable, due to the death of predators [reaction (1c)]. The species Y will eventually die out, unless there is an influx of prey (X) into the region. As the density p of barriers increases, such an influx thought to become impossible, and Y goes extinct. However, this argument is not completely correct, since the steady-state density of prey increases with the increase of p. More refined theories and arguments are When the species Y becomes extinct ($p > p_0$), our system (1) is represented only by reaction (1b). It is therefore thought that the prey (X) occupies the whole lattice points. Nevertheless, this argument is not true:

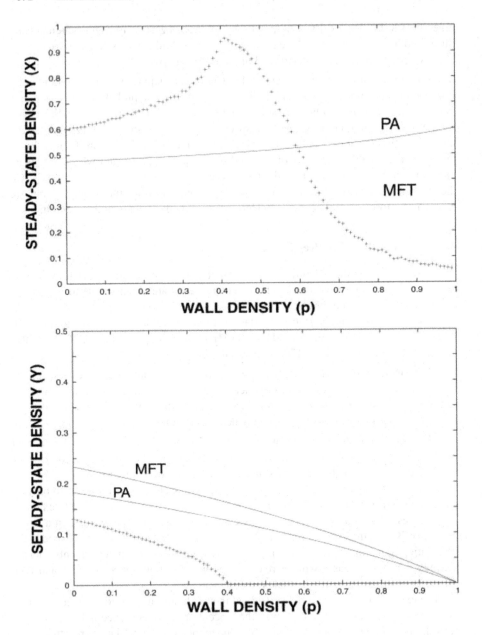

Fig. 6. The steady-state densities of species X and Y are plotted against the wall (barrier) density p ($r = 0.5$ and $d = 0.6$). The theoretical results of the random collision model (mean-field theory: MFT) and pair approximation (PA) are also shown.

X cannot increases, since the fragmentation of habitat of X becomes severe for a large value of p. In particular, when p exceeds the percolation transition p_c ($p_c = 1/2$), the prey X is enclosed in small segments. Hence, the prey density decreases with increasing p (Fig. 6). In both theories (MFT and PA), the effect of fragmentation is not taken into account.

The extinction of predator (Y) observed in simulation thought to be understood by the following argument: The only way that species Y may reproduce is by consuming X. A domain containing only Y is unstable, due to the death of predators [reaction (1c)]. The species Y will eventually die out, unless there is an influx of prey (X) into the region. As the density p of barriers increases, such an influx thought to become impossible, and Y goes extinct. However, this argument is not completely correct, since the steady-state density of prey increases with the increase of p. More refined theories and arguments are

When the species Y becomes extinct ($p > p_0$), our system (1) is represented only by reaction (1b). It is therefore thought that the prey (X) occupies the whole lattice points. Nevertheless, this argument is not true: X cannot increases, since the fragmentation of habitat of X becomes severe for a large value of p. In particular, when p exceeds the percolation transition p_c ($p_c = 1/2$), the prey X is enclosed in small segments. Hence, the prey density decreases with increasing p (Fig. 6). In both theories (MFT and PA), the effect of fragmentation is not taken into account.

The extinction of predator (Y) observed in simulation thought to be understood by the following argument: The only way that species Y may reproduce is by consuming X. A domain containing only Y is unstable, due to the death of predators [reaction (1c)]. The species Y will eventually die out, unless there is an influx of prey (X) into the region. As the density p of barriers increases, such an influx thought to become impossible, and Y goes extinct. However, this argument is not completely correct, since the steady-state density of prey increases with the increase of p. More refined theories and arguments are necessary to explain the extinction of Y.

So far, we considered the press perturbation that the barrier density is jumped from zero to a nonzero value of p, and we obtain the steady-state densities (Fig. 6). Now we can consider more general cases; namely, p is increased from p_1 to p_2. If $p_1 < p_0 < p_2$ is satisfied, then the species Y becomes extinct; no matter how the difference $p_2 - p_1$ is small, the extinction occurs. When there is an endangered species, it may become extinct by a slight perturbation to its habitat. We deal with a very simple prey-predator model, so that it is expected that in many web models, the similar extinction due to indirect effect takes place.

5   Concluding Remarks

In summary, the dynamics of Rock-Paper-Scissors (RPS) system is represented in Table 1. From this table, qualitative understanding may be sufficient, while the dynamics for lattice system is not fully found quantitatively. Especially, the asymptotical profile on two dimensional lattice ($d = 2$) cannot be explained by the first (MFT) and second (PA) approximations. It is still unknown which order of approximation explains the asymptotical stability.

In voter and predator-prey models, I illustrate the phase transitions caused by paradoxical reasons. We find that such cases are not special. By computer simulation, we perform various kinds of press perturbation, in which a parameter of a target species is altered and held at higher or lower levels. It is found that the system usually satisfies a parity law [28, 29], where the parity is defined by whether the system contains even or odd number of species. When the parity is even, or when the system has even number of states, then the long-term response of density of the target species is similar to the short-term one. Both short- and long-term responses are intuitively predicted. On the other hand, when the parity is odd, the long- and short-term responses are just opposite. The parity law is useful for model systems, but it is still suspicious whether this law does hold in real ecosystems. In summary, the dynamics of Rock-Paper-Scissors (RPS) system is represented in Table 1. From this table, qualitative understanding may be sufficient, while the dynamics for lattice system is not fully found quantitatively. Especially, the asymptotical profile on two dimensional lattice ($d = 2$) cannot be explained by the first (MFT) and second (PA) approximations. It is still unknown which order of approximation explains the asymptotical stability.

In voter and predator-prey models, I illustrate the phase transitions caused by paradoxical reasons. We find that such cases are not special. By computer simulation, we perform various kinds of press perturbation, in which a parameter of a target species is altered and held at higher or lower levels. It is found that the system usually satisfies a parity law [28, 29], where the parity is defined by whether the system contains even or odd number of species. When the parity is even, or when the system has even number of states, then the long-term response of density of the target species is similar to the short-term one. Both short- and long- term responses are intuitively predicted. On the other hand, when the parity is odd, the long- and short-term responses are just opposite. The parity law is useful for model systems, but it is still suspicious whether this law does hold in real ecosystems.

Table 1. Dynamics in lattice mode

Spatial dimension	Dynamics
$d = 1$	Decrease of kinks
$d = 2$	Asymptotically stable

Finally, we emphasize that the approaches of modeling and simulation are useful to study real problems. (i) in the case of theory or simulation, we can know the true cause of phase transition. In voter model, nobody cannot believe such a cause: the candidate 2 disappears (phase transition) by the news of the candidate 3. In predator-prey case, Y goes extinct by the increase of habitat destruction p. In real fields, however, no one may believe such a cause, since the prey population increases; there is no causal relation. (ii) in real systems, the long-term response is very hard to know, because there are a lot of miner factors. (iii) we cannot carry out real experiments of election and extinction.

References

1. M. Ono, T. Igarashi, E. Ohno and M. Sasaki, Unusual thermal defence by a honeybee against mass attack by hornets. Nature, 377, 334-336 (1995).
2. B. Sinervo and C. M. Lively, The rock-paper-scissors game and the evolution of alternative male strategies. Nature, 380, 240-243 (1996).
3. R. Aelrod, The Complexity of Cooperation. (Basic Books, New York, 1997).
4. R. Aelrod and W. D. Hamilton, The evolution of cooperation. Science 211, 1390-1396 (1981).
5. D. Kraines and V. Kraines, Learning to cooperate with Pavlov: an adaptive strategy for the iterated Prisoner's Dilemma game. Theory Decision 35, 107-150 (1993).
6. M. A. Nowak and K. Sigmund, A strategy of win-stay, lose-shift that outperforms tit-for tat in the Prisoner's Dilemma game. Nature 364, 56-58 (1993).
7. K. Tainaka, Lattice model for the Lotka-Volterra system. J. Phys. Soc. Jpn. 57, 2588-2590 (1988).
8. Y. Itoh, On a ruin problem with interaction. Ann. Instit. Statst. Math. 25, 635-641 (1973).
9. Y. Itoh, Integrals of a Lotka-Volterra system of odd number of variables. Prog. Theor. Phys. 78, 507-510 (1987).
10. M. Bramson and D. Griffeath, Flux and fixation in cyclic particle systems, Ann. Probability, 17, 26-45 (1989).
11. L. Frachebourg, P. L. Krapivsky and E. Ben-Naim, Segregation in a one-dimensional of interacting species. Phys. Rev. Lett. 77, 2125-2128 (1996).
12. K. Tainaka, Stationary pattern of vortices or strings in biological systems: lattice version of the Lotka-Volterra model. Phys. Rev. Lett. 63, 2688-2691 (1989).
13. K. Tainaka, Topological phase transition in biological ecosystems. Europhys. Lett. 15, 399-404 (1991).
14. K. Tainaka and Y. Itoh, Apparent selforganized criticality. Phys. Lett. A 220 58-62 (1996).
15. K. Tainaka, Paradoxical effect in a 3-candidates voter model. Phys. Lett. A 176, 303-306 (1993).
16. K. Tainaka, Indirect effect in cyclic voter models. Phys. Lett. A 207 53-57 (1995).
17. T. E. Harris, Contact interaction on a lattice. Ann. Prob. 2, 969-988 (1974).
18. T. M. Liggett, Interacting Particle Systems. (Springer-Verlag, New York, 1985).
19. J. Marro and R. Dickman, Nonequilibrium Phase Transition in Lattice Models (Cambridge University Press, Cambridge, 1999).
20. N. Nakagiri and K. Tainaka, Indirect relation between species extinction and habitat destruction. To be published in Ecol. Model.
21. Y. Takeuchi, Global Dynamical Properties of Lotka-Volterra System (World Scientific, Singapore, 1996).
22. K. Tainaka and N. Araki, Press perturbation in lattice ecosystems: parity law and optimum strategy. J. Theor. Biol. 197, 1-13. (1999).
23. J. E. Satulovsky and T. Tome, Phys. Rev. E 49, 5073 (1994).
24. J. Hofbauer and K. Sigmund, The Theory of Evolution and Dynamical Systems (Cambridge University Press, Cambridge, 1988).
25. K. Tainaka, Intrinsic uncertainty in ecological catastrophe. J. Theor. Biol. 166, 91-99 (1994).
26. D. Stauffer, Introduction to Percolation Theory (Taylor & Francis, London, 1985).
27. M. Sahimi, Applications of Percolation Theory (Taylor & Francis, London, 1993).
28. K. Kobayashi and K. Tainaka, Critical phenomena in cyclic ecosystems: parity law and self-structuring extinction pattern. J. Phys. Soc. Jpn. 66, 38-41 (1997).
29. K. Tainaka and T. Sakata, Perturbation experiment and parity law in a cyclic ecosystem. J. Phys. Soc. Jpn. 68, 1055-1056 (1999)

Review: Computer Language Games

Michael L. Littman

AT&T Labs–Research
180 Park Ave. Room A275
Florham Park, NJ 07932-0971 USA
mlittman@research.att.com

Abstract. In language games, word meanings play a central role. This means that closed world assumptions that are so powerful in many games cannot be depended on. This paper describes implemented systems that play the language games of American-style crosswords, cryptic crosswords, Trivial Pursuit[TM], and punning riddles. These early efforts illustrate the challenge of language games, demonstrate some emerging ideas for tackling language games by computer, and indicate fruitful areas for future research.

Keywords: word meanings, word games, hangman, Boggle, crosswords, cryptics, Trivial Pursuit, puns.

1 Introduction

The rules of checkers and Rubik's cube[TM] are extremely simple. Nonetheless, the games are fun and challenging for people to play because the richness of the interactions of the rules mean an enormous number of configurations must be considered to find good moves. Powerful ideas such as alpha-beta search and pattern databases have been developed and applied to create computer programs that achieve human, and sometimes super-human, performance in these games [10,8].

Games like these satisfy the closed-world assumption—relationships not explicitly stated to hold do not hold [5]. To put it another way, if it's not written in the rules, it can't happen. Even though the games admit mathematically definable correct answers, judgment is often important to combat the computational intractability of searching for an optimal move. I define *language games* as games involving natural language in which word meanings play an important role. Because natural language can be used to describe the full range of human experiences, language games are inconsistent with the closed world assumption—no fixed set of rules will be sufficient to define game play. Thus, judgment is needed for language games not just for computational reasons, but even to interpret the rules themselves. This makes language games an exciting testbed for creating programs that display various types of judgment in decision making.

This paper describes several attempts to develop programs for playing language games. Section 3 describes American-style crosswords, Section 4 describes cryptic crosswords, Section 5 Trivial Pursuit[TM], and Section 6 punning riddles. For all these games, the programs designed to play them demonstrate general techniques for building

T.A. Marsland and I. Frank (Eds.): CG 2000, LNCS 2063, pp. 396–404, 2001.
© Springer-Verlag Berlin Heidelberg 2001

language-centered applications. Each includes databases of general linguistic information along with task-specific information, although they differ in the precise form these databases take.

To help clarify the notion of language games, the next section describes several games that involve words, but are not language games because the meanings of words are not important.

2 Word Games

Words are a popular feature of many games. Often, words are simply a large and familiar set of legal patterns. The fact that the words refer to things in the real world is not relevant to the game.

Scrabble^{TM} is an example of a word game. Here, players take turns placing letters in a grid to form words. Letters and grid positions have point values associated with them and a player's object is to score more points than the opponent. Maven, the top Scrabble^{TM}-playing program [12], plays a nearly perfect game.

In Boggle^{TM}, a full 5×5 grid of letters is randomly generated and players must construct as many words as possible from sequences of contiguous letters in the grid within a fixed time limit (typically 3 minutes). Points are scored based on finding words that other players did not find, with bonuses for finding long words. A Boggle^{TM} board can be considered "solved" if every legal word in it has been identified. Like "word find" puzzles, this task is easily handled by a computer. In fact, Boyan [4] attacked the problem of "Boggle Board Setup"—search through the space of boards to find one whose solution yields the highest possible score. Solving this optimization problem involves solving a hundred thousand individual Boggle^{TM} boards, each of which took his program approximately a millisecond.

Efforts such as these indicate that we have techniques that are very well suited to attacking pure word games. To illustrate how language games differ from word games, I'd like to contrast the games of hangman and Wheel of Fortune.

In hangman, a player must identify a hidden word. Initially, the player is told the number of letters in the word via a series of blanks, one for each letter. The player then guesses a letter of the alphabet. If that letter appears in the hidden word, it is written into the corresponding blank(s). If the letter does not appear in the word, it is added to the "missed letters" list. If the player discovers the identity of the word before missing six letters, the player wins.

As an example, imagine the hidden word is "indecipherable". After the player guesses 'a', 'e', 'i', 'o', 'u', 'r', 'l', 's', 't', and 'n', the board looks like:

word	missed letters
in_e_i__era_le	o u s t

It is not too hard to write a program that plays hangman at a super-human level. At each stage, the program can guess the letter that appears in the most legal words consistent with the current board. Justin Boyan and I tested this scheme on the 18,814 9-letter words in a version of the Scrabble^{TM}-player's dictionary. It solved all but 5 words with five or fewer misses and averaged 1.35 misses per game. The word "indecipherable" is uniquely identified in the list of 1,774 14-letter words with no misses after guessing only "i" and "a".

The Wheel of Fortune game is similar, with three players competing to be the first to solve a hangman-like puzzle. There are two significant differences in the puzzle, however, that suggest that Wheel of Fortune should be classified as a language game and hangman a word game. First, Wheel of Fortune answers are often names of people or short phrases (i.e., there is no official list of legal answers). Second, players are told the category of the answer, which provides a hint to the meaning of the answer phrase.

A simple word-list-based approach would fail for Wheel of Fortune because (a) the answer will likely not appear on any finite-size list, and (b) it would not take advantage of the information in the category hint. Dealing successfully with these issues is central to many language games, including several described in this paper.

First, a brief note about anagrams and cryptograms. In an anagram, the letters in a word or phrase (e.g., "computer language games") are rearranged to form a new word or phrase (e.g., "Let me gauge magna corpus" or "Gem puns gag ace emulator"). Creating nonsense anagrams ("mop crag league augments") and single-word anagrams (from "umsblje" to "jumbles") are word games. Generating phrase-level anagrams, like above the examples, whose meanings are suggestive of the phrase they are anagramming is a language game (see http://www.anagramgenius.com/server.html for Anagram Genius by William Tunstall-Pedoe).

In cryptograms, a short passage is encoded by replacing each letter consistently with some other letter. To solve a cryptogram, one must recover the original phrase (e.g., "Gtszyda scrrcj gipgryrirytd winnscg yg h eydm to shdaihac ahkc." to "Solving letter substitution puzzles is a kind of language game."). I would classify this as a language game because the meaning of the words matters. However, they can often be solved by ignoring this and simply making substitutions that maximize the number of valid words (see the Oneacross Cryptogram tool at http://oneacross.com/cryptograms/ by Noam Shazeer).

3   American-Style Crosswords

PROVERB [7] was designed and built as part of a seminar class at Duke University. It solves the language game of American-style crosswords, such as those that appear in The New York Times, TV Guide, or USA Today. Over a set of 370 such puzzles, PROVERB averaged 95% words correct.

A crossword puzzle is specified via two sets of clues (across and down) and a grid of interlocking slots into which the answers to the clues must be written. In American-style puzzles, no grid cells are "unchecked": each is part of both a down answer and an across answer. In part because of the constraints of the grid, puzzle designers are often quite liberal with what they consider to be a legal answer, especially for the "theme" clues that appear in many puzzles. This can also add to the fun for human solvers, as designers include novel puns and names from current events in their puzzles. One of my favorites is the clue "Simian singer" with answer "barbaramandrill", blending the name of the singer "Barbara Mandrell" with the ape "mandrill".

The architecture of PROVERB reflects both its group development and the eclectic nature of crossword clues. In particular, a set of 30 separate expert modules attack each of the puzzle's clues. These expert modules differ a great deal in their dependence on the exact form of the clues. For example, the "movie" module converts clues into queries against a detailed television-movie database. The "exact match" module looks for a clue in a large database of previously published crossword clues. Both of these modules take the form of the clue very literally. On the other hand, there is a set of modules that completely *ignore* their clues and simply return words or short phrases from huge word lists. The output of all the different modules is merged into a unified list in a way that assigns results from more precise modules higher confidence.

Including a collection of expert modules with varying degrees of precision helps make PROVERB robust in the face of the complexity of natural language. When possible, precise modules provide accurate answers. When accurate answers are not found, less precise modules provide a "backstop" that makes the system prefer answers that are known to be real words to those that are suspected to be nonsense. A sophisticated probability-based grid-filling algorithm [11] is then used to select a final answer for each clue based on a combination of how well it fits with its clue (prior) and how well it fits in the grid with answers from other clues (posterior).

Although PROVERB contains dozens of different kinds of databases (e.g., a list of colors, a list of famous authors, several thesauri, an encyclopedia, huge word lists), experiments indicated that the single most important database is its list of 350,000 previously published crossword clues with their answers. This database serves as a useful source of background knowledge on words and their associations. It is also application specific, as it can be used very directly to solve clues by exact match or simple transformations.

There are many other ways in which computers are used in the context of crossword puzzles. Creating a crossword puzzle given a grid and a word list is a standard testbed for constraint satisfaction and search methods [6]. Many professional crossword constructors use computers to help organize their favorite grids, clues, and answers, and PC programs such as "Crossword Compiler" (http://www.x-word.com/) are used to help generate high quality fills. I am not aware of any substantial work in machine generation of clues.

Many programs are available to help humans solve crossword puzzles by, for one, indexing letter patterns. For example, a program can quickly point out that the pattern "_n_g_a___" only fits with one word in the Scrabble^{TM}-player's dictionary ("enigmatic"). The web site www.oneacross.com goes a major step further by integrating this type of pattern matching with actual clue solving. It does this using a version of several of the expert modules from the PROVERB project.

4   Cryptic Crosswords

Cryptic, or UK-style, crosswords differ from their American counterparts in both cluing and grid styles. A cryptic clue often consists of two parts—the definition (similar to an American-style crossword clue), and a subsidiary indication. The subsidiary indication can be thought of as a set of instructions for creating a sequence of letters that is the answer to the definition. Because cryptic clues are usually self-reinforcing (definition and subsidiary indication must match, so a solver can have high confidence in the uniqueness of an answer), the grids in cryptic crosswords generally include "unchecked" grid cells.

Crossword Maestro (http://www.genius2000.com/cm.html) by William Tunstall-Pedoe is a commercial program that can solve cryptic crosswords. Because it is a commercial product, technical information about how it works is not available. However, behaviorally, it is clear that, like PROVERB, it uses confidence scores, has a rich database of answer words and a representation of their meanings, and has a sophisticated "back off" strategy.

As an example, here is the output produced by Crossword Maestro for a cryptic clue from a published puzzle:

Clue: A clue like this about Japanese religion (6)
Suggested answer: SHINTO
SHINTO **Confidence**: 100%
'japanese religion' is the **definition**. I am not sure about the 'japanese' bit but 'Shinto' can be an answer for 'religion'.
'a clue like this about' is the **subsidiary indication**. 'a clue' becomes 'hint', 'like this' becomes 'so', 'about' means one lot of letters goes inside another. 'hint' placed inside 'so' is 'shinto'.

There are several interesting things to note about this example. First, the program successfully breaks the clue into its definition part and its subsidiary indication part. Based on this split, it decides that "SHINTO" is the correct answer. Its database indicates that Shinto is a religion, but not, apparently, that it is a Japanese religion. However, its back-off strategy is to assume that a "Japanese religion" is a kind of religion, so it has a match. The lexicon appears to have a rich hierarchy of meanings, so back-offs can get more and more tenuous as the need arises. Categories I have observed include "manmade objects", "competing", "communicating", "bodily activities", "natural acts", "motion", "contact", and more generally "singular nouns", "verbs in base form", "adjectives", etc.

In the subsidiary indication, the program's thesaurus maps "a clue" to "hint" and "like this" to "so". Crossword Maestro also has a list of cryptic indicators that tells it that the word "about" can mean that one set of letters is placed inside another. Following the instructions, when "hint" is placed inside of "so", the string "shinto" is formed. It is clear that the program must have considered a huge number of possibilities before it found one that fit so perfectly.

The program's online advertising material claims that the system has a success rate of about 75% on individual clues. This seems right to me, at least in the context of solving an entire puzzle (i.e., using crossing information from other clues). The documentation also claims that Crossword Maestro can solve American-style clues. However, on a small random sample of New York Times puzzles, I found that it scored under 5% words correct. I think this is because the program is extremely strong in lexical knowledge (word meanings), but much weaker in handling the kinds of cultural references that are very common in American-style puzzles.

All in all, Crossword Maestro is one of the most sophisticated language-game-playing programs that has been written, successfully combining a detailed lexicon, puzzle-specific knowledge, search, and even natural language explanations into a single system.

5 Trivial Pursuit

Question answering is a problem domain that is receiving a great deal of attention by text retrieval and natural language processing researchers [13]. As in text retrieval, the idea is to find passages in a text database relevant to a query. But, instead of returning the passage containing the answer, the word or phrase representing the answer is extracted and presented to the user. So, for a question like "Who invented the compact disk?" a question-answering system returns "James Russell" instead of a link to

```
http://inventors.about.com/science/inventors/library/
                inventors/blcomputer_peripherals.htm.
```

Question answering is a key competence in many language games such as College Bowl, Who Wants to Be a Millionaire, Jeopardy^{TM} and Trivial Pursuit^{TM}. As a proof of concept, I created a system called *wigwam* to play Trivial Pursuit^{TM} (teepee).

The game of Trivial Pursuit^{TM} is played by two to six players on a wheel-shaped game board with seventy-three spaces on it. Each space is labeled with one of six colors representing six different categories (G: Geography, E: Entertainment, H: History, AL: Art & Literature, SN: Science & Nature, SL: Sports & Leisure) or "roll again." Players take turns rolling a die, moving their tokens around the board, and answering questions in the category of the space their tokens land on. The first time a player answers a question correctly in a category's designated "headquarters" space, the player collects a "wedge" for that category. Whenever a player answers a question wrong, the turn is forfeited to the next player. The first player to collect all six wedges, reach the center of the board, and answer a question correctly there wins the game.

A Trivial Pursuit^{TM} player consists of two basic components: question answering and move selection. In *wigwam*, I used AT&T's question answering system, AQUA [1], for question answering. Like most state-of-the-art broad-domain question answering systems, AQUA works by searching in a large text database for short passages that contain many of the same words as the question. It then extracts and labels each of the "entities" in the retrieved passages as being people, places, organizations, money, dates, durations, sizes or distances, quantities, or other names. AQUA determines what kind of entity the question is asking for ("Who" usually means "person", "How many" refers to quantities, etc.). Finally an entity of the correct type that appeared in a sufficient number of high-scoring passages is returned as the top answer.

AQUA also demonstrates a kind of graceful back off: If the question appears verbatim in the text collection with its answer, the corresponding answer is likely to be returned. If no exact match is found, the system will suggest answers that are associated with (nearby) words in the question. So, in response to "When was Amelia Earhart's last flight?", AQUA correctly returns "1937" first, but it also gives high weight to "1997", the year of a recreation of her flight and also the year in which a fictitious book about her won a literary award.

The move-selection component of *wigwam* was built with the help of Fan Jiang. Based on an estimate of the probability of answering questions correctly in each of the categories, the system builds a Markov decision process whose solution is the move-selection policy that results in the minimum expected number of turns until game completion.

Table 1. Estimated question-answering accuracy for *wigwam*, two human subjects, and the web database used by AQUA. Estimates are based on a fixed set of 10 questions. The starred (*) estimate is based on 40 questions. The last column indicates the expected number of turns a player would take to complete a game following an optimal move-selection strategy.

category	G	E	H	AL	SN	SL	est. turns
wigwam	0.1	0.3	0.1	0.3	0.2	0.025*	414.2
person A	0.2	0.3	0.2	0.6	0.7	0.6	47.8
person B	0.4	0.5	0.5	0.6	0.7	0.7	21.8
web data	0.9	0.9	0.9	0.9	0.7	0.4	8.2

To evaluate *wigwam*, I asked it ten questions from each of the six categories. Table 1 gives the fraction of correct answers given by the system. In the Sports & Leisure category, the system returned zero correct answers in the first ten, so I expanded the question set in this category to 20, 30, and then 40 questions before finally finding a question it could answer.

Based on these probabilities, I used the Markov decision process approach to compute the average number of turns the optimal move-selection strategy would take to complete a game. The result is given in the final column of Table 1. For comparison, the second and third rows of the table give comparable results for two human subjects; the second person was recommended to me as a trivia buff and is estimated to finish a game in a nineteenth the number of turns of *wigwam*.

AQUA can only answer questions that are covered in its large text database (approximately 100 gigabytes of downloaded web pages). To understand whether *wigwam*'s sub-par performance is due to a lack of coverage or an ineffectiveness at extracting information from its database, I searched through the database looking for passages or pairs of passages that support the correct answer to each of the sample questions. I found that 47 of the 60 answers could be justified based on the text in the database (with Sports & Leisure again being the weakest category). The resulting performance, given in the final row of Table 1, is probably close to that of expert human players. It is an open problem, however, to develop extraction software that could perform this well. We are currently exploring more precise representations to help with this problem.

6 Punning Riddles

Question: How do two periodic curves communicate?
Answer: Sine language!

This type of joke, which takes the form of a question and the humor is derived from a play on words, is called a punning riddle. Generating and answering punning riddles is a very widespread type of language play, although it is rarely played as an actual "game" (people generally don't keep score).

Binsted and Ritchie [2] studied computer-generation of punning riddles. Their program, JAPE-1, focused on jokes in which the punchline was a common noun phrase with a homonym substitution. Later developments included other types of jokes, a much larger lexicon, and even a Japanese version [3].

The algorithm employed in JAPE-1 carried out the following steps:

1. Start with a word or phrase A in a lexicon. Ex.: *sign language* is a form of communication between hearing-impaired people.
2. Find a shorter word or phrase B in the lexicon that is phonologically similar to part of A. Ex.: *sine* is a type of periodic curve.
3. Create a fake word or phrase C by substituting B into A using one of a number of schema. Ex.: *sine language* is form of communication between periodic curves.
4. Create a question for which C is the answer via one of a number of templates.
5. Check to make sure the result seems to be a "legal" joke.

JAPE-1 makes use of a number of databases to carry out these steps. The *lexicon* can be thought of as a list of words and noun phrases along with pronunciations and definitions. Note that this resource is not application specific, similar to the lexicon in Crossword Maestro. The humor-specific parts of JAPE-1 are a set of six *schemata* that give strategies for generating the meaning of a constructed noun phrase, a set of *templates* that give the surface form of the question from the meaning, and some *post-production checks* that make sure that the system didn't inadvertently create an existing noun phrase or accidentally use part of the answer in the question. Even these last few databases are arguably somewhat humor-independent, as defining novel noun phrases and making up questions are useful tasks beyond punning riddles.

In evaluations with human judges, JAPE-1 produces punning riddles that are of "comparable quality" to those published in joke books. Unfortunately, this is in part due to the fact that many of the judges didn't think the published jokes were funny either. Nonetheless, JAPE-1 successfully plays its language game.

7   Discussion and Conclusion

This paper defined language games as those in which word meanings are important. It described four computer programs that play language games. They combined large expandable task-independent language databases with smaller task-specific databases to address the breadth of natural language. They also included various "back-off" strategies to help contend with the open-ended nature of language.

The form of the natural language background knowledge differs in each of the systems. In *wigwam* and PROVERB, large collections of text are used to establish associations between words. This type of database is easily expandable (just add more text) and can be quite powerful. Landauer and Dumais [9] showed that the associations-through-text approach could be used to answer synonym questions from the Test of English as a Foreign Language (TOEFL) standardized exam at a passing level. Nevertheless, association information can be imprecise; knowing that iris and flower are related is less useful than knowing that an iris is a *type* of flower. Crossword Maestro and JAPE-1 use lexicons with more highly structured relationships, but these lexicons are difficult to expand and create and are often inconsistent in their coverage.

An important area for future work, therefore, is developing methods for automatically deriving lexicons with structured semantic representations (synonyms, antonyms, type-of and part-of relations, etc.). This type of database would be helpful in all the language games described in this paper and would have many other uses.

Games like chess and backgammon have helped advance the state of the art in artificial intelligence. However, if we intend to create programs that can collaborate with us on real-world problems, we must find techniques that allow the closed world assumption underlying these games to be relaxed. Language games are an excellent testbed for developing new ideas for tackling open-ended applications.

Acknowledgement

I received valuable help from Justin Boyan, Peter Stone, William Tunstall-Pedoe, David Reiter, Lisa Littman, Jennifer Davis, Giles Davis, Kim Binsted, Amit Singhal, Steve Abney, Satinder Singh, and Jonathan Schaeffer during the preparation of this paper.

References

1. Steven Abney, Michael Collins, and Amit Singhal. Answer extraction. In *Proceedings of the 6th Applied Natural Language Processing Conference*, pages 296–301, 2000.
2. Kim Binsted and Graeme Ritchie. An implemented model of punning riddles. In *Proceedings of the Twelfth National Conference on Artificial Intelligence (AAAI-94)*, 1994.
3. Kim Binsted and Osamu Takizawa. A model of Japanese puns and its implementation. In *Proceedings of the 4th conference of the Australian Cognitive Science Society*, 1998.
4. Justin Andrew Boyan. *Learning Evaluation Functions for Global Optimization*. PhD thesis, School of Computer Science, Carnegie Mellon University, 1998.
5. Paul R. Cohen and Edward A. Feigenbaum, editors. *The Handbook of Artificial Intelligence: Volume III*. William Kaufmann Inc., 1982.
6. M. L. Ginsberg, M. Frank, M. P. Halpin, and M. C. Torrance. Search lessons learned from crossword puzzles. In *Proceedings of the Eighth National Conference on Artificial Intelligence*, pages 210–215, 1990.
7. Greg A. Keim, Noam Shazeer, Michael L. Littman, Sushant Agarwal, Catherine M. Cheves, Joseph Fitzgerald, Jason Grosland, Fan Jiang, Shannon Pollard, and Karl Weinmeister. Proverb: The probabilistic cruciverbalist. In *Proceedings of the Sixteenth National Conference on Artificial Intelligence*, pages 710–717, 1999.
8. Richard E. Korf. Finding optimal solutions to Rubik's cube using pattern databases. In *Proceedings of the Fourteenth National Conference on Artificial Intelligence*, pages 700–705. AAAI Press/The MIT Press, 1997.
9. Thomas K. Landauer and Susan T. Dumais. A solution to Plato's problem: The latent semantic analysis theory of acquisition, induction and representation of knowledge. *Psychological Review*, 104(2):211–240, 1997.
10. Jonathan Schaeffer. *One Jump Ahead: Challenging Human Supremacy in Checkers*. Springer-Verlag, 1997.
11. Noam M. Shazeer, Michael L. Littman, and Greg A. Keim. Solving crossword puzzles as probabilistic constraint satisfaction. In *Proceedings of the Sixteenth National Conference on Artificial Intelligence*, pages 156–162, 1999.
12. Brian Sheppard. Computers and language games. *IEEE Intelligent Systems*, 14(6):15–16, November/December 1999. In Trends and Controversies, Playing with AI (Haym Hirsch, editor).
13. Ellen M. Voorhees and Dawn M. Tice. Building a question answering test collection. In *SIGIR 2000: Proceedings of the 23rd Annual International ACM SIGIR COnference on Research and Development in Information Retrieval*, pages 200–207. ACM Press, 2000.

Review: Computer Go 1984–2000

Martin Müller

Department of Computing Science
University of Alberta
Edmonton, Canada T6G 2E1
mmueller@cs.ualberta.ca

Abstract. Computer Go is maybe the biggest challenge faced by game programmers. Despite considerable work and much progress in solving specific technical problems, overall playing strength of Go programs lags far behind most other games. This review summarizes the development of computer Go in recent years and points out some areas for future research.

Keywords: Computer Go, Go programs

1 Introduction

The introduction briefly describes the rules of the game. Section 2 summarizes the history and current state of computer Go, and Section 3 contains three sample games to illustrate the progress made from 1988 until today. The final Section 4 poses some challenge problems for further research in the field.

1.1 The Game of Go

Go is played between two players Black and White, who alternatingly place a stone of their own color on an empty intersection on a Go board, with Black playing first. The standard board size is 19×19, but smaller sizes such as 9×9 and 13×13 are also used. The goal of the game is to control a larger area than the opponent. Figure 1 shows the opening phase of a typical game.

The capturing rule states that if stones of one color have been completely surrounded by the opponent, so that no adjacent empty point remains, they are removed from the board. Figure 2 shows two white stones with a single adjacent empty point (liberty) at 'a'. If Black plays there, the two white stones are captured and removed from the board. If White plays on the same point first, it will now require Black three moves at 'a', 'b' and 'c' to capture the three stones. Capturing and recapturing stones can potentially lead to the infinite repetition of positions. The *ko rule* forbids such a repetition. A basic ko is shown in Figure 3. After Black captures a single White stone, White cannot immediately take back at 'a' in the diagram on the right side. A large number of rule sets exist, but the variations between them rarely affect the outcome of games. The main differences occur in the evaluation of coexistence or *seki* positions and in the treatment of rare, complex cases of position repetition.

Players can pass at any time; consecutive passes end the game. Differences in playing strength can be balanced by a handicap system, which allows Black to place several

T.A. Marsland and I. Frank (Eds.): CG 2000, LNCS 2063, pp. 405–413, 2001.

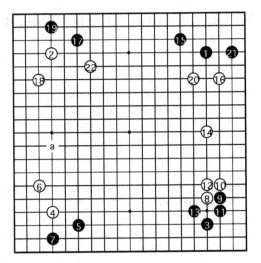

Fig. 1. The game of Go

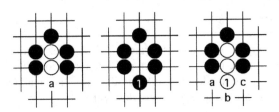

Fig. 2. The capturing rule

stones in a row at the start of the game. For detailed information about rules and many other aspects of Go, see [1].

2   The Development and Current State of Computer Go

Of all games of skill, Go is second only to chess in terms of research and programming efforts spent. Yet in playing strength, Go programs lag far behind their counterparts in almost any other game. While Go programs have advanced considerably in the last 10-15 years, they can still be beaten easily by human players of moderate skill. An exact ranking of Go programs is difficult, but Figure 4 shows the rough interval on the human ranking scale in which current programs can be found.

2.1   Overview Literature about Computer Go

The development of computer Go has been documented in a number of surveys. Wilcox [7] has written extensively about the early US-based Go programs in the seventies and eighties developed by Zobrist, Ryder, and by Reitman and Wilcox. Important early papers

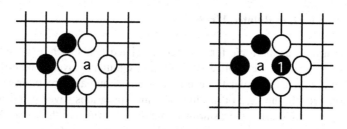

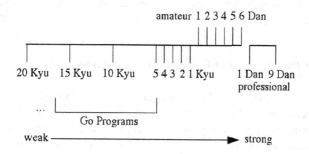

Fig. 3. Ko

Fig. 4. Go programs on the human ranking scale

on computer Go are contained in Levy's collection [5]. Kierulf's Ph.D. thesis [4] contains references for most of the programs that participated in the computer Go tournaments of 1985-1989. Erbach [3] gives a good overview of the state of the art in the early nineties. Burmeister and Wiles have published detailed descriptions and comparisons of several modern Go programs [2].

Information about computer Go programs, tournaments and literature is available from many interconnected web sites. Ph.D. theses about computer Go started appearing 30 years ago, and are recently published at a rate of about one per year. See my forthcoming detailed survey [6] and its web companion mentioned at the end of this report for an extensive bibliography and web links.

2.2   Specialized Research in Subtopics of Computer Go

The most tangible progress in the field of computer Go has not been in programs that play a complete game, but rather in programs that address one specific problem. For many subproblems of Go, specialized methods have been developed which achieve a much greater heuristic accuracy than general methods, or can even solve a subproblem precisely. Thomas Wolf's *GoTools* has reached the level of strong amateur players [8] in solving Life and Death puzzles in a small, completely enclosed region. Methods have been developed for proving the safety of stones and territory, fighting semeai and solving difficult endgame puzzles [6].

2.3   Go Research in Related Fields

Go has been used as the topic for research in related fields such as cognitive science and machine learning. In cognitive science studies, human players with different levels of Go skill are tested in order to develop models of human perception and problem-solving behavior. In the field of machine learning, programs have been developed that can pick up basic Go principles, starting from only the rules of the game. A promising approach is the integration of a priori knowledge from expert Go modules into neural networks. There are many books, papers and theses in the field of combinatorial game theory related to computer Go, especially in the areas of endgame and *ko* evaluation. Again, please see [6] or its online companion for further references on these topics.

2.4   Recent Development of Computer Go Programs

While Go programming started in the late sixties, it got a big boost in the mid eighties, with the appearance of affordable PC's on one hand, and of tournament sponsors such as the Ing foundation on the other hand. In early tournaments, Taiwanese programs such as *Dragon* were successful. From 1989-91, Mark Boon's *Goliath* dominated all tournaments, followed by Ken Chen's *Go Intellect* and Chen Zhixing's programs *Handtalk* and *Goemate*. In recent years, *Go4++* by Michael Reiss, David Fotland's *Many Faces of Go* and the controversial North Korean program *KCC Igo* have also won major tournaments. In total there are about 10 top class programs, including *Haruka, Wulu, FunGo, Star of Poland* and *Jimmy*. *Goemate* and *Go4++* seem to be slightly ahead of the rest. A step behind the top 10 is a set of about 30 medium-strength programs. An interesting recent phenomenon is the appearance of good open source programs such as the new *GnuGo*. The total size of the computer Go community can be estimated at about 200 programmers, and is growing steadily.

Several milestones have been reached in the short history of computer Go: In 1991, *Goliath* won a yearly playoff with three strong young human players, taking a handicap of 17 stones. *Handtalk* won the 15 and 13 stone matches in 1995, and the 11 stone match in 1997. Programs such as *Handtalk* and *Go4++* have achieved some success in even games against human players close to *dan* level strength. However, experienced human players can still beat all current programs on much more than 11 stones. *Handtalk* was successively awarded 5, 4 and 3 *kyu* diplomas by the Japanese Go Association *Nihon Kiin* after winning the 1995-97 FOST cups, and *KCC Igo* received a 2 kyu diploma in 1999.

In Japan, in recent years there has been an enormous increase in the number of Go software packages. There are more than two dozen Go-related titles on the market, with prices ranging from 5 − 100$.

At this time, the future of big, world championship caliber events is uncertain. The FOST cup has been cancelled for lack of funding this year, and the traditional Ing tournament will stop altogether. However, tournaments are likely to continue. Small-scale events continue to be held in Asia, Europe, North America and on the internet. The Computer Olympiad has recently been revived in the context of the Mind Sports Olympiad, and efforts are underway to organize a new large-scale tournament in Japan.

Table 1. Results of International Computer Go Tournaments. NP = number of participants, ICGC = International Computer Go Congress (Ing Cup), FOST = FOST Cup, CO = Computer Olympiad, MSO = Mind Sports Olympiad, ICOT = ICOT tournament, CGF = Computer Go Forum (CGF) Computer Go Tournament, SOP = Star of Poland.

Event	Location	NP	Winner	2nd place	3rd place
1985 ICGC	Taipei	?	Dragon, 3:0	?	?
1986 ICGC	Taipei	10	(Author: Du), 4:0	Dragon, 3:1	Nemesis, 3:1
1987 ICGC	Taipei	18	Friday, 4:0	Dragon, 3:1	Peanut, 3:1
1988 ICGC	Taipei	16	Codan, 4:0	Dragon, 3:1	Goliath, 3:1
1989 CO	London	10	Explorer, 8:1	Goliath, 7:2	SOP, 6:3
1989 ICGC	Taipei	14	Goliath, 4:0	Nemesis, 3:1	Go Intellect, 3:1
1990 CO	London	3	Go Intellect, 4:0	Explorer, 2:2	Go 4++, 0:4
1990 ICGC	Beijing	10	Goliath, 5:1	Go Intellect, 5:1	SOP, 4:2
1991 CO	Maastricht	?	Goliath	?	?
1991 ICGC	Singapore	15	Goliath, 6:0	Go Intellect, 5:1	Dragon, 4:2
1991 ICOT	Tokyo	8	Goliath, 5:0	Intellect, 4:1	SOP, 3:2
1992 CO	London	?	Go Intellect	?	?
1992 ICGC	Tokyo	10	Go Intellect, 5:1	Handtalk, 4:2	Goliath, 4:2
1993 ICGC	Chengdu	13	Handtalk, 6:0	SOP, 5:1	Go Intellect, 4:2
1994 ICGC	Taipei	9	Go Intellect, 5:1	Many Faces, 5:1	Handtalk, 5:1
1995 FOST	Tokyo	14	Handtalk, 7:0	Go 4++, 6:1	Many Faces, 5:2
1995 ICGC	Seoul	10	Handtalk, 5:0	Go 4++, 4:1	Go Intellect, 3:2
1996 FOST	Tokyo	19	Handtalk, 8:1	Go 4++, 7:2	Many Faces, 7:2
1996 ICGC	Guangzhou	12	Handtalk, 6:0	Go Intellect, 5:1	Stone, 4:2
1997 FOST	Nagoya	38	Handtalk, 9:1	Go 4++, 8:2	Go Intellect, 8:2
1997 ICGC	San Francisco	10	Handtalk, 8:1	Go 4++, 8:1	Go Intellect, 7:2
1998 FOST	Tokyo	38	Silver Igo, 6:0	Goemate, 4:2	Go 4++, 4:2
1998 ICGC	London	17	Many Faces, 6:1	Wulu, 6:1	Go 4++, 5:2
1999 CGF	Tsukuba	28	Go 4++, 8:1	Haruka, 7:2	Goemate, 7:2
1999 FOST	Tokyo	16	KCC Igo, 7:1	Go 4++, 7:1	Many Faces, 6:2
1999 ICGC	Shanghai	16	Go 4++, 6:0	Goemate, 5:1	KCC Igo, 4:2
2000 MSO	London	6	Goemate, 10:0	Go 4++, 8:2	Aya, 5:5

2.5 Computer Go Tournaments

This section lists computer Go tournaments in three separate tables. Table 1 contains major international tournaments, Table 2 North American championships, and Table 3 European championships. For a complete list of computer Go tournaments see the web page http://www.usgo.org/computer.

3 Computer Go 1988, 1994 and 2000: Three Sample Games

The following three games illustrate the performance of the top Go programs in 1988, 1994 and 2000. The first game, shown in Figure 5, was the final of the International Computer Go Congress (Ing Cup), played in Taipei, Taiwan on November 11, 1988. The komi was 8 points, as in all Ing-sponsored competitions. *Codan* by Kazuyoshi Hayashi playing White won by 7 points against Liu Dong-Yue's *Dragon*. Overall, both programs

Table 2. North American Go Tournament. G2 was an early version of Many Faces. In 1990, Goliath was not eligible for the title because it is not an American program. The best American programs in 1990 were Go Intellect, Nemesis and Many Faces. In 1994, third place was shared by Contender and RisciGo.

Year	Location	NP	Winner	2nd place	3rd place
1984	USENIX	4	Nemesis, 4:1	Goanna, 3:2	Ogo, 2:3
1985	USENIX	?	Og	?	?
1986	USENIX	?	Og	?	?
1987	USENIX	4	Golem, 5:1	Og, 3:3	Codan, 2:4
1988	USENIX	5	G2*	Goo	Goanna
1988	Berkeley	5	Many Faces, 3:0	Nemesis, 1:2	Infinity Go, 1:2
1989	New Brunswick	10	Go Intellect, 4:0	MicroGo 2, 3:1	Many Faces, 3:1
1990	Denver	7	Goliath*, 6:0	Go Intellect, 5:1	Nemesis, 4:2
1991	Rochester	6	Many Faces, 5:0	Go Intellect, 4:1	Stone, 3:2
1992	Salem	7	Many Faces, 5:1	Go Intellect, 5:1	Nemesis, 4:2
1993	South Hadley	7	Stone, 6:0	Go Intellect, 5:1	Prototype, 4:2
1994	Arlington	4	Go Intellect, 6:0	Many Faces, 4:2	2 programs*, 1:5
1995	Seattle	5	Many Faces, 3:1	Explorer, 3:1	Poka, 3:1
1996	Cleveland	5	Many Faces, 4:0	Explorer, 3:1	Poka, 2:2
1997	Lancaster	2	Many Faces, 2:0	TeamGo, 0:2	-
1998	Santa Fe	5	Many Faces, 4:0	Smart Go, 3:1	Explorer, 2:2
1999	San Francisco	4	Many Faces, 3:0	Gnu Go, 2:1	Smart Go, 1:2
2000	Denver	3	Many Faces, 2:0	Smart Go, 1:1	Poka, 0:2

Table 3. European Computer Go Championship. SOP = Star of Poland. In 1993, there was a three-way tie for second to fourth place between Modgo, TurboGo and Gogelaar.

Year	Location	NP	Winner	2nd place	3rd place
1987	Grenoble	7	SOP*, 6:0	Microgo 2, 5:1	Goliath, 4:2
1988	Hamburg	10	Goliath, 8:1	SOP, 7:2	Progo, 7:2
1989	Nis	6	Goliath, 5:0	Microgo 2, 4:1	SOP, 3:2
1990	Vienna	7	SOP, 6:0	Explorer, 5:1	Nemesis, 4:2
1991	Namur	7	Goliath, ?:?	SOP, ?:?	Progo, ?:?
1992	Canterbury	7	GO 4.3, 6:0	Modgo, 5:1	Progo, 4:2
1993	Prague	6	Progo, 5:0	3 programs*, 3:2	3 programs*, 3:2
1994	Maastricht	7	SOP, 6:0	Imago, 5:1	?
1995	Tuchola	6	SOP, 5:0	TurboGo, 3:2	Argus, 3:2
1997	Marseille	6	SOP, 5:0	GoAhead, 4:1	TurboGo, 3:2
1999	Podbanske	4	TurboGo, 3:0	The Turtle, 2:1	Alpha, 1:2
2000	Strausberg	5	GoAhead, 4:0	TS-Go, 3:1	TurboGo, 2:2

play a very solid, territory-oriented game. *Codan* loses a group on the top edge, but gives it up early enough to avoid disaster, and wins the game by making slightly more efficient moves on average. Much of the game consists of simple, boundary-settling local sequences of play.

Six years later, Ken Chen's *Go Intellect* won the same event on a tiebreak with five wins to one loss (see Table 1). The game shown in Figure 6 was played in Taipei, Taiwan on November 17, 1994 between the two other programs with five wins, Chen Zhixing's

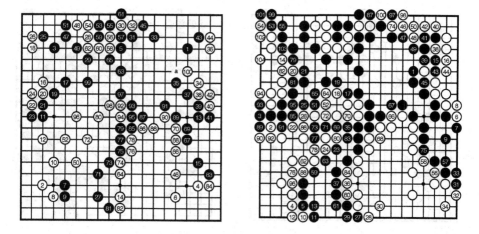

Fig. 5. ICGC 1988: Dragon (B) - Codan (W)

Handtalk and David Fotland's *Many Faces of Go*. This game is one of *Handtalk's* rare losses in the time from 1993 to 1997, when it dominated the computer Go scene. In contrast to the 1988 game, this game is characterized by intense fighting, with the focus on the attack and defense of weak groups. The ability to cut and connect weak groups, and the amount of knowledge about Life and Death, play the dominant roles. In this game, *Handtalk* makes two decisive group-related mistakes: move 99 loses the black group at the bottom. It should be played below 96, after which Black could either capture the stones 94 and 96 by playing to the left of 96, or capture the other cutting stone 98 by a move at 100. The second mistake is 141 (41 in the diagram on the right side), which lets White lead out the previously dead group in the center. After these two reversals, White has a big lead. Finally, White blunders at 158, which should be at 160 immediately, but Black misses the chance to turn the corner into *ko* and lets White repair the damage one move later.

As an example of the current state of the art, Figure 7 shows the deciding game of the computer Go tournament at the Mind Sports Olympiad, played in London on August 22, 2000. Playing Black and giving a *komi* of 6.5 points, *Go4++* by Michael Reiss lost by just half a point to *Goemate*, developed by Chen Zhixing as the successor program of *Handtalk*. This game develops in a tight territorial fashion typical of most current top programs, with little fighting going on. Up to 26, standard opening sequences are played out to occupy all corners and sides. The moves from 27 to 33 are also a standard sequence. After that, both programs continue to surround territory, with Black simply giving up stones such as 25 and 35/47 rather than risking a big fight by running out. While the play of both programs in this game is rather simple and safe, their overall performance is very respectable. There may still a large number of less-than-optimal moves, but there are few really big mistakes. Both programs demonstrate an understanding of many aspects of Go. For example, they can build safe territory as well as large frameworks, and can react early to reduce an opponent's sphere of influence. Programs are careful to avoid getting weak groups, and play a reasonable endgame.

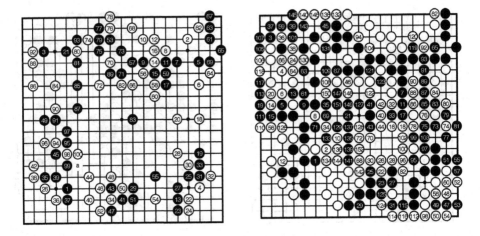

Fig. 6. ICGC 1994: Handtalk (B) - Many Faces of Go (W)

Comparing these three games, computer Go seems to have come full circle. Early programs knew little more than simple rules and patterns to surround territory and capture stones. The next generation, lead by *Goliath*, *Go Intellect* and *Handtalk*, dominated their opponents through a better knowledge of attack and defense. A typical game from this period was decided by a large margin, with the stronger program saving more of its own groups and killing more opponent groups. *Go4++* lead the revolution leading to the current state, by demonstrating that a program which is not so strong in fighting but very efficient in taking territory can win a large percentage of games, even when giving up a few small groups along the way. In recent years, these two extreme approaches have converged to the point where it is hard to distinguish between playing styles. Current programs are stronger than fifteen years ago in judging group strength, life and death, and tactics, but most prefer to play a peaceful game where these strengths are not so apparent. Their style of play hides much of the inherent complexity of Go.

4   Challenges for Computer Go Research

Develop a search-bound Go program In contrast to most other games, in Go there has not yet been a clear demonstration of correlation between deeper search and playing strength. Develop such a Go program that can automatically take advantage of greater processing power.

Comprehensive local analysis Develop a search architecture that can integrate all aspects of local fighting and evaluation.

Threats and forcing moves Develop a system that can systematically detect threats and use them for double threats, ko threats, or for forcing moves, while avoiding bad forcing moves, which have unexpected side effects.

Test suite Develop a comprehensive public domain computer Go test suite.

Computer Go source code library Provide a highly (re-)usable library of common functions.

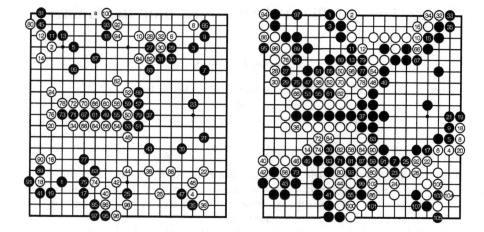

Fig. 7. Mind Sports Olympiad 2000: Go4++ (B) - Goemate (W)

Sure-win program for high handicaps Build a program that can demonstrably win all games on a high handicap, then successively reduce the handicap.

Integrate exact modules in heuristic program Solve the problems of interfacing specialized modules, which handle one aspect of the game very well, with a general Go program.

Solve Go on small boards Human players have analyzed Go on many small rectangular boards, but there are few exact proofs. Solve Go on small board sizes such as 5×5 or 7×7.

Acknowledgement and Web Reference

Parts of this review are based on a forthcoming detailed survey paper *Computer Go* [6], to appear in a special issue of Artificial Intelligence Journal on the state of the art in computer game-playing. The web page http://www.cs.ualberta.ca/~mmueller/cgo/survey contains further information.

References

1. R. Bozulich. *The Go Players Almanac*. Ishi Press, Tokyo, 1992.
2. J. Burmeister and J. Wiles. AI techniques used in computer Go. In *Fourth Conference of the Australasian Cognitive Science Society*, Newcastle, 1997.
3. D. W. Erbach. Computers and Go. In Richard Bozulich, editor, *The Go Player's Almanac*. The Ishi Press, 1992. Chapter 11.
4. A. Kierulf. *Smart Game Board: a Workbench for Game-Playing Programs, with Go and Othello as Case Studies*. PhD thesis, ETH Zürich, 1990.
5. D. Levy, editor. *Computer Games I+II*. Springer Verlag, 1988.
6. M. Müller. Computer Go, 2000. Artificial Intelligence, 2001. To appear.
7. B. Wilcox. Computer Go. In D.N.L. Levy, editor, *Computer Games*, volume 2, pages 94–135. Springer-Verlag, 1988.
8. T. Wolf. Forward pruning and other heuristic search techniques in tsume go. *Information Sciences*, 122:59–76, 2000.

Review: Intelligent Agents for Computer Games

Alexander Nareyek

GMD FIRST, Kekuléstr. 7, D-12489 Berlin, Germany
alex@ai-center.com
http://www.ai-center.com/home/alex/

Abstract. In modern computer games – like action, adventure, role-playing, strategy, simulation and sports games – artificial intelligence (AI) techniques play an important role. However, the requirements of such games are very different from those of the games normally studied in AI.

This article discusses which approaches and fields of research are relevant to achieve a sophisticated goal-directed behavior for modern computer games' non-player characters. It also presents a classification of approaches for autonomous agents and gives an overview of a solution developed in the EXCALIBUR project.

Keywords: commercial games, action planning, real time, dynamics, incomplete knowledge, resources.

1 Modern Computer Games

The use of game applications has a long tradition in artificial intelligence. Games provide high variability and scalability for problem definitions, are processed in a restricted domain and the results are generally easy to evaluate. But there is also a great deal of interest on the commercial side, the "AI inside" feature being a high-priority task [6,7,26] in the fast-growing, multi-billion-dollar electronic gaming market (the revenue from PC games software alone is already as big as that of box office movies [15]).

Many "traditional" games, such as card/board/puzzle games like Go-Moku [1] and the Nine Men's Morris [12], have recently been solved by AI techniques. Deep Blue's victory over Kasparov was another milestone event here. However, it is highly questionable whether and to what extend the techniques used in this field of research can be applied to today's "modern" computer games. Of the techniques used in this field, A* [14] (an improved version of Dijkstra's shortest-path algorithm [8]) and its variants/extensions are practically the only ones employed in modern computer games (see also [28]).

Such games pose problems for AI that are infinitely more complex than those of traditional games. Modern computer games feature:

- Real Time: There is only very limited time for reasoning.
- Dynamics: Computer games provide a highly dynamic environment.
- Incomplete Knowledge: A game character has only incomplete knowledge of the world.
- Resources: The game character's/world's resources may be restricted.

T.A. Marsland and I. Frank (Eds.): CG 2000, LNCS 2063, pp. 414–422, 2001.

Techniques from the AI fields of autonomous agents, planning, scheduling, robotics and learning would appear to be much more important than those from traditional games.

AI techniques can be applied to a variety of tasks in modern computer games. A game that uses probabilistic networks to predict the player's next move in order to speed up graphics may be on a high AI level. But although AI must not always be personified, the notion of artificial intelligence in computer games is primarily related to characters. These characters can be seen as *agents*, their properties perfectly fitting the AI agent concept.

But how does the player of a computer game perceive the intelligence of a game agent/character? This question is dealt with neatly in [16]. Important dimensions include physical characteristics, language cues, behaviors and social skills. Physical characteristics like attractiveness are more a matter for psychologists and visual artists (e.g., see [11]). Language skills are not normally needed by game agents and are ignored here too.

The most important question when judging an agent's intelligence is the goal-directed component, which we look at in the rest of this paper. The standard procedure followed in modern computer games to implement a goal-directed behavior is to use predetermined behavior patterns. This is normally done using simple if-then rules. In more sophisticated approaches using neural networks, behavior becomes adaptive, but the purely reactive property has still not been overcome.

Many computer games circumvent the problem of applying sophisticated AI techniques by allowing computer-guided agents to cheat. But the credibility of an environment featuring cheating agents is very hard to ensure, given the constant growth of the complexity and variability in computer-game environments. Consider a situation in which a player destroys a communication vehicle in an enemy convoy in order to stop the enemy communicating with its headquarters. If the game cheats in order to avoid a realistic simulation of the characters' behavior, directly accessing the game's internal map information, the enemy's headquarters may nonetheless be aware of the player's subsequent attack on the convoy.

2   A Classification of Autonomous Agents

Wooldridge and Jennings [29] provide a useful starting point by defining autonomy, social ability, re-activity and proactiveness as essential properties of an agent. Agent research is a wide area covering a variety of topics. These include:

- *Distributed Problem Solving (DPS)*
  The agent concept can be used to simplify the solution of large problems by distributing them to a number of collaborating problem-solving units. DPS is not considered here because computer games' agents should normally act fully autonomously: Each agent has individual goals.
- *Multi-Agent System (MAS)*
  MAS research deals with appropriate ways of organizing agents. These include general organizational concepts, the distribution of management tasks, dynamic organizational changes like team formation and underlying communication mechanisms.

Fig. 1. Interaction with the Environment

- *Autonomous Agents*

    Research on autonomous agents is primarily concerned with the realization of a single agent. This includes topics like sensing, models of emotion, motivation, personality, and action selection and planning. This field is our main focus here.

    An agent has goals (stay alive, catch player's avatar, ...), can sense certain properties of its environment (see objects, hear noises, ...), and can execute specific actions (walk northward, eat apple, ...). There are some special senses and actions dedicated to communicating with other agents.

    The following sections classify different agent architectures according to their trade-off between computation time and the realization of sophisticated goal-directed behavior.

2.1   Reactive Agents

Reactive agents work in a hard-wired stimulus-response manner. Systems like Joseph Weizenbaum's Eliza [27] and Agre and Chapman's Pengi [2] are examples of this kind of approach. For certain sensor information, a specific action is executed. This can be implemented by simple if-then rules.

The agent's goals are only implicitly represented by the rules, and it is hard to ensure the desired behavior. Each and every situation must be considered in advance. For example, a situation in which a helicopter is to follow another helicopter can be realized by corresponding rules. One of the rules might look like this:

```
IF (leading_helicopter == left) THEN
   turn_left
ENDIF
```

But if the programmer fails to foresee all possible events, he may forget an additional rule designed to stop the pursuit if the leading helicopter crashes. Reactive systems in more complex environments often contain hundreds of rules, which makes it very costly to encode these systems and keep track of their behavior.

The nice thing about reactive agents is their ability to react very fast. But their reactive nature deprives them of the possibility of longer-term reasoning. The agent is doomed if a mere sequence of actions can cause a desired effect and one of the actions is different from what would normally be executed in the corresponding situation.

2.2 Triggering Agents

Triggering agents introduce internal states. Past information can thus be utilized by the rules, and sequences of actions can be executed to attain longer-term goals. A possible rule might look like this:

```
IF (distribution_mode) AND (leading_helicopter == left)
THEN
   turn_right
   trigger_acceleration_mode
ENDIF
```

Popular Alife agent systems like CyberLife's Creatures [13], P.F. Magic's Virtual Petz [25] and Brooks' subsumption architecture [5] are examples of this category. Indeed, nearly all of today's computer games apply this approach, using finite state machines to implement it.

These agents can react as fast as reactive agents and also have the ability to attain longer-term goals. But they are still based on hard-wired rules and cannot react appropriately to situations that were not foreseen by the programmers or have not been previously learned by the agents (e.g., by neural networks).

2.3 Deliberative Agents

Deliberative agents constitute a fundamentally different approach. The goals and a world model containing information about the application requirements and consequences of actions are represented explicitly. An internal refinement-based planning system uses the world model's information to build a plan that achieves the agent's goals. Planning systems are often identified with the agents themselves. A great deal of research has been done on planning, and a wide range of planning systems have been developed, e.g., STRIPS [9], UCPOP [24], Graphplan [3] and SATPLAN [17,18].

The basic planning problem is given by an initial world description, a partial description of the goal world, and a set of actions/operators that map a partial world description to another partial world description. A solution is a sequence of actions leading from the initial world description to the goal world description and is called a *plan*. The problem can be enriched by including further aspects, like temporal or uncertainty issues, or by requiring the optimization of certain properties[1].

Deliberative agents have no problem attaining longer-term goals. Also, the encoding of all the special rules can be dispensed with because the planning system can establish goal-directed action plans on its own. When an agent is called to execute its next action, it applies an internal planning system:

```
IF (current_plan_is_not_applicable_anymore) THEN
   recompute_plan
ENDIF
execute_plan's_next_action
```

---

[1] Note that the use of the term *planning* in AI is different from that expected by people in the operations research (OR) community (e.g., *scheduling*).

Even unforeseen situations can be handled in an appropriate manner, general reasoning methods being applied. The problem with deliberative agents is their lack of speed. Every time the situation is different from that anticipated by the agent's planning process, the plan must be recomputed. Computing plans can be very time-consuming, and considering real-time requirements in a complex environment is mostly out of the question.

2.4 Hybrid Agents

Hybrid agents such as the 3T robot architecture [4], the New Millennium Remote Agent [23] or the characters by Funge et al. [10] apply a traditional off-line deliberative planner for higher-level planning and leave decisions about minor refinement alternatives of single plan steps to a reactive component.

```
IF (current_plan-step_refinement_is_not_applicable_anymore)
THEN

    WHILE (no_plan-step_refinement_is_possible) DO
      recompute_high-level_plan
    ENDWHILE
    use_hard-wired_rules_for_plan-step_refinement

ENDIF
execute_plan-step_refinement's_next_action
```

There is a clear boundary between higher-level planning and hard-wired reaction, the latter being fast while the former is still computed off-line. For complex and fast-changing environments like computer games, this approach is not appropriate because the off-line planning is still too slow and would – given enough computation time – come up with plans for situations that have already changed.

2.5 Anytime Agents

What we need is a continuous transition from reaction to planning. No matter how much the agent has already computed, there must always be a plan available. This can be achieved by improving the plan iteratively. When an agent is called to execute its next action, it improves its current plan until its computation time limit is reached and then executes the action:

```
WHILE (computation_time_available) DO
  improve_current_plan
ENDWHILE
execute_plan's_next_action
```

For short-term computation horizons, only very primitive plans (reactions) are available, longer computation times being used to improve and optimize the agent's plan. The more time is available for the agent's computations, the more intelligent the behavior will become. Furthermore, the iterative improvement enables the planning process to easily adapt the plan to changed or unexpected situations. This class of agents is very important for computer games applications. An architecture for anytime agents was realized within the EXCALIBUR project, which is described in Section 4.

3   Other Requirements

The previous sections pointed to the need for *anytime planning* capable of *easily adapting to a changing environment*. However, these are only two aspects relating to the planning of an agent's behavior. The expressiveness of planning is also very important:

- **Temporal planning:** The planning mechanisms must take a rich temporal annotation into account. Actions have a duration, the game world's agents can act simultaneously, and actions may have synergetic effects.
- **Resources:** Resources such as food, hit points or magical power are standard features of today's computer games. Computer-guided characters often have goals that are related to these resources and must take this into account in planning their behavior.
- **Incomplete knowledge:** An agent can only know about the things that it has sensed via its sensors. The type and number of the world's objects is not known in advance and can change without any action on the part of the agent.
- **Domain-specific guidance:** The ability to make use of domain-specific knowledge often makes the difference between a planning's quick success and miserable failure. It is critically important that this knowledge can be incorporated in a simple and appropriate manner.

There are many planning approaches that address one of these problems, but there are virtually none (except the EXCALIBUR agent's planning system) that try to address all of them.

4   The EXCALIBUR Project

To illustrate how computer games can implement anytime agents that also meet the requirements formulated in the previous section, we give a brief overview of the approaches applied in the EXCALIBUR project. In this project, a generic architecture was developed for autonomously operating agents, like computer-guided characters/mobiles/items, that can act in a complex computer-game environment.

The planning of an agent's behavior is realized in a constraint programming (CP) framework, which makes the approach highly declarative. Furthermore, CP methods have proved very successful in handling problems that involve resources and scheduling and are thus highly suitable as an underlying paradigm for the agents. Problems are formulated in a framework of variables, domains and constraints. The constraint satisfaction problem (CSP) consists of

- a set of variables $x = \{x_1, \ldots, x_n\}$
- where each variable is associated with a domain $d_1, \ldots, d_n$
- and a set of constraints $c = \{c_1, \ldots, c_m\}$ over these variables.

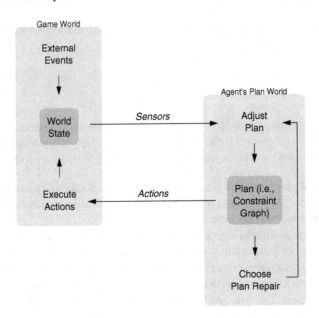

Fig. 2. Interleaving Sensing, Planning and Execution

The domains can be symbols as well as numbers, continuous or discrete (e.g., "door", "13", "6.5"). Constraints are relations between variables (e.g., "x_a is a friend of x_b", "$x_a < x_b \times x_c$") that restrict the possible value assignments. Constraint satisfaction is the search for a variable assignment that satisfies the given constraints. Constraint optimization requires an additional function that assigns a quality value to a solution and tries to find a solution that maximizes this value.

However, if the CP framework is used to compute an agent's behavior, several extensions are needed to achieve the required efficiency and expressiveness. These include structural constraint satisfaction [19,20], which enhances conventional constraint satisfaction by allowing problems to be formulated in which the constraints and variables must not be given in advance. The search for the correct structure of the CSP is part of the satisfaction process here. This is necessary for the task of generating an agent's behavior plan because multiple CSP structures are possible, the number/type of actions for the plan being unknown a priori.

Another important extension is the combination of constraint programming and local search [22]. This combination is a key feature, providing the agent with efficient anytime reasoning and allowing the uncomplicated handling of the environment's dynamics. Local-search approaches perform a search by iteratively improving an initial state/plan. In the applied approach, domain-dependent knowledge can be encapsulated in the constraints, which choose advantageous changes to the current plan by themselves. Figure 2 shows how sensing, planning and execution are interleaved.

The resource focus is reflected in the architecture's planning model [21] and allows us to use and optimize temporal, spatial and all other kinds of resources. Furthermore, the model is expressive enough to handle incomplete knowledge and information gathering.

The system was tested on several sample problems and yielded good results for specialized domain-dependent solutions as well as for domain-independent tasks. Detailed information on the EXCALIBUR project is available at:
`http://www.ai-center.com/projects/excalibur/`

5 Conclusion

Given the fast-growing complexity of modern computer games, approaches used for traditional games can no longer be directly applied to the modern games. Techniques from fields like autonomous agents and planning would appear to be more appropriate. However, even these fields have to be extended by mechanisms designed to satisfy the sophisticated requirements of modern computer games – such as real time, dynamics, resources and incomplete knowledge. The EXCALIBUR agents illustrate how these requirements can be tackled.

To conclude, I wish to draw attention to the setting up of an Artificial Intelligence Special Interest Group within the International Game Developers Association. Technology transfer from academia is one of the main issues, and anyone interested is invited to join the SIG and present/discuss their work. More information is available at:
`http://www.igda.org/SIGs/game_ai.htm`

Acknowledgments

The EXCALIBUR project is supported by the German Research Foundation (DFG), Conitec Datensysteme GmbH, Cross Platform Research Germany (CPR) and NICOSIO. Thanks to Reijer Grimbergen for his suggestions on how to improve this paper.

References

1. L. V. Allis, H. J. van den Herik, and M. P. H. Huntjens. Go-Moku Solved by New Search Techniques. In *Proceedings of the 1993 AAAI Fall Symposium on Games: Planning and Learning*, 1993.
2. P. Agre and D. Chapman. PENGI: An Implementation of a Theory of Activity. In *Proceedings of the Sixth National Conference on Artificial Intelligence* (AAAI-87), 268–272, 1987.
3. A. L. Blum and M. L. Furst. Fast Planning Through Planning Graph Analysis. *Artificial Intelligence* 90: 281–300, 1997.
4. R. P. Bonasso, R. J. Firby, E. Gat, D. Kortenkamp, D. P. Miller, M. G. Slack. Experiences with an Architecture for Intelligent, Reactive Agents. *Journal of Experimental and Theoretical Artificial Intelligence* 9(1), 1997.
5. R. A. Brooks. A Robust Layered Control System for a Mobile Robot. *IEEE Journal of Robotics and Automation RA-2* (1): 14–23, 1986.
6. C. Charla. Mind Games: the Rise and Rise of Artificial Intelligence. *Next Generation* November, 1995.
7. D. Coco. Creating Intelligent Creatures. *Computer Graphics World*, July, 1997.
8. E. W. Dijkstra. A note on two problems in connexion with graphs. *Numerische Mathematik*, 1: 269–271, 1959.
9. R. E. Fikes and N. Nilsson. STRIPS: A New Approach to the Application of Theorem Proving to Problem Solving. *Artificial Intelligence* 5(2): 189–208, 1971.

10. J. Funge, X. Tu, and D. Terzopoulos. Cognitive Modeling: Knowledge, Reasoning and Planning for Intelligent Characters. In *Proceedings of the International Conference on Computer Graphics and Interactive Techniques* (SIGGRAPH'99), 29–38, 1999.

11. T. Gard. Building Character. *Gamasutra*,
http://www.gamasutra.com/features/20000720/gard_01.htm,
June, 2000.

12. R. Gasser. Solving Nine Men's Morris. *Computational Intelligence* 12(1): 24–41, 1996.

13. S. Grand, D. Cliff, and A. Malhotra. Creatures: Artificial Life Autonomous Software Agents for Home Entertainment. In Proceedings of the First International Conference on Autonomous Agents (Agents'97), 22–29, 1997.

14. P. E. Hart, N. J. Nilsson, and B. Raphael. A formal basis for the heuristic determination of minimum cost paths. *IEEE Transactions on Systems Science and Cybernetics* 4(2): 100–107, 1968.

15. K. Hause. What to Play Next: Gaming Forecast, 1999-2003. Report #W21056, International Data Corporation, Framingham, Massachusetts, 1999.

16. K. Isbister. Perceived Intelligence and the Design of Computer Characters. *Lifelike Computer Characters* (LCC'95), Snowbird, Utah, 1995.

17. H. Kautz and B. Selman. Planning as Satisfiability. In *Proceedings of the Tenth European Conference on Artificial Intelligence* (ECAI-92), 359–363, 1992.

18. H. Kautz and B. Selman. Pushing the Envelope: Planning, Propositional Logic, and Stochastic Search. In *Proceedings of the Thirteenth National Conference on Artificial Intelligence* (AAAI-96), 1194–1201, 1996.

19. A. Nareyek. Structural Constraint Satisfaction. In *Papers from the 1999 AAAI Workshop on Configuration*, Technical Report, WS-99-05, 76–82. AAAI Press, Menlo Park, California, 1999.

20. A. Nareyek. Applying Local Search to Structural Constraint Satisfaction. In *Proceedings of the IJCAI-99 Workshop on Intelligent Workflow and Process Management: The New Frontier for AI in Business*, 1999.

21. A. Nareyek. Open World Planning as SCSP. In *Papers from the AAAI-2000 Workshop on Constraints and AI Planning*, Technical Report, WS-00-02, 35–46. AAAI Press, Menlo Park, California, 2000.

22. A. Nareyek. Using Global Constraints for Local Search. In E. C. Freuder and R. J. Wallace (eds.), *Constraint Programming and Large Scale Discrete Optimization*, American Mathematical Society Publications, DIMACS Volume 57, 9–28, 2001

23. B. Pell, D. E. Bernard, S. A. Chien, E. Gat, N. Muscettola, P. P. Nayak, M. D. Wagner, and B. C. Williams. A Remote Agent Prototype for Spacecraft Autonomy. In *Proceedings of the SPIE Conference on Optical Science, Engineering, and Instrumentation*, 1996.

24. J. S. Penberthy and D. S. Weld. UCPOP: A Sound, Complete, Partial Order Planner for ADL. In *Proceedings of the Third International Conference on Principles of Knowledge Representation and Reasoning* (KR'92), 102–114, 1992.

25. A. Stern, A. Frank, and B. Resner. Virtual Petz: A Hybrid Approach to Creating Autonomous Lifelike Dogz and Catz. In *Proceedings of the Second International Conference on Autonomous Agents* (AGENTS98), 334–335, 1998.

26. A. Stern. AI Beyond Computer Games. AAAI Symposium on Computer Games and Artificial Intelligence, 1999.

27. J. Weizenbaum. ELIZA – A Computer Program for the Study of Natural Language Communication between Man and Machine. *Communications of the ACM* 9(1): 36–45, 1966.

28. S. Woodcock. Game AI: The State of the Industry. *Game Developer*, August, 2000.

29. M. Wooldridge, and N. R. Jennings. Intelligent Agents: Theory and Practice. *The Knowledge Engineering Review* 10(2): 115–152, 1995.

Review: RoboCup through 2000

Itsuki Noda and Ian Frank

Electrotechnical Laboratory, Tsukuba Ibaraki 305, Japan
noda@etl.go.jp, ianf@fun.ac.jp

Abstract. Can a team of robots beat the human world champions at soccer? That is the 50-year grand challenge at the heart of the Robotic Soccer World Cup (RoboCup) initiative. Every year, researchers from around the world gather at the RoboCup tournaments to test their teams of software and hardware soccer players against each other. We report here on the first three of these tournaments, which were held in 1997 (Nagoya), 1998 (Paris) and 1999 (Stockholm). We summarise the game results, the practical and scientific lessons learned, and the progress towards that grand challenge goal.

Keywords: soccer, multi-agent systems, learning, opponent-modelling.

1 Introduction: What Is RoboCup?

A robotic soccer team capable of beating the human world champions may seem like a pipe dream. But this is the stated goal of the Robot Soccer World Cup (RoboCup) Games and Conferences. Since the early 1990's, researchers have been working towards establishing soccer as an effective multi-disciplinary testbed for evaluating competing theories, algorithms, and architectures for robotics, learning and multi-agent systems. Most prominently, this effort has led to the establishment of an annual World Cup of soccer for computers [5].

The Robot Soccer World Cups began in 1997, when RoboCup-97 was staged in conjunction with IJCAI-97 (the International Joint Conference on Artificial Intelligence), in Nagoya, Japan. In this inaugural event, 39 teams participated in three leagues (two robotic leagues and one simulation league), and a parallel workshop featured over 20 paper presentations. One year later, RoboCup-98 was held in Paris at the same time as ICMAS-98 (the International Conference on Multi-agent Systems) and the France98 FIFA Soccer World Cup. Most recently, RoboCup-99 was staged in conjunction with IJCAI-99 in Stockholm, Sweden. By 1999, the tournament had grown to include 90 teams, and also included a new "legged robot league" based on Sony's AIBO pet dog robots.

Let us give a brief overview of the basic physical conditions of each league:

- **Simulation League**. The simulation is realised by the *Soccer Server* [11,10], see http://ci.etl.go.jp/~noda/soccer/server.html), which models a virtual soccer field populated by two teams of 11 players. In a game, the actions of each player are controlled by separate "client" programs that connect independently to the Soccer Server. The server carries out the commands issued by the clients

T.A. Marsland and I. Frank (Eds.): CG 2000, LNCS 2063, pp. 423–432, 2001.

Fig. 1. View of the middle-sized robot league final at RoboCup-99

and monitors the game for fouls, free kicks, *etc.* It also sends the clients relevant sensor information such as stamina, vision, and auditory information. Vision and communication are limited by the physical constraints of the game (*e.g.*, players cannot see behind themselves, and cannot hear their teammates' shouts at long distances).

- **Small Robot league.** The real robot games are divided into three leagues according to the physical properties of the robots. In the smallest league the footprint of the robots must be contained in a circle of 15cm diameter. Teams may have up to five robots, and play is carried out on a pitch the size of a table-tennis table. A camera above the pitch provides global vision that any team can make use of to guide the robots with remote computers. The ball in the small robot league is a golf ball.
- **Middle-sized Robot League.** Again, the teams may have up to five robots, but their size can now be up to 50cm in diameter. The playing field is the size of a 3 × 3 arrangement of table-tennis tables. Global vision is prohibited, so each robot must sense the field status using local sensors (such as camera, infrared, or sonar). Remote computers and built-in computers can be used to control robots. The ball is a regulation size-4 ball. (See Figure 1).
- **Legged Robot League.** This uses a prototype of Sony's AIBO robots that can be programmed in an object-oriented programming language. The AIBO must use only their on-board camera and sensors to interpret the game, and play in teams of four.

Soccer has a number of unique features that distinguish it from most of the other games featured in these proceedings. First, it is a team game calling for the ability to model opponents, to work constructively with team-mates, and often also for the ability to explain the play at a high level (*e.g.*, a coach can shout directions during stoppages

Table 1. Scientific Challenge Awards, 1997–1999

Year	Awardees	Award Citation
1997	Univ. Maryland, USA	*for demonstrating the utility of an evolutionary approach by co-evolving soccer teams in the simulator league.*
1998	ETL, Japan Sony CSLI, Japan DFKI, Germany	*for the development of fully automatic commentator systems for the RoboCup simulator league.*
1999	Univ. Southern California, USA ETL, Japan Chubu University, Japan	*for the development of automated and statistical game analysis systems and methodologies.*

that help a team improve their play when the game resumes). Second, each player only has access to incomplete information on the game state, yet they must still tackle 'real-world' problems such as interpreting and controlling the movements of objects in three dimensions. This combination of teams consisting of a medium-sized number of agents each with only local information forms an example of what Casti [2] characterises as a *complex system*. We have discussed in [10] how RoboCup represents a promising testbed for researchers to investigate the properties of such domains.

In this review for Computers and Games, we first give an overview of the results of the first three RoboCups (§2), and then highlight some of the research lessons learned to date (§3). We concentrate here on the progress in the simulation league because the absence of low-level issues (such as robot motor control) has allowed researchers in this league to focus on the higher-level questions of most interest to Computers and Games researchers: teamwork, learning and adaptation, and opponent modelling. Additionally, we review the infrastructure developments and challenges that have accompanied the growth in RoboCup (§4) before concluding (§5).

2   Basic Results

Table 1 and Table 2 summarise the results of the first three RoboCups. In each RoboCup a "scientific challenge award" is presented, to encourage scientific and engineering innovations. We describe the research of the award winners (see Table 1) in the body of the paper. The actual winning teams are shown in Table 2. We also describe the research of some of these teams in our review of the progress in the simulation league. Note that from 1999 the simulation league incorporated a selection process to reduce the number of teams participating in the actual event.

3   Progress of Strategy in the Simulation League

The simulation teams have improved year by year. This is easily demonstrated by playing the winners of successive tournaments against each other. An important reason for the yearly improvement in RoboCup is the open-source policy: according to the competition rules, all participants are required to make their code open to the public

Table 2. Finalists in the RoboCup tournaments, 1997 – 1999

RoboCup-97
Simulation League (33 teams)
1: AT-Humboldt (Humboldt Univ., Germany)
2: Andhill (Tokyo Institute of Technology, Japan)
Small Size League (4 teams)
1: CMUnited (CMU, USA)
2: NAIST (Nara Advanced Institute of Science and Technology, Japan)
Middle Size League (5 teams)
1: Dreamteam (ISI/USC, USA)
1: Trakies (Osaka Univ., Japan)

RoboCup-98
Simulation League (34 teams)
1: CMUnited (CMU, USA)
2: AT-Humboldt'98 (Humboldt Univ, Germany)
Small Size League (12 teams)
1: CMUnited-98 (CMU, USA)
2: Roboroos (Univ. Queensland, Australia)
Middle Size League (16 teams)
1: CS-Freiburg (Germany)
2: Tubingen Univ. (Germany)

RoboCup-99
Simulation League (38 teams)
1: CMUnited-99 (CMU, USA)
2: MagmaFreiburg (Albert-Ludwigs-Univ. Freiburg, Germany)
Small Size League (18 teams)
1: The Big Red (Cornell University, USA)
2: FU-Fighters (Free University of Berlin, Germany)
Middle Size League (20 teams)
1: CS Sharif (Sharif University of Technology, Iran)
2: Azzurra Robot Team (RoboCup Italia, Italy)

after each competition. A source code archive is maintained at the official web site, `http://medialab.di.unipi.it/Project/Robocup/pub/`, so that all researchers have instant access to even the code of the best previous teams. In practice, this means that good solutions for the problems of low-level control (less significant from the viewpoint of AI research) can be found in the past work of others, and new research can therefore be concentrated on the more significant problems of high-level control.

Progress in these high-level control techniques is measured not just by performance in RoboCup tournaments, but also by the way that teams answer the specific challenges of *teamwork*, *learning*, and *opponent modelling* laid out in [6]. In the following sections we discuss each of these challenges in turn.

3.1   Teamwork

In the first RoboCup, most teams used inflexible teamwork strategies, in which each player had a fixed home position and covering area. Basically, players chased the ball only when it was inside their designated areas, and otherwise stayed in their "home" positions. However, such teams suffered in 1998 with the introduction of the offside rule. Teams that could not coordinate their movements to catch their opponents in an offside trap, or avoid being trapped offside themselves, were at a distinct disadvantage.

The sharing of source code and the resulting improvement in basic skills has also had a major impact on teamwork, by expanding the basic "unit" of a possible player action from the simple kicking and running provided by the Server environment. It is now feasible for designers to let their programs select between candidate plays such as "dribble", "short pass", and "long pass". Observing the winning teams in the two most recent RoboCups (CMUnited [18], and magmaFreiburg [3]) clearly shows that they can perform a wide variety of plays in well-formed combinations.

However, the teams with the best competitive performance are typically those that have spent time hand-tuning their teamwork systems specifically to the soccer domain. In contrast, several teams have been investigating how to fit soccer into a more complete framework of general teamwork. For example, Scerri and Ydren [15] have developed a system to compile high-level team strategies into behaviour-based agents. In this system, a user designs several strategies by specifying with diagrams what constitutes a good pass in a range of situations. The system then decomposes these diagrams into conditions that invoke behaviours of agents automatically. Because they use behaviour-based agents, it is easy to integrate multiple behaviours into an agent.

Another example of a more general approach is that of Stolzenburg et al. [17], who have proposed a system for decomposing team-level strategies into rules for individual agents. They use logical forms of BDI scripts to define strategies, and translate them into constraint logic programs to control each agent. Also, ISIS [8] have used STEAM to design team and individual planning trees. Originally, STEAM was developed to design team plans for military combat domains. ISIS showed the generality of STEAM's inference rules for executing team strategy by applying most of the rules of the combat domain to RoboCup.

These high-level approaches have not yet had time to mature, and so have not led to cup-winning teams. However, we believe that as teamwork issues become more critical in the future, such kinds of system will become increasingly competitive. One interesting twist to the problem of teamwork is being explored by Team OZ (Open Zeng) [9]. In this project, eleven different research groups (from universities and institutes all over Japan) collaborate to produce a single team by designing just one player each. Each research group has independent research goals and techniques for developing their agent, and the development is also carried out largely independently. The resulting team is then truly an example of a heterogeneous system. Of course, it is impossible to develop well organised multi-agent systems without *any* common assumptions between the agents, but Team OZ is designed to investigate exactly which assumptions are the most important in generating teamwork.

3.2 Learning and Adaptation

More than half the teams participating in the simulation league use learning to some degree. Most commonly, a learning technique is used to tackle single tasks (such as "passing" or "shooting"), but Stone and Veloso [19,18] have used a hierarchical machine learning technique, called 'layered learning', in which lower-level machine learning modules create the action or input spaces of higher-level learning modules. They used the following three types of machine learning modules in the RoboCup domain:

- Neural networks, to train individual players to learn how to intercept a moving ball.
- Decision trees (C4.5), to learn to estimate pass success/fail ratios.
- Reinforcement learning, to choose an action policy (a direction to carry the ball) from among several options.

Layered learning does not automate the selection of the actual hierarchical learning layers or the methods, but with the appropriate task decompositions and learning methods for soccer, it proved very effective within RoboCup.

Evolutionary methods have also been applied to RoboCup. Notably, Luke et. al. [7] were awarded the 1997 Scientific Challenge Award for their application of genetic programming (GP) to developing a team of players. Starting with just randomly generated programs, they played teams against each other and used GP techniques to produce the next generation of programs. Remarkably, the evolution of their programs went through a number of recognisable phases. Initially, there was no discernible strategy, or indeed any goals. But after several generations, the players started to gather around the ball and kick it towards the opponents' goal. In the next stage, some players stayed on their own side as defenders. Finally, all the team's players became distributed over the field and moved the ball up the field by passing it to teammates. The evolution did not progress further, as it did not model the methods of communication that would be necessary for evolving higher-level strategies. However, the experiment provided impressive evidence of the feasibility of learning techniques, even with few assumptions and little built-in knowledge.

In general, reinforcement learning has the following open problems in multi-agent domains:

- One agent's changes of behaviour caused by learning can affect all other agents. Because of exploitation in reinforcement learning, the learning of a multi-agent system tends to become uneven.
- It is difficult to decide how to share reward for learning among agents. Especially in the case of multi-agent systems without top-down control, we must introduce a mechanism to transfer reward between agents.

In the domain of RoboCup, Ohta and Ando have proposed a method that tackles these problems by learning without communication [13,12]. The basic principles of this method are that agents assign rewards to behaviours only when they observe them themselves, (the knowledge to recognise good behaviour is given by the designers), and they also monitor the rewards received by other agents. This system has the following advantages:

- Since the reward given by observation is small, behaviour changes gradually when the agent only uses exploration rather than exploitation in reinforcement learning. This means that change of behaviours does not affect other agents' learning drastically.

- The local minima into which reinforcement learning tends to fall can be avoided because observation can cover a wide area of the state space of the agent.

This learning system is not yet proven within RoboCup, but since it requires no explicit communication among agents, it is a very promising technique for coping with domains containing unknown agents. One of the key results to come from RoboCup research may well be knowledge of how to establish cooperative behaviours between multiple agents that have a minimum of shared knowledge.

3.3 Opponent Modelling

To date, evidence of genuine opponent modelling in RoboCup has been rather sparse. Since each player has only a local view of the field, the prime consideration has been to orient the players, give them good low-level skills, and produce cooperation with team-mates. However, this situation should change significantly in the coming RoboCup2000 with the introduction of the *coach client*, which allows each team to obtain from the Soccer Server a high-quality real-time log of the game from a bird's eye, global perspective. Of course, it is forbidden for the teams to directly feed this global information to the players. Rather, the receiving program (the "coach") is only allowed to shout information to the player clients during breaks in the play (*e.g.*, for throw-ins, or goal kicks). This communication is unrestricted, but must cease once play is under way again. The teams that benefit from this new ability will clearly be those that can model teamwork at a high enough level to represent tactics and strategy with limited communication abilities. They will also be those whose representations of the game allow them to model their opponents actions and reason about the best counter-strategies.

4 Infrastructure: Viewers, Commentators, Analysis

The Soccer Server renders its simulation of a soccer game with a full-pitch, overhead 2-D view. This is very different to the way that the majority of soccer fans experience the majority of their soccer: on television. The TV coverage of soccer typically incorporates a huge array of enhancements such as slow motion replays, multiple camera angles, mobile cameras on the touchlines, play-by-play commentaries, expert analysis, and on-screen statistics. Here we look at some of the infrastructure research supporting RoboCup that addresses these questions.

4.1 Viewers

There are now a number of 3-dimensional viewer systems for the Soccer Server, but the first 3-D viewer was produced by Shinjoh and Yoshida [16]. This system, called SPACE, takes the raw output of Soccer Server and represents it as an isometric 3-dimensional scene. The early versions of SPACE simply used five fixed cameras that could 'pan' to capture the movement of the ball. There was also a simple set of rules for switching between these cameras. But, it was soon found that this basic control structure did not produce good results. For example, [16, Page 43] reports that when watching the output of SPACE "most audiences... were bored", and that "because switching of the camera was done frequently, the audience was easily confused". More recent work on SPACE has concentrated on dealing with these problems by developing an Intelligent Navigation System (INS) to control more carefully the selection of the displayed views. One of the principles of the INS control system is the incorporation of meta-rules such as 'the reduction of audience confusion by the minimisation of camera movements'. The important feature of this rule is that it takes into account not just what happens in the virtual world but also assesses the information that spectators should receive, or may have received already.

4.2 Commentary

Another example where it is critical to consider the knowledge of the spectators is soccer commentary. To date, RoboCup has produced two commentators: MIKE [21] and ROCCO [1]. Each of these generates real-time commentary for Soccer Server games in the form of a stream of (spoken or textual) utterances. In commentary, just as in the presentation of a 3-dimensional visualisation, the appropriate information to convey next will depend on what the audience has been told already. In general, the flexibility to present a game to different audiences will therefore call for investigation of issues such as Grice's cooperative principles of conversation [4] (*e.g.*, the maxim of quantity, that your contribution should be no more and no less informative than is required). Rules of communication become especially significant if a commentary *team* is being modeled, for example with an announcer following the ball-by-ball action and an expert providing higher-level analysis. Real-time commentary is a significant challenge for natural language, calling for reactive control in the face of quickly-changing events. The commentary systems in RoboCup were selected to receive the Scientific Challenge award at RoboCup98.

4.3 Automated Analysis

The automated analysis of RoboCup games was highlighted as another significant area of progress in RoboCup by the 1999 Scientific Challenge Award. One of the systems receiving this award was ISAAC [14], which analyses game logs using C4.5 techniques to identify features of teams. It can generate natural language descriptions of these features and predict results of unseen games, thus helping developers analyse and improve the performance of their teams.

Another recipient of the 1999 award was the statistical analysis of soccer carried out by [20]. The main focus of this research was the analysis of past RoboCups to demonstrate which areas showed most improvement (for instance the measurement of "compactness" showed a significant improvement in teams' players moving together over the three RoboCups). However, this analysis has now been extended to produce a *statistics proxy server* that can produce on demand any of a range of over 30 different statistics. In addition to being a useful development tool for team designers, this server promises to combine powerfully with the coach client discussed above. For the first time, teams will have a significant ability to track the game, to cover their own weaknesses, and exploit the shortcomings of the opponents.

5   Conclusions

One of the main challenges in automating soccer is the difficulty of decomposing the overall game into independent sub-goals that can be solved by an agent that has only local information. RoboCup has shown that initial progress is possible by simply providing each agent with good low-level skills, but that soon 'teamwork' and 'agent modelling' become the only way to increase an agent's knowledge about other parts of the environment. In its first three years, RoboCup has produced promising results both in play (using learning and hard-coded techniques for representing teamwork) and in the *focusing* of domain events (through 3D viewers, commentator systems, and automated analysis systems).

References

1. E. André, G. Herzog, and T. Rist. Generating multimedia presentations for RoboCup soccer games. In H. Kitano, editor, *RoboCup-97: Robot Soccer World Cup I*, volume 1395 of *LNAI*, pages 200–215. Springer, 1998.
2. J. L. Casti. *Would-be Worlds: how simulation is changing the frontiers of science.* John Wiley and Sons, Inc., 1997a.
3. Klaus Dorer. Behavior networks for continuous domains using situation-dependent motivations. In *Proc. of Sixteenth International Joint Conference on Artificial Intelligence*, pages 1233–1238, Aug. 1999.
4. H. P Grice. Logic and conversation. In P. Cole and J. L. Morgan, editors, *Syntax and Semantics: Speech Acts*, volume 3. Academic Press, 1975.
5. Hiroaki Kitano, Minoru Asada, Yasuo Kuniyoshi, Itsuki Noda, Eiichi Osawa, and Hitoshi Matsubara. RoboCup – a challenge problem for AI –. *AI Magazine*, 18(1):73–85, Spring 1997.
6. Hiroaki Kitano, Milind Tambe, Peter Stone, Manuela Veloso, Silvia Coradeschi, Eiichi Osawa, Hitoshi Matsubara, Itsuki Noda, and Minoru Asada. The robocup synthetic agent challenge 97. In Hiroaki Kitano, editor, *Proc. of The First International Workshop on RoboCup*, pages 45–50, Aug. 1997.
7. Sean Luke, Charles Horn, Jonathan Farris, Gary Jackson, and James Hendler. Co-evolving soccer softbot team coordination with genetic programming. In Hiroaki Kitano, editor, *Proc. of The First International Workshop on RoboCup*, pages 115–118, Aug. 1997.
8. Stacy Marsella, Jafar Adibi, Yaser Al-Onaizan, Ali Erdem, Randall Hill, Gal A. Kaminka, Zhun Qiu, and Milind Tambe. Using an explicit teamwork model and learning in robocup. In Minoru Asada, editor, *Proc. of the second RoboCup Workshop*, pages 195–203, 1998.
9. Junji Nishino, Takuya Morishita, Shuuhei Kinoshita, Takashi Suzuki, Igarashi, Tetsuhiko Kofuji, and Hiroki Shimora. The dream team oz for the robocup simulation league (in japanese). In *JSAI Technical Report SIG-HOT/PPAI-9909*, pages 16–19. JSAI, March 2000. JSAI Technical Report, SIG-Challenge-9909-4(3/21).
10. Itsuki Noda and Ian Frank. Investigating the complex with virtual soccer. In J.-C. Heudin, editor, *Virtual Worlds*, pages 241–253. Springer Verlag, Sep. 1998.
11. Itsuki Noda and Hitoshi Matsubara. Soccer server and researches on multi-agent systems. In Hiroaki Kitano, editor, *Proceedings of IROS-96 Workshop on RoboCup*, pages 1–7, Nov. 1996.
12. M. Ohta and T. Ando. Cooperative reward in reinforcement learning (in japanese). In *Proc. of 3rd JSAI RoboMech Symposia*, pages 7–11, Apr. 1998.

432 Itsuki Noda and Ian Frank

13. Masayuki Ohta. Gemini – learning cooperative behaviors without communicating. In Silvia Coradeschi, Tucker Balch, Gerhard Kraetzschmar, and Peter Stone, editors, *RoboCup-99 Team Descriptions: Simulation League*, pages 36–39. Linköping University Electronic Press, Aug. 1999.

14. T. Raines, M. Tambe, and S. Marsella. Towards automated team analysis: A machine learning approach. In *Proc. of the RoboCup Workshop at the International Joint Conference on Artificial Intelligence(IJCAI'99)*, Aug. 1999.

15. Paul Scerri and Johan Ydren. End user specification of robocup teams. In Manuela M. Veloso, editor, *Proc. of The Third International Workshop on RoboCup*, pages 187–192, Aug. 1999.

16. A. Shinjoh and S. Yoshida. The intelligent three-dimensional viewer system for robocup. In *Proceedings of the Second International Workshop on RoboCup*, pages 37–46, 1998.

17. Frieder Stolzenburg, Oliver Obst, Jan Murray, and Björn Bremer. Spatial agents implemented in a logical expressible language. In Manuela M. Veloso, editor, *Proc. of The Third International Workshop on RoboCup*, pages 205–210, Aug. 1999.

18. Peter Stone and Manuela Veloso. Layered learning and flexible teamwork in robocup simulation agents. In Manuela M. Veloso, editor, *Proc. of The Third International Workshop on RoboCup*, pages 211–216, Aug. 1999.

19. Peter Stone, Manuela Veloso, and Patrick Riley. Cmunited-98 simulation team. *AI Magazine*, 21(1):20–28, Spring 2000.

20. K. Tanaka-Ishii, I. Frank, I. Noda, and H. Matsubara. A statistical perspective on the robocup simulator league: Progress and prospects. In *Proceedings of the Third International Workshop on RoboCup*, 1999.

21. K. Tanaka-Ishii, I. Noda, I. Frank, H. Nakashima, K. Hasida, and H. Matsubara. MIKE: An automatic commentary system for soccer. In *Proceedings of ICMAS-98*, pages 285–292, 1998.

Review: Computer Shogi through 2000

Takenobu Takizawa[1] and Reijer Grimbergen[2]

[1] School of Political Science and Economics, Waseda University
1-6-1 Nishi-Waseda, Shinjuku-ku, Tokyo, Japan 169-8050
takizawa@mse.waseda.ac.jp
[2] Complex Games Lab, Electrotechnical Laboratory
1-1-4 Umezono, Tsukuba-shi, Ibaraki-ken, Japan 305-8568
grimbergen@fu.is.saga-u.ac.jp

Abstract. Since the first computer shogi program was developed by the first author in 1974, more than a quarter century has passed. During that time, shogi
programming has attracted both researchers and commercial programmers and
playing strength has improved steadily. Currently, the best programs have a level
that is comparable to that of a strong amateur player (about 4-dan), but the level of
experts is still beyond the horizon. The basic structure of strong shogi programs is
similar to chess programs. However, the differences between chess and shogi have
led to the development of some shogi-specific methods. In this paper we will give
an overview of the computer shogi history, summarise the most successful techniques and give some ideas for the future directions of research in computer shogi.

Keywords: shogi, computer shogi history, evaluation function, plausible move
generation, SUPER SOMA, tsume shogi, tesuji search.

1   Introduction

Shogi, or Japanese chess, is a two-player complete information game similar to chess.
As in chess, the goal of the game is to capture the opponent's king. However, there are a
number of differences between chess and shogi: the shogi board is slightly bigger than
the chess board (9x9 instead of 8x8), there are different pieces that are relatively weak
compared to the pieces in chess (no queens, but gold generals, silver generals and lances)
and the number of pieces in shogi is larger (40 instead of 32).

But the most important difference between chess and shogi is the possibility to re-use
captured pieces. A piece captured from the opponent becomes a *piece in hand* and at
any move a player can *drop* a piece that was captured earlier on a vacant square instead
of moving a piece on the board. For a more detailed description of the rules of shogi,
see [9].

There are only a few restrictions on where a piece can be dropped. Creating a double
pawn by a drop is not allowed, but dropping a piece with check or mate is perfectly legal
(with one exception: mate with a pawn drop is not allowed). As a result of these drop
moves, the number of legal moves in shogi is on average much larger than for chess. In
chess the average number of legal moves is estimated at 35, while for shogi the figure is
about 80 [11]. In the endgame, the situation is even worse, as in most endgame positions

T.A. Marsland and I. Frank (Eds.): CG 2000, LNCS 2063, pp. 433–442, 2001.

there will be various pieces in hand and the number of legal moves can easily grow to over 200 [3].

Since shogi is similar to chess, the techniques that have proven effective in chess also have been the foundation of most shogi programs. However, to deal with the high number of legal moves in shogi, shogi-specific methods have to be developed because deep search is more difficult than in chess. Shogi could therefore be considered as an intermediate step between chess and Go [12].

The improvement of shogi programs over the last couple of years has been impressive. When Kasparov lost to DEEP BLUE in 1997, computer shogi programs could only compete with average amateur players (about 2-dan). After the latest computer shogi tournament, Hiroyuki Iida (a professional shogi player and an associate professor at Shizuoka University) estimated the strength of the top programs at about 4-dan. Even though on the Japanese grading scale this is only two grades difference, the actual difference in playing strength is quite large. Even for strong amateurs, losing against the computer (especially in quick games) is no longer an embarrassment.

In this review paper we will start with a brief overview of the history of computer shogi in Section 2. In Section 3, we will then give an overview of the techniques that have been most successful. In Section 4 we will explain about some of the recent developments in computer shogi. We will end with some conclusions and ideas about the future direction of computer shogi in Section 5.

2   A Brief History of Computer Shogi

The first shogi program was developed by the research group of the first author in November 1974. At that time, shogi seemed so complex that the prediction was that it would take about 50 years for shogi programs to reach the level of a 1-dan amateur[1]. However, this prediction was way off, as Iida estimated that the strength of the top programs in 1995 was already strong 1-dan.

One of the reasons for this has been the establishment of the *Computer Shogi Association* (CSA) in 1986 by Yoshiyuki Kotani and the first author. This organisation started organising computer shogi tournaments in 1990, tournaments that were also supported by the *Nihon Shogi Renmei*, the Japanese shogi organisation.

2.1   The Computer Shogi Championships (1990–1999)

The first computer shogi championship was held on December 2nd 1990. Table 1 gives an overview of all the computer shogi tournaments that have been held since then. It is clear that computer shogi tournaments have become big events. The number of participants in the CSA tournaments has increased every year, from 6 participants in the first tournament (including 2 invited programs) to 45 entries in the last event. Since 1996, the finals of

---

[1] 1-dan is an interesting psychological barrier in shogi and most other Japanese sports. The Japanese ranking system in shogi starts at 15-kyu, going down to 1-kyu and then going up from 1-dan to 6-dan. Therefore, the difference between 1-kyu and 1-dan is only one grade, but the psychological difference is considerable. Unlike players with a kyu grade, players with a dan grade are considered strong players.

Table 1. Results of the computer shogi tournaments 1990-2000. CSA = CSA Tournament; SGP = Computer Shogi Grand Prix; MSO = Mind Sports Olympiad.

	Date	Entries	Winner	2nd	3rd
CSA1	Dec 2 1990	6	EISEI MEIJIN	KAKINOKI	MORITA
CSA2	Dec 1 1991	9	MORITA	KIWAME	EISEI MEIJIN
CSA3	Dec 6 1992	10	KIWAME	KAKINOKI	MORITA
CSA4	Dec 5 1993	14	KIWAME	KAKINOKI	MORITA
CSA5	Dec 4 1994	22	KIWAME	MORITA	YSS
CSA6	Jan 20-21 1996	25	KANAZAWA	KAKINOKI	MORITA
CSA7	Feb 8-9 1997	33	YSS	KANAZAWA	KAKINOKI
CSA8	Feb 12-13 1998	35	IS	KANAZAWA	SHOTEST
CSA9	Mar 18-19 1999	40	KANAZAWA	YSS	SHOTEST
SGP1	Jun 19-20 1999	8	KAKINOKI,YSS		IS, KCC
CSA10	Mar 8-10 2000	45	IS	YSS	KAWABATA
MSO1	Aug 21-25 2000	3	YSS	SHOTEST	TACOS

the tournament have been a round robin of eight programs. These eight programs are decided by taking the five strongest programs from the first day(s) of the tournament, and adding the three best programs from the previous year's contest.

The first CSA tournament was won by EISEI MEIJIN (made by Yoshimura), but even though this program is still strong and has participated in most tournaments since 1990, its only other top three finish was in the 1991 tournament. Since 1992, the CSA tournament has been dominated by KIWAME/KANAZAWA, two programs written by Shinichiro Kanazawa. Kanazawa's programs have won the CSA tournament five times and have been runner-up three times.

Other past winners are MORITA SHOGI (written by Morita), YSS (Yamashita) and IS SHOGI (Tanase, Goto, Kishimoto and Nagai), which is the reigning champion. In recent years, there have also been a number of foreign entries. The best results have been achieved by Jeff Rollason from England with his program SHOTEST, which entered for the first time in 1997 and finished third in both the 1998 and the 1999 tournaments. Other foreign entries have been KCC from Korea, SHOCKY by Pauli Misikangas from Finland, and SPEAR, a Dutch program by the second author. KCC managed to qualify for the finals of the 1999 CSA tournament and SHOCKY did the same in the 2000 CSA tournament.

Until recently, the CSA tournament was the only computer shogi tournament, but in 1999 the first *Computer Shogi Grand Prix* was organised as part of an international shogi festival called the *Shogi Forum*. The Computer Shogi Grand Prix was an invitation tournament for the best 8 programs of the previous CSA tournament and was won by KAKINOKI SHOGI (Kakinoki) and YSS. Unfortunately, KANAZAWA SHOGI had to pull out of this tournament because of problems with running the program on the single platform hardware that was used. Finally, there was also a (very small) computer shogi tournament held as part of the Computer Olympiad at the 2000 Mind Sports Olympiad. This tournament was won by YSS.

Table 2. Results of the finals of the CSA tournament 2000.

No	Program Name	1	2	3	4	5	6	7	Pt	SB
1	IS SHOGI	6+	8+	3-	2+	5+	4-	7+	5	15
2	YSS 10	8+	5+	6+	1-	3-	7+	4+	5	14
3	KAWABATA SHOGI	4-	7-	1+	5+	2+	8+	6-	4	16
4	KANAZAWA SHOGI	3+	6-	5-	8+	7+	1+	2-	4	13
5	KAKINOKI SHOGI	7+	2-	4+	3-	1-	6+	8+	4	10
6	KFEND	1-	4+	2-	7-	8-	5-	3+	2	8
7	SHOTEST 4.0	5-	3+	8-	6+	4-	2-	1-	2	6
8	SHOCKY 3	2-	1-	7+	4-	6+	3-	5-	2	4

2.2   The Millennium CSA Computer Shogi Championship

Now let's look at the results of the latest CSA Computer Shogi Championship in a little more detail. The millennium tournament was held from March 8th to March 10th, 2000. This tournament had a record number of 45 participants and was also the first tournament that was played over three days: two days of qualification tournaments were played with the Swiss tournament system and the finals were then held on a single day.

The detailed results of the finals are given in Table 2. The finalists are an interesting mix between programs that have participated for a long time (IS SHOGI: 4 years, YSS: 9 years, KANAZAWA SHOGI: 9 years, KAKINOKI SHOGI: 10 years), some relatively new entries (KAWABATA SHOGI (Kawabata) and KFEND (Arioka)) and non-Japanese entries (SHOTEST and SHOCKY).

A second observation is that the tournament was very close. Before the start of the final round there were four programs that could still win the tournament. IS SHOGI in the end was the lucky one, as it won in the final round and two other games ended in its favour. The fourth place of KANAZAWA SHOGI can be called a surprise, but if the program would have won in the final round against YSS, it would have won the tournament: another illustration of how close the tournament was.

This year the foreign entries did not do very well, as SHOTEST and SHOCKY ended at the bottom of the table. Still, it was only SHOCKY's second appearance in this tournament and SHOTEST has proven its strength in the two previous CSA tournaments. Furthermore, the Korean program KCC just missed the final eight, even though it beat IS SHOGI in the qualification tournament. Although computer contests are still dominated by the Japanese programs, this is an indication that computer shogi is becoming more and more international. More details about the 10th CSA tournament can be found in [4].

3   Techniques Used in Computer Shogi Programs

As pointed out earlier, shogi is similar to chess, and strong shogi programs have a structure that is similar to that of chess programs. Typically, mini-max game trees are built that are explored by an iterative alpha-beta search. Shogi programs also make use of common refinements of this scheme such as PVS-search, quiescence search, aspiration search, null-move pruning, history heuristic and killer moves.

Despite these similarities, however, the specific features of shogi have made it necessary to explore other methods. In this section we will look at the following elements that need to be handled differently in shogi, or are shogi specific: *data structures, the evaluation function, hash tables, plausible move generation* and *tsume shogi*.

3.1   Data Structures

Probably the best English overview of the data structures used in shogi is given by Yamashita on his webpage [19]. The most important extra data structure that is used in shogi seems to be the *kiki table* (or piece attack table). In the kiki table, information about which piece attacks which square is stored. The table distinguishes between pieces that can move to a square directly and pieces that can only move there indirectly (for example, a rook behind another rook). In shogi the kiki table is very important as it is accessed by a number of other modules in the program such as the evaluation function, plausible move generation and mating search.

3.2   The Evaluation Function

The evaluation function of chess contains many different features, but material is the dominant component. For example, it is almost impossible to have enough positional compensation for the loss of a queen. On the other hand, in shogi the balance between material and king safety is the most important part of the evaluation function [7,18].

As pointed out, in shogi captured pieces do not disappear from the game, so a game of shogi almost always ends with one of the kings being mated (there are some possibilities of a draw in shogi, but less than 1% of the professional games end in a draw). Therefore, the endgame in shogi is usually a mating race where the speed of the attack has the highest priority. Losing a strong piece such as a rook often leads to disaster in the opening and middle game, but can be completely insignificant in the endgame.

Strong shogi programs therefore need an understanding of the stage of the game (opening, middle game or endgame). The weights of the features of the evaluation function can change dramatically based on this game stage. A game of shogi usually has a slow build-up, so in the opening there is almost no weight given to the king safety. The most important features are material balance, castle formations, the mobility of major pieces (in shogi the rook and bishop are the strongest pieces) and the control of squares near the centre. In the middle game increasing weight is given to king safety, even though material still is the most important feature. In the endgame the king safety takes priority over material considerations. The best shogi programs can handle these transitions quite well and are able to accurately adjust their evaluation function during the game [20].

3.3   Hash Tables

The possibility of having pieces in hand also changes the way in which hash tables are used in shogi. In chess, only the pieces on the board matter, so hash tables are only used for transpositions. Transpositions are the same for chess and shogi: if two board positions are identical and the same player is to move, then there is a transposition if the pieces in hand for both players are identical. However, in shogi it is also possible to have *domination* of positions:

Definition 1 *A position* P *is* dominating *a position* Q *if the pieces on the board in* P *and* Q *are identical, the same player is to move in* P *and* Q *and the pieces in hand for the player to move in* P *are a superset of the pieces in hand of the player to move in* Q.

Search can be stopped in these types of positions as this is a cycle with a material advantage (or disadvantage) for the player to move.

Hash tables can be used to handle domination as well. In shogi, hashing of positions is only done on the pieces on the board and an extra hashcode is stored in the hash table for the pieces in hand [17].

3.4 Plausible Move Generation

To deal with the large number of legal moves in shogi, most shogi programs use plausible move generation to reduce the number of candidate moves that need to be searched. Some of the cuts are very safe. For example, promotion of shogi pieces is almost always optional. It is possible to construct positions where not promoting a rook, bishop or a pawn is the only way to win, even though the moves of these pieces in their promoted state are a superset of the moves of the non-promoted ones. However, in practice the non-promotion moves for these pieces are so rare that they can be safely removed from the list of legal moves.

These almost safe cuts are only a small percentage of the total number of legal moves, so most shogi programs also perform a number of more speculative, unsafe cuts (for more details, see [3]). The general approach is to have a set of categories with tactical moves, ordered by urgency. Common categories are *Capture material, Attack material, Move attacked piece, Defend attacked piece, Attack king, Defend king, Discovered attack, Promote piece, Threaten promotion* and *Defend against promotion threat*. For example, IS SHOGI has 10 different categories [18]; KAKINOKI SHOGI uses 8 general move categories [6] and YSS uses a very detailed move categorisation with no less than 30 move categories [20]. In YSS the move categories also depend on the search depth. For example, moves that attack a piece are not generated if the remaining search depth is 1, since the position evaluation will not change much if the attacked piece moves to a safe square.

3.5 Tsume Shogi

Tsume shogi, or checkmating problems in shogi, has been an independent research domain for many years. Unlike the mating problems in chess, each move by the attacker in a tsume shogi problem must be a check, finally leading to mate in all variations. The aim of the defender in a tsume shogi problem is to prolong the mate as long as possible and the solution of a tsume shogi problem is the longest variation that leads to mate.

The best tsume shogi solver has been developed by Seo [16]. While previous tsume shogi solvers used iterative alpha-beta search, Seo's solver used conspiracy number search [13] and this worked much better than previous approaches. After further improvements of his program (for example using proof number search [1] instead of conspiracy number search), Seo was able to solve the longest tsume shogi problem to date called *Microcosmos*. This problem was composed by Koji Hashimoto in 1986 and has a solution of 1525 ply. Seo's tsume shogi solver solved Microcosms in about 30 hours in April 1997. Currently, tsume shogi is the only area in shogi where computer programs outperform world class shogi players. More details about the history and design of tsume shogi solvers can be found in [2].

Most strong shogi programs use a tsume shogi solver to find mate in the final stages of the game. However, because of the rule that each move by the attacker must be a check, the use of the tsume shogi solver is limited. Seo showed that the branching factor of an average search tree in tsume shogi is only about 5 [16], which is very different from the branching factor of the search tree in general endgame positions.

4   Recent Developments in Computer Shogi

Of course, all methods currently used in computer shogi are still being revised and improved. However, there are also some new ideas that have been developed recently, which we summarise here.

4.1   Extended Use of the Tsume Shogi Solver

As pointed out above, tsume shogi solvers outperform human experts, so a logical step is to use the tsume shogi solver during normal tree search. IS SHOGI seems to be the program that is most advanced in this respect [18]. First of all, if the opponent is threatening mate, the moves in the mating sequence are added to the list of killer moves. A more sophisticated use of the tsume shogi solver is to store the mating sequence that was found for future use. A problem in shogi is that by dropping pieces it is possible to play long checking sequences that push the winning variation over the search horizon. IS SHOGI uses the tsume shogi solver to find mating threats in the endgame. The sequences that lead to mate are then stored and the tree node is marked as being a mating threat. If the opponent then starts a sequence of checks that continue until the search horizon is reached, the stored mating sequences are retrieved and it is tested if these also work in the position at the end of the variation that was searched.

4.2   Artistic Evaluation of Tsume Shogi

An interesting related research domain is the artistic evaluation of tsume shogi problems. As with mating problems in chess, there are big differences between the artistic impression that a tsume shogi problem makes to a human solver. Koyama et al. have done some interesting work in this area [8]. Their ideas were pursued by Noshita [14] and Hirose et al. [5], who developed programs that could automatically generate tsume shogi problems using retrograde analysis. With Koyama's rules for artistic evaluation it was possible to distinguish good tsume shogi problems from bad ones. Some of the problems made by Hirose's program were sent under a pseudonym to monthly magazines that publish tsume shogi problems and were accepted for publication.

4.3   Tesuji Search

Another interesting idea used in IS SHOGI is *tesuji search* [18]. In the middle game in shogi there are often sequences of moves that are played in a particular order to reach a certain goal. In shogi these standard move sequences are called *tesuji*. For a search algorithm it is hard to recognise tesuji, as their goal can be beyond the search horizon. In IS SHOGI a small number of 3-ply move sequences are implemented. If such a sequence is found in the search tree, then the search is extended by two ply. At this point, IS SHOGI implements only 6 different move sequences to deal with certain blind spots of the program, but this method seems more generally applicable. Even though we do not know of any other published methods, we have observed that most programs have some way to implement standard move sequences, especially in the opening.

4.4 SUPER SOMA

In Jeff Rollason's SHOTEST, the horizon effect is dealt with in a way that is very different from other programs. In SHOTEST a static tactical move analysis called *SUPER SOMA* is used to guide the search [15]. SUPER SOMA is an extension of the SOMA algorithm for static analysis of capture sequences [10]. SUPER SOMA can not only statically analyse captures, but also pins, ties to material as well as ties to mate and promotion (if a defending piece moves, material will be lost or mate will be possible), discovered attacks, forks and defensive moves. SHOTEST uses the results of this static tactical evaluation to guide the search.

SHOTEST's static tactical evaluator slows down the program considerably (it is the program that has the lowest node per second search rate among the top programs). Also, the tactical evaluator just returns one line of play, generating only the top move according to the tactical evaluator at each depth. However, SHOTEST's tournament performance show that this approach is a good alternative to the fast evaluation used in most other programs.

5 Conclusions and Expectations for the Future

Shogi programs have improved significantly in the past few years and are a good match for most strong amateur players. Unfortunately, there is no tradition of playing shogi programs against human players under tournament conditions, so it is not completely clear how strong shogi programs actually are.

We feel that the near future will be very important for computer shogi as the latest CSA tournament was the closest ever and the differences in strength between the programs seem to be getting smaller. It will therefore be interesting to see if there is some kind of limit to the methods that are now being used in computer shogi, or if this is the start of a combined effort of a large number of programs towards the ultimate goal of beating the best players in the world.

As for this ultimate goal, the best human player Yoshiharu Habu is one of the few professionals who recognises how much shogi programs have improved. When asked in 1996 when he thought a computer would beat him, his clear answer was: "2015". This sounds like a reasonable estimate, but there is still a lot of work to do, as Habu (already a living legend) will be only 45 years old by then and very much at the peak of his abilities.

To beat Habu, we might need the help of special purpose hardware like the chess chip that was used in DEEP BLUE. Feng-hsiung Hsu of the DEEP BLUE team has shown interest in designing such a chip for shogi, but so far there have been no concrete steps taken to design one.

Acknowledgement

The authors are grateful to the members of the CSA for their kind help.

References

1. L.V. Allis, M. van der Meulen, and H.J. van den Herik. Proof-number search. *Artificial Intelligence*, 66:91–124, 1994.
2. R. Grimbergen. A Survey of Tsume-Shogi Programs Using Variable-Depth Search. In H.J. van den Herik and H.Iida, editors, *Computers and Games: Proceedings CG'98. LNCS 1558*, pages 300–317. Springer Verlag, Berlin, 1999.
3. R. Grimbergen. Plausible Move Generation Using Move Merit Analysis with Cut-Off Thresholds in Shogi. In *Computers and Games: Proceedings CG2000.*, Hamamatsu, Japan, 2000.
4. R. Grimbergen. Report on the 10th CSA Computer-Shogi Championship. *ICGA Journal*, 23(2):115–120, June 2000.
5. M. Hirose, H. Matsubara, and T. Ito. The composition of Tsume-Shogi problems. In *Advances in Computer Chess 8*, pages 299–319, Maastricht, Holland, 1996. ISBN 9062162347.
6. Y. Kakinoki. The Search Algorithm of the Shogi Program K3.0. In H. Matsubara, editor, *Computer Shogi Progress*, pages 1–23. Tokyo: Kyoritsu Shuppan Co, 1996. ISBN 4-320-02799-X. (In Japanese).
7. S. Kanazawa. The Kanazawa Shogi Algorithm. In H. Matsubara, editor, *Computer Shogi Progress 3*, pages 15–26. Tokyo: Kyoritsu Shuppan Co, 2000. ISBN 4-320-02956-9. (In Japanese).
8. K. Koyama and Y. Kawano. Analysis of Favorable Impression Factors on Masterpieces of Tsume-shogi (Mating Problems of Japanese Chess). In *Game Programming Workshop in Japan '94*, pages 12–21, Kanagawa, Japan, 1994. (In Japanese).
9. T. Leggett. *Shogi: Japan's Game of Strategy*. Charles E. Tuttle Company, England, 1966. ISBN 0-8048-1903-3.
10. D. Levy and M. Newborn. *How Computers Play Chess*. Computer Science Press, 1991. ISBN 0-7167-8121-2.
11. H. Matsubara and K. Handa. Some Properties of Shogi as a Game. *Proceedings of Artificial Intelligence*, 96(3):21–30, 1994. (In Japanese).
12. H. Matsubara, H. Iida, and R. Grimbergen. Natural developments in game research: From Chess to Shogi to Go. *ICCA Journal*, 19(2):103–112, June 1996.
13. D.A. McAllester. Conspiracy Numbers for Min-Max search. *Artificial Intelligence*, 35:287–310, 1988.
14. K. Noshita. An Algorithm for the Automatic Generation of Tsume Shogi Problems. In H. Matsubara, editor, *Computer Shogi Progress 2*, pages 32–46. Tokyo: Kyoritsu Shuppan Co, 1998. ISBN 4-320-02799-X. (In Japanese).
15. J. Rollason. SUPER SOMA - Solving Tactical Exchanges in Shogi without Tree Searching. In *Computers and Games: Proceedings CG2000.*, Hamamatsu, Japan, 2000.
16. M. Seo. The C* Algorithm for AND/OR Tree Search and its Application to a Tsume-Shogi Program. Master's thesis, Faculty of Science, University of Tokyo, 1995.
17. M. Seo. On Effective Utilization of Dominance Relations in Tsume-Shogi Solving Algorithms. In *Game Programming Workshop in Japan '99*, pages 129–136, Kanagawa, Japan, 1999. (In Japanese).

18. Y. Tanase. The IS Shogi Algorithm. In H. Matsubara, editor, *Computer Shogi Progress 3*, pages 1–14. Tokyo: Kyoritsu Shuppan Co, 2000. ISBN 4-320-02956-9. (In Japanese).

19. H. Yamashita. `http://plaza15.mbn.or.jp/~yss/book_e.html`.

20. H. Yamashita. YSS: About its Datastructures and Algorithm. In H. Matsubara, editor, *Computer Shogi Progress 2*, pages 112–142. Tokyo: Kyoritsu Shuppan Co, 1998. ISBN 4-320-02799-X. (In Japanese).

Author Index

Lecture Notes in Computer Science

For information about Vols. 1–2179
please contact your bookseller or Springer-Verlag